Terahertz Astronomy

Terahertz Astronomy

Christopher K. Walker

CRC Press
Taylor & Francis Group
Boca Raton London New York

CRC Press is an imprint of the
Taylor & Francis Group, an **informa** business

CRC Press
Taylor & Francis Group
6000 Broken Sound Parkway NW, Suite 300
Boca Raton, FL 33487-2742

First issued in paperback 2019

ISBN-13: 978-1-4665-7042-9 (hbk)
ISBN-13: 978-1-138-89464-8 (pbk)

Library of Congress Cataloging-in-Publication Data

Walker, Christopher K. (Christopher Kidd)
Terahertz astronomy / Christopher K. Walker.
pages cm
"A CRC title."
Includes bibliographical references and index.
ISBN 978-1-4665-7042-9 (alk. paper)
1. Radio astronomy. 2. Radio astronomy--Problems, exercises, etc. I. Title.

QB476.5.W35 2016
522'.682--dc23 2015013733AA

Visit the Taylor & Francis Web site at
http://www.taylorandfrancis.com

and the CRC Press Web site at
http://www.crcpress.com

To my Family, who lit the path to the writing of this book

CONTENTS

PREFACE

For now we see through a glass, darkly....

1 Corinthians 13:12

The Universe is large—so large, in fact, that normal Earth-bound distance scales lose their significance. We express the distance to even the nearest stars in terms of light years (i.e., the distance light travels in a year, ~6 trillion miles). Such distances would require more than 100 human lifetimes to reach even the nearest stars with our current, fastest spacecraft. Except for high energy particles believed to originate primarily from supernova explosions and, perhaps, active galactic nuclei, all we have to probe the space beyond the boundaries of our solar system is light. Being children of a yellow, G-type star, evolution has built into us a natural prejudice for "optical" light. It is in this narrow spectral regime that our star puts out most of its radiant energy, and to which our eyes are highly tuned. From the dawn of civilization until the invention of the telescope in the early 1600s, all astronomy was performed with the unaided human eye. With its large lenses and mirrors, the telescope greatly increased the light-gathering area of the eye, extending the realm and details of the cosmos observable by humans. A further, and similarly important, leap occurred in the nineteenth century, when the newly invented technology of photography was quickly applied to astronomical observations. But still, all astronomy occurred only in the narrow frequency range associated with optical light.

In the latter half of the nineteenth century, James Clerk Maxwell derived his now famous electromagnetic field equations. David Edward Hughes, Heinrich Hertz, and others, soon after, demonstrated the existence of nonoptical (radio) light. With both a theoretical underpinning and successful experiments in place, physicists and engineers were able to enhance the sensitivity of early radios. But unlike the relatively rapid application of the breakthrough technology of photography to astronomy, no one thought of trying to use this new radio technology to observe the cosmos; it was simply beyond the human experience. It was quite by accident that a Bell Telephone Laboratories physicist (Karl Jansky) discovered 160 MHz radio emission from beyond Earth (it came from the very center of our Milky Way!) in January, 1932. Even then, the discovery was not embraced by the astronomical community. It was the continued pursuit and persistence of an amateur astronomer/radio engineer (Grote Reber) that broke the ground for "nonoptical" astronomy in the late 1930s and early 1940s. The origin of these new extraterrestrial radio emissions was unknown, or misunderstood, by astronomers and physicists of the time, many of who believed the radio light to result from thermal black-body radiation. The advent of World War II and the need for high frequency radar greatly accelerated the evolution of radio technology. The great increase in radiosensitivity, combined with breakthroughs in quantum mechanics, led to the prediction and discovery of extraterrestrial radio emission from neutral atomic hydrogen at 1420 MHz (1.42 GHz) in the early 1950s. For more than 60 years, the sensitivity and frequency range of radios has continued to increase until now radios can be used to probe the emission and absorption of light by atoms and molecules between ~0.3 and 10 terahertz (THz). This is the realm of *THz astronomy*. THz observations of interstellar atoms, molecules, and dust can serve as powerful probes of the conditions within the interstellar medium that permeates our galaxy and provide insights into the origins of stars, planets, galaxies, and the Universe itself.

The goal of this book is to introduce the student and researcher to THz astrophysics and the technologies that make this rapidly evolving field possible. Throughout the book, our philosophy is to "follow the light." The book is written to be cross-disciplinary, making this exciting subject accessible to physics, astronomy, and engineering students alike. Chapters 1 through 4 discuss the origin and interpretation of THz light in astrophysical sources. Chapters 5 through 9 provide an introduction and overview of the technologies used to collect and detect THz light. Each chapter contains worked-out example problems, in addition to the ones provided for student practice at the end of the chapters. Care is taken to present each new topic area as intuitively as possible, together with the equations needed to apply this intuition to real-life astrophysical situations. Figures and diagrams are used extensively to reinforce important concepts. The appendices include lists of useful physical/astronomical constants, important THz transitions of atoms and molecules, and commonly used expressions in the design and operation of THz instrumentation. The author has used the information contained in this volume to interpret THz observations and design, build, and deploy THz astronomical instrumentation on high mountain tops, the Antarctic Plateau, and high-altitude balloons. However, there are a number of areas that are only touched upon or not discussed at all. For this, the author apologizes. Our field is broad and ever growing, making it difficult to cover it in a single text.

Historically, THz technology has been almost exclusively developed and used by a relatively small group (~100) of astronomers, engineers, and physicists to probe conditions around astrophysical objects. However, a new interest in terrestrial remote sensing at THz frequencies has led to an explosion of activity, with the number of active THz researchers increasing by more than an order of magnitude in just a few years. Much of the background science of Chapters 1 through 4 and almost all of the THz technology discussed in Chapters 5 through 9 are directly applicable to this new area of interest.

The author wishes to thank friends, colleagues, and students who proofread early versions of the text and contributed figures. These include Peter Ade, John Bally, John Bieging, Matt Bradford, Greg Engargiola, J. R. Gao, Jason Glenn, Paul Goldsmith, Dathon Golish, Karl Gordon, Chris Groppi, Jeffrey Hesler, David Hollenbach, Casey Honniball, Jenna Kloosterman, Craig Kulesa, Charlie Lada (who first showed me how to compute column density), David Lesser, Phil Maloney, Tony Marston, John Mather, Ben Mazin, Imran Mehdi, Gary Melnick, Harvey Moseley, Desika Narayanan, Gopal Narayanan, Bill Peters, Dick Plambeck, George Rieke, Gordon Stacey, Brandon Swift, and Mark Wolfire—writing a textbook is a community effort! The author also wishes to thank Buell Jannuzi and the Astronomy Department staff of the University of Arizona for their support through the writing process. The style of the book is inspired by the classic textbook, *Radio Astronomy*, by John D. Kraus, the author's Master's thesis advisor.

Finally, the author thanks his family members for their encouragement, and wishes the readers all the best in pursuing their research into THz astronomy.

Christopher K. Walker

AUTHOR

Professor Christopher Walker has over 30 years of experience designing, building, and using state-of-the-art receiver systems for THz astronomy. He is a professor of astronomy and an associate professor of optical sciences and electrical and computer engineering at the University of Arizona (UofA). He has published numerous papers on star formation and protostellar evolution. He has served as dissertation director for a number of PhD students and is a topical editor for *IEEE Transactions on TeraHertz Science and Technology*. He has worked in industry (TRW Aerospace and Jet Propulsion Laboratory (JPL)) as well as academia. As a Millikan Fellow in Physics at the California Institute of Technology, he worked on the development of low-noise, superconductor-insulator-superconductor (SIS) waveguide receivers above 400 GHz, and explored techniques for etching waveguide out of silicon. On joining the UofA faculty in 1991, he began the Steward Observatory Radio Astronomy Lab (SORAL), which has become a world leader in developing THz receiver systems for astronomy and other remote sensing applications. These instruments are multi-institutional efforts, with key components coming from JPL, several universities, and a number of industrial partners. Instruments developed by his team have served as primary facility instruments at the Heinrich Hertz Telescope on Mt. Graham, Arizona, and the Antarctic Submillimeter Telescope/Remote Observatory (AST/RO) telescope at the South Pole. He led the effort to design and build the world's largest (64 pixels) submillimeter-wave heterodyne array receiver, called SuperCam. He is principal investigator of the long duration balloon project "The Stratospheric THz Observatory (STO)" funded by NASA (National Aeronautics and Space Administration, USA). In 2013, he was selected as a NASA Innovative Advanced Concepts (NIAC) Fellow for his proposed work on the "10 Meter Suborbital Large Balloon Reflector (LBR)" project. He was named a Galileo Circle Fellow in 2015 by the College of Science of the University of Arizona.

ABOUT THE COVER

The top image shows the region around the Horsehead Nebula in the constellation of Orion as seen in the light emitted by CO molecules at a frequency of 0.346 THz. The Horsehead Nebula is about 1500 light years from Earth. The full image covers 0.5×1.1 degrees or, equivalently, $\sim 13 \times 29$ light years of interstellar space. The bottom image shows the same region as seen in optical light. In the optical image, the Horsehead appears as a quiescent dark cloud silhouetted against the pinkish glow of hydrogen gas. In contrast, the CO image reveals the turbulent nature of the cold, dense gas from which the Horsehead and its parent cloud are made. The color-coding of the CO image has been chosen to reflect the motions of the gas in the cloud: blue/red portions have velocities directed more toward/away from the observer, respectively (by about 1 km/s), compared to the average velocity of the cloud (emission coded in green). The data was taken in December 2014 with the SuperCam, 64 pixel, heterodyne array receiver mounted on the APEX telescope, located on the Chajnantor Plateau (5100 m elevation) in northern Chile.

The CO image is courtesy of Thomas Stanke (European Southern Observatory).

The optical image is courtesy of Adam Block (Steward Observatory).

1

THE INTERSTELLAR MEDIUM (ISM) AT TERAHERTZ (THz) FREQUENCIES

PROLOGUE

The terahertz (THz) portion of the electromagnetic spectrum ($\nu \sim$ 1–10 THz; $\lambda \sim$ 300–30 μm) provides us with a powerful window into cosmic evolution. THz photons arriving at Earth can yield valuable insights into everything from the birth and death of stars to the cataclysmic events associated with the origin of galaxies and the Universe itself. Most of the THz photons we observe are emitted by the gas and dust between the stars, that is, the interstellar medium (ISM). At THz frequencies we can observe photons associated with the ISM of our own galaxy, the Milky Way, as well as from the ISMs of distant galaxies. In this chapter, we will examine the ISM and introduce some of the THz observational probes that can be used in discerning its nature.

1.1 INTRODUCTION

Today the interstellar medium (or ISM) accounts for only ~1% of the total gravitational mass of the Milky Way, with the balance being in stars (~6%) and dark matter (~93%). However, in the beginning, most of the galaxy's baryonic mass (everything made from protons and/or neutrons) was in ISM. The mass and energetics of the ISM were and continue to be the principal drivers of galactic evolution. Indeed, we owe our very existence to violent processes occurring in the local ISM ~4.6 billion years ago. These violent events are what led to the formation of our solar system. Before then, every atom that makes up the Earth and us (!) was part of the ISM, floating in space, the ash of ~1000 generations of stellar birth and death.

Energy sources within the ISM include gravity, cosmic rays (high energy particles), stellar winds, magnetic fields, and electromagnetic (EM) radiation. Sources of EM radiation include the cosmic background (i.e., Big Bang, peaking at microwave/GHz or gigahertz frequencies), photospheres of stars (mostly at ultraviolet (UV)/optical/infrared (IR) wavelengths), thermal dust emission (peaking in the far-IR/THz frequencies), emission lines of atoms and molecules (found across the entire spectrum), and supernovae (emitting from gamma ray through radio frequencies).

The evolution of THz astronomy has been driven largely by two factors: (1) atmospheric absorption of THz light and (2) the availability of detector technology. Water vapor in the Earth's atmosphere is a very efficient absorber of THz photons (see Figure 1.1). Therefore, THz observations are best conducted from space-based telescopes, balloon-borne telescopes, airborne observatories, or telescopes at high, dry, and cold sites on Earth.

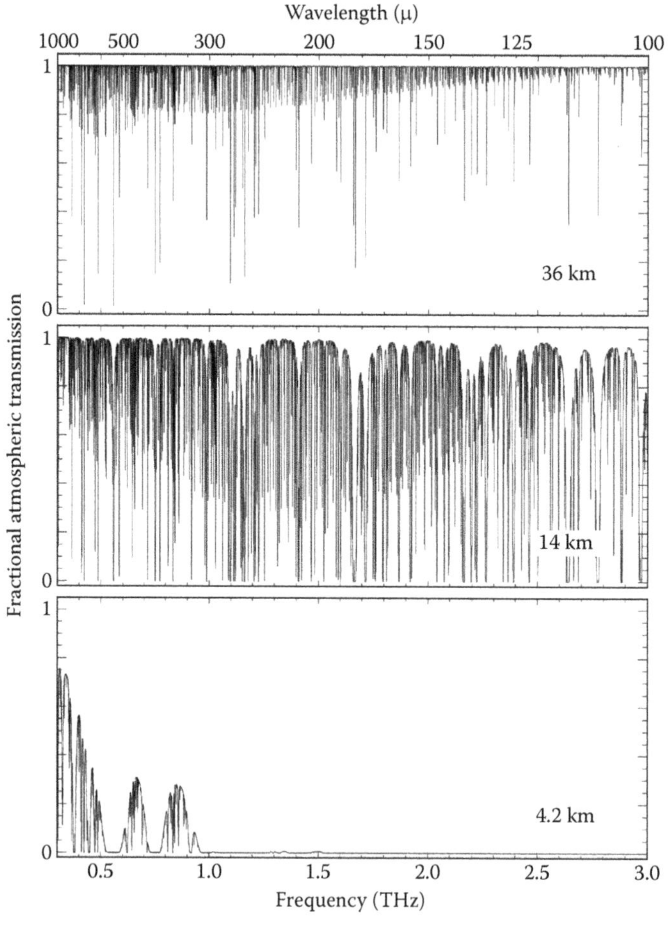

FIGURE 1.1 THz atmospheric transmission for typical weather conditions from mountaintop site (4.2 km), airborne altitudes (14 km), and balloon-borne altitudes (32 km). (Atmospheric model courtesy G. Melnick; see also Melnick (1988).)

The history of THz detector technology (see Appendix A) can be traced back to the nineteenth and early twentieth century, with the work of Maxwell, Hughes, Hertz, and Bose (among others), and from the latter half of the twentieth century, with the development of the gallium doped germanium bolometer (Low, 1961), the millimeter-wave Schottky diode mixer (Burrus, 1963; Young and Irvin, 1965; Wilson et al., 1970), the superconductor–insulator–superconductor (SIS) mixer (Dolan et al., 1979; Richards et al., 1979), first narrow-band then wide-bandwidth hot electron bolometer (HEB) mixer (Phillips and Jefferts, 1973; Gershenzon et al., 1990; Gol'tsman et al., 1995), the transition edge sensor (TES) (Irwin et al., 1995), and, more recently, the microwave kinetic inductance detector (MKID) (Day et al., 2003). Mixers and MKIDs require low-noise amplifiers and spectrometers to read them out. Breakthroughs in these areas came with the development of the first digital autocorrelator (Weinreb, 1961) and low-noise, solid-state amplifiers (Weinreb, 1980). Research in device, spectrometers, and local oscillator (LO) technologies has continued and accelerated in recent years, leading to the realization of a new generation of powerful detector systems with which to explore the THz universe.

1.2 ISM COMPONENTS OF THE MILKY WAY

Our Milky Way is a barred spiral galaxy (Class SBc) with a disk approximately 100,000 light years (ly) in diameter and ~1000 ly thick, containing $\approx 8.5 \times 10^{11}$ solar masses of material (Peñarrubia et al., 2014). The central 10,000 ly forms a bulge containing a central bar in the plane of the disk. The solar system is located about 27,000 ly from the galactic center on the inner edge of the Orion–Cygnus Arm (see Figure 1.2a). The Milky Way contains ~400 billion stars, many of which are believed to harbor one or more planets. The rotation curve of the galaxy is flat outside the bulge, with an orbital velocity of 220 km/s. The non-Keplerian nature of the rotation curve was one of the first indications that "dark matter," in the form of an extended halo, makes up the majority of the Milky Way's mass. It takes the Sun 240 million years to complete one orbit of the Milky Way.

Figure 1.2b is an optical photograph of the Milky Way taken on a clear night from the summit of Mt. Graham, Arizona. The Milky Way is seen edge-on as a broad diffuse band of light about 30° wide, stretching from horizon to horizon. The imaginary arc that cuts through the center of the band is the galactic plane. It has an inclination of ~60° to the plane of the solar system (i.e., the ecliptic plane). In order to facilitate the study of the galaxy, a system of galactic coordinates is defined, with l representing galactic longitude and b galactic latitude. In this system, the galactic center is located at (l,b) of (0,0).

All stars seen in Figure 1.2b are within our Milky Way. The stars outside the galactic plane are in the foreground between us and the more distant parts of the galaxy. The dust component of the ISM can be seen as dark clouds of heavy extinction within the galactic plane. The brightest part of the Milky Way is toward the galactic center in the top-left of the photograph. The band of dark, overlapping clouds trailing off diagonally down to the right from the galactic center is referred to as the "Great Rift." These clouds are approximately 300 ly away and contain ~1 million solar masses of material. As we will discuss in Chapter 3, the dark clouds mark the location of large amounts of dust and gas, with the gas mass surpassing the dust mass by a ratio of ~100 to 1. These clouds of gas and dust are often referred to as giant molecular clouds (GMCs) and are believed to have formed from the interaction

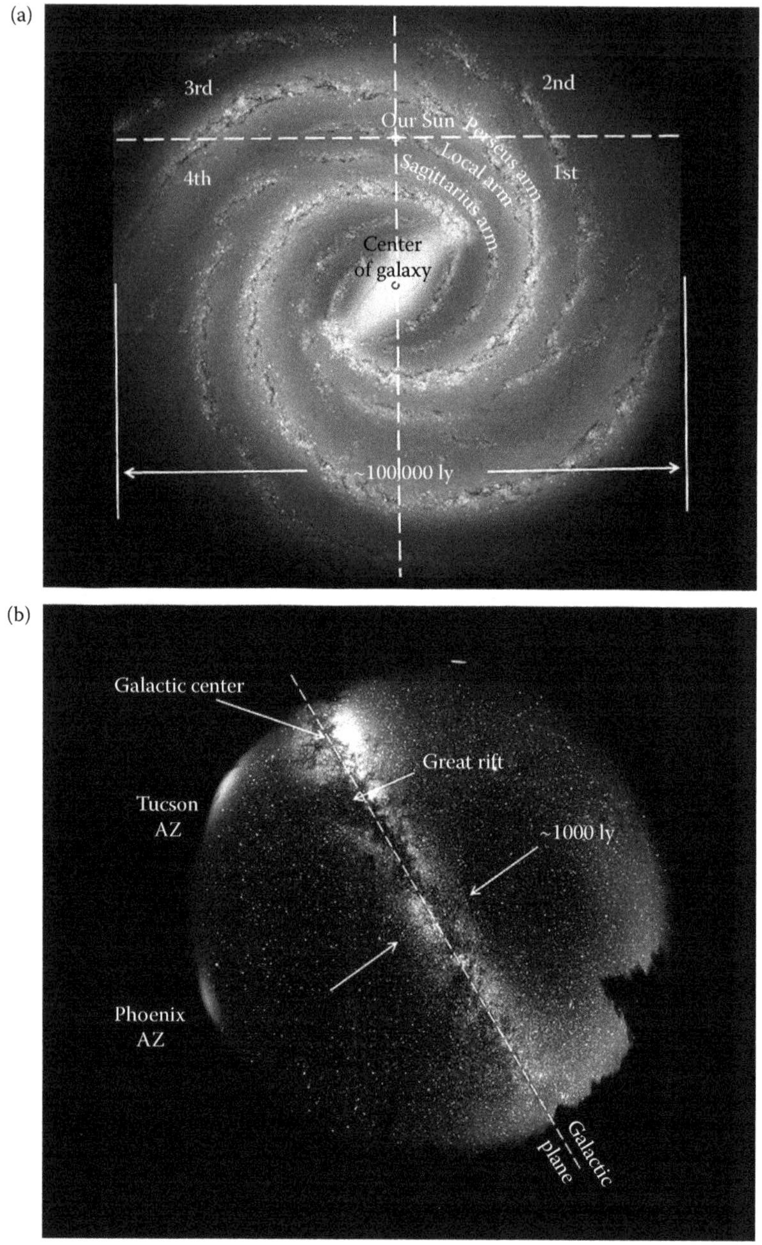

FIGURE 1.2 (a) Face-on view of the Milky Way. Our galaxy is a Class SBc, barred-spiral galaxy approximately 100,000 ly in diameter, containing ~400 billion stars. The Sun is located ~27,000 ly from the center. The central bulge or nucleus of the galaxy containing the bar is ~10,000 ly in diameter. The galaxy is divided into four quadrants, with our solar system at the origin. (Underlying Milky Way image credit: NASA https://solarsystem.nasa.gov/multimedia/display.cfm?IM_ID=8083: 19 Jan 2015.) (b) Edge-on view of a portion of the Milky Way as seen from the summit of Mt. Graham, Arizona. The galactic plane appears ~30° in width, and is the home of numerous, overlapping interstellar dust clouds (e.g., the Great Rift), clear evidence of an extensive interstellar medium (ISM). The galactic center is along a line of sight toward the brightest region of the Milky Way. (Underlying Milky Way photo: Copyright 1997, Steward Observatory, The University of Arizona. With permission.)

of shock waves from supernovae explosions (Mac Low and Klessen, 2004). GMCs can be ~15 to 600 ly across and can contain 10^3 to 10^7 $M_\odot$ (i.e., solar masses) of gas and dust. GMCs are the most massive objects in the galaxy. The dust component of the clouds consists of a staggering number of micron-size carbon and/or silicon grains belched out from the atmospheres of dying stars. These dust grains are what produce the visual extinction and also the thermal emission seen toward GMCs. The gas is predominantly molecular hydrogen, H_2. Unfortunately, due to its symmetric nature, H_2 does not have observable transitions at THz frequencies. However, GMCs also contain trace amounts of carbon, oxygen, nitrogen, sulfur, and other gases. Deep within the GMCs the column density of H_2 and dust is sufficient to block the far ultraviolet (FUV) light (6 eV < $h\nu$ < 13.6 eV) from background stars, and permit the formation of a variety of molecules; some simple, like CO, O_2, and H_2O, and some complex, like ethanol and formaldehyde. These atomic and molecular gases play a key role in cooling the ISM, and are the primary tools used by THz astronomers to probe physical conditions within the ISM. Regions of the ISM where FUV photons dominate the composition and energetics are called PhotoDissociation Regions (PDRs). All of the atomic and most of the molecular gas in the Milky Way are in PDRs (Hollenbach and Tielens, 1997). For some other galaxies, this might not be the case. In practice, the regions associated with diffuse, atomic, neutral hydrogen HI emission within the ISM are rarely referred to as PDRs.

The ISM is a violent, dynamic place (see Figure 1.3). However, its composition and energetics leads to the establishment of characteristic phases or submediums (Wolfire et al., 2003, 2010; Draine, 2011). These submediums include the: hot ionized medium (HIM); warm neutral medium (WNM); warm ionized medium (WIM); and the cool neutral medium (CNM).

1.2.1 HOT IONIZED MEDIUM (HIM)

The HIM is very low-density (~0.01 to 0.001 proton/cm^3) ionized gas superheated by supernova blast waves to temperatures ≥$10^{5.5}$ K. It constitutes ~50% (or more) of the ISM's volume and fills in the voids between denser components. The HIM forms an extended "atmosphere" above and below the galactic plane that gradually merges with the even lower density intergalactic medium (IGM).

1.2.2 WARM NEUTRAL MEDIUM (WNM)

The WNM is predominately composed of neutral atomic hydrogen with a temperature and density of $T_H \sim 10^{3.7}$ K and $n_H \sim 0.5$ cm^{-3}, respectively. It constitutes approximately 40% of the ISM volume.

1.2.3 WARM IONIZED MEDIUM (WIM)

The WIM is gas photoionized by the extreme ultraviolet (EUV) and soft x-ray (13.6 eV < $h\nu$ < 100 eV) photons from massive O and B stars. Often, the ionizing stars are found in or next to the dense molecular clouds from which they are formed. The FUV photons from the stars photodissociate nearby molecular gas, and the EUV photons ionize the atomic hydrogen, leading to the formation of a "blister HII (singly ionized hydrogen) region" and dense PDR (e.g., the Orion Nebula). The ionizing EUV photons are free to stream through interstellar space in directions not bounded by a cloud, and create extended,

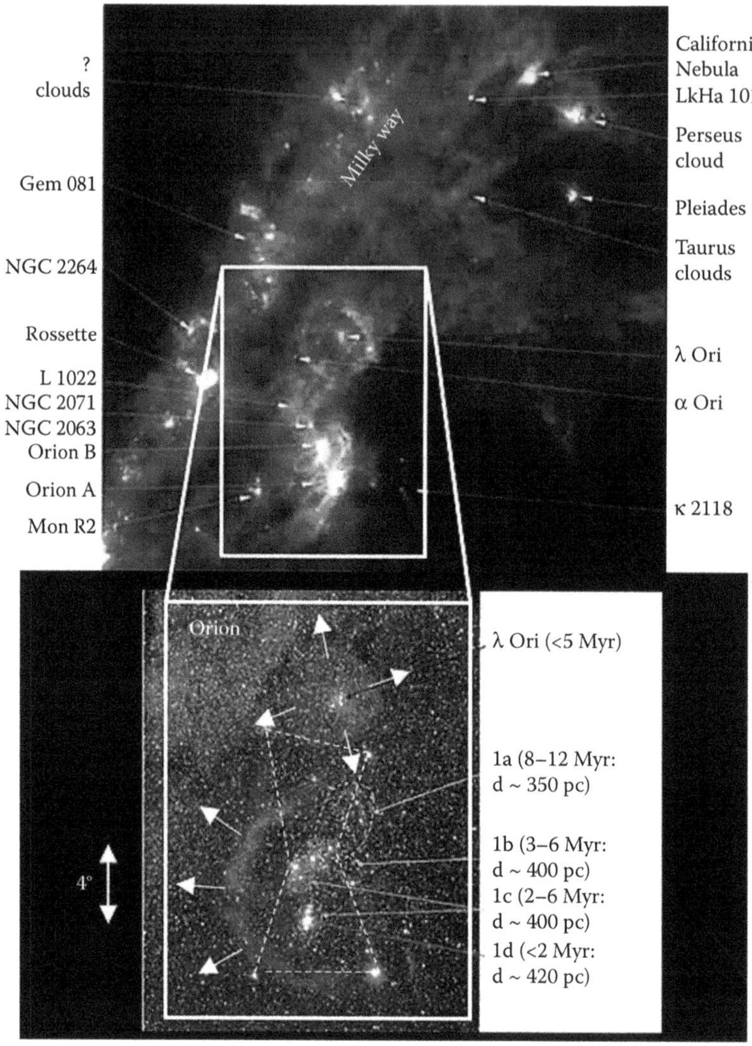

FIGURE 1.3 The dynamic ISM. (Top) Far-infrared (~3 THz) image of Milky Way thermal dust emission taken in the vicinity of Orion by the IRAS satellite. (Bottom) The Orion complex (shown here at optical wavelengths), at a distance of ~390 pc, has given birth to $>10^4$ stars over the past 12 Myr. The Orion Nebula itself (corresponding to 1d) has a diameter of ~24 ly. The chaotic nature of the region is due to the combined effects of UV radiation, stellar winds, and past supernovae. Expansion velocities are ~5 km/s. (Adapted from Bally, J., 2008, *Handbook of Star Forming Regions Vol. I* ASP Conference Series, Bo Reipurth. With permission.)

low-density ionized regions. These low-density ionized regions are estimated to take up ~10% of the Milky Way's volume.

1.2.4 COOL NEUTRAL MEDIUM (CNM)

The vast majority of the CNM is in the form of neutral atomic hydrogen clouds with a characteristic temperature and density of $T_H \sim 10^2$ K and $n_H \sim 30$ cm^{-3}. These are the clouds often observed in the 1420 MHz (megahertz) ($\lambda = 21$ cm) fine structure emission line. The density in some regions of these clouds may be sufficiently high to form diffuse molecular hydrogen, H_2, clouds. The CNM accounts for just ~1% of the volume of the Milky Way, with the diffuse H_2 clouds accounting for only 0.1%.

1.2.5 COLD DENSE MOLECULAR CLOUDS (CDM)

The CDM represents molecular cloud clumps or cores with temperatures, $10 \leq T_{H2} \leq 50$ K and densities, $10^3 \leq n_{H2} \leq 10^6$ cm^{-3}, respectively. It is in these regions that stars are formed. GMCs typically contain one or more CDM clumps. The interclump medium within a GMC consists largely of lower density (~100 cm^{-3}) H_2. The CDM accounts for just 0.01% of the volume of the Milky Way.

1.2.6 RELATIONSHIP BETWEEN ISM PHASES

The ISM phases are listed in Table 1.1 along with their galactic volumetric filling factors, f_V, characteristic densities, n_H, and phase temperatures, T_P. Figure 1.4 is a simplified representation of the Milky Way ISM, with each phase represented roughly to scale. An image of M33, showing the interplay between the WNM (as traced by HI) and the CNM/CDM (as traced by CO) is shown in Figure 1.5.

The observed phase temperatures are the end product of radiative, collisional, and magnetic interactions within the ISM. Therefore, we can gain an understanding of the relative energy content of the phases by computing the thermal energy, E_P^{th}, within each. Let $\langle n_P \rangle$ represent the average galactic density within a given phase, such that

$$\langle n_P \rangle \propto f_V n_H \tag{1.1}$$

The total number of gas particles within each phase, N_P, is

$$N_P = \langle n_P \rangle V_G$$

where
V_G = the volume of the Milky Way (cm^3).

The corresponding thermal energy within each phase is then

$$E_P^{th} \approx \frac{\alpha}{2} N_P k T_P \tag{1.2}$$

TABLE 1.1 ISM Phase Properties

Phase	f_V	n_H (cm^{-3})	T_P (K)	% H Mass	% Thermal Energy in Each Phase
Coronal (HIM)	0.5	0.004	$\geq 10^{5.5}$	~0.24	~34
HII (WIM)	0.1	0.3–10^4	10^4	~2.4	~11
Warm HI (WNM)	0.4	0.6	~5000	~24	~53
Cool HI (CNM)	0.01	30	~100	~37	~2
Diffuse H_2 (CNM)	0.001	100	~50	~12	~0.3
Dense H_2 (CDM)	0.0001	10^3–10^6	10–50	~24	~0.4

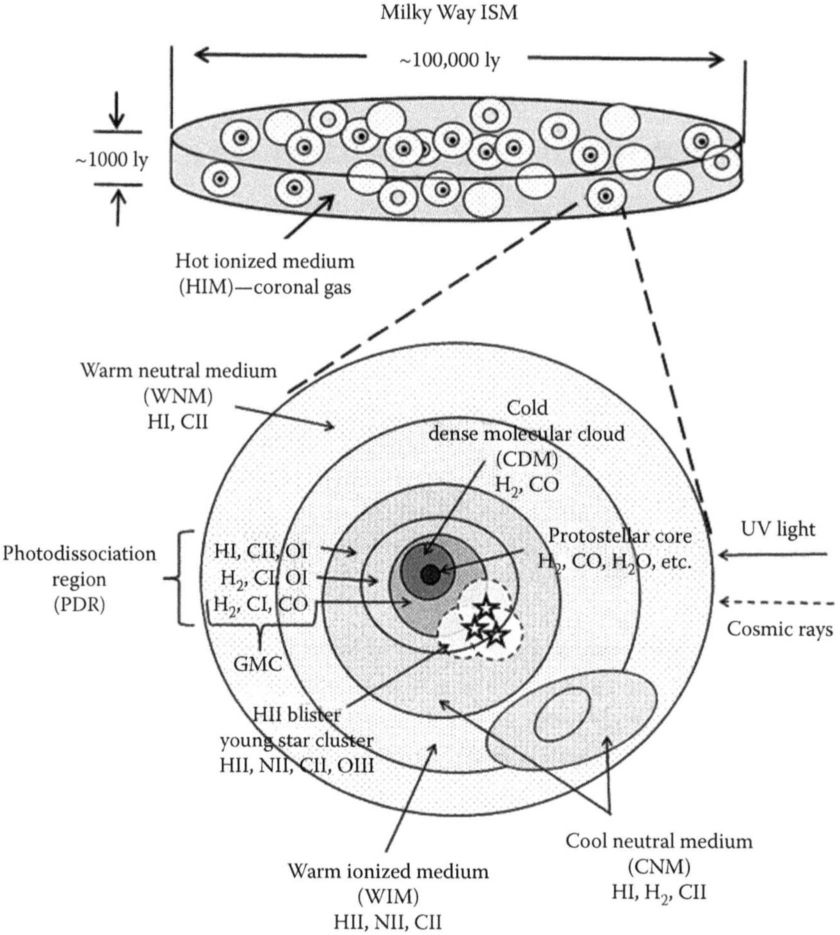

FIGURE 1.4 Schematic of the Milky Way interstellar medium (ISM). The composition and energetics of the ISM lead to the establishment of several phases or submediums, some in equilibrium, and others in a constant flux. The phases are drawn in approximate proportion to their volumetric filling factor. The hot ionized medium (HIM—coronal gas) takes up half or more of the Milky Way's volume, followed in relative volume by the warm neutral medium (WNM—warm HI gas), warm ionized medium (WIM—which includes dense and diffuse PDRs), cool neutral medium (CNM—cool HI clouds), and last, but not least, the cold dense medium (CDM—within GMCs), from which all stars are born.

where

E_P^{th} = thermal energy in a phase (ergs)
N_P = number of particles in each phase
k = Boltzmann's constant = 1.38×10^{-16} (ergs K^{-1})
T_P = phase temperature (K)
α = number of degrees of freedom (3 for monatomic and 5 for diatomic)

All but the CDM phase are dominated by atomic hydrogen, and will have a value of $\alpha \sim 3$. The CDM phase, for which the constituent particles are primarily H_2, will have an

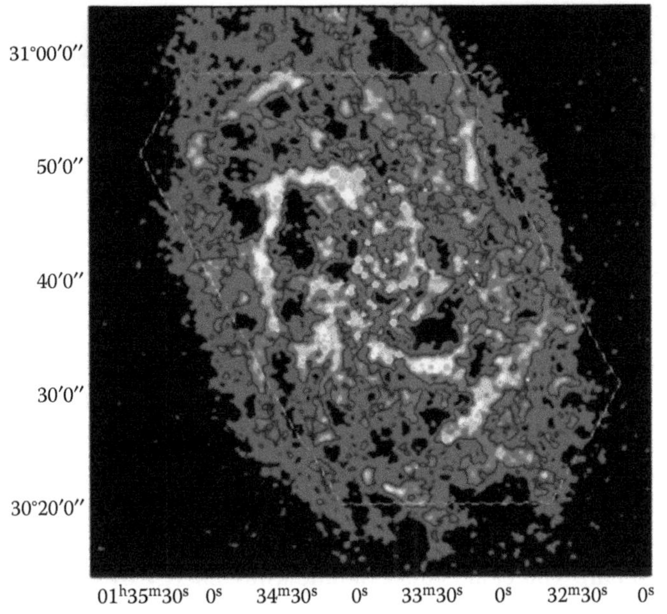

FIGURE 1.5 Relationship between the WNM, CNM, and CDM in M33. The extended emission is the atomic WNM as traced by HI (Deul and van der Hulst, 1987). The circles are regions of molecular emission as traced by CO (see analogous structures in Figure 1.4). CO traces CNM clouds and CDM clumps that form in regions of HI overdensity. (From Engargiola, G. et al., 2003, *Ap. JS.*, 149, 343. With permission.)

$\alpha \sim 5$. The last column of Table 1.1 lists the relative percentage of energy within each ISM phase, E_P^{th}, as derived using Equation 1.2.

Representative "pie-charts" of the ISM thermal energy and baryonic mass partitions within the Milky Way are shown in Figure 1.6. Examination of the figure reveals that cool HI (CNM) dominates the baryonic mass, while the warm HI dominates the energy. It is interesting to note that the coronal gas (HIM) that takes up most of the galaxy, contributes only a tiny bit to the overall mass, but is second only to the WNM in terms of energy content.

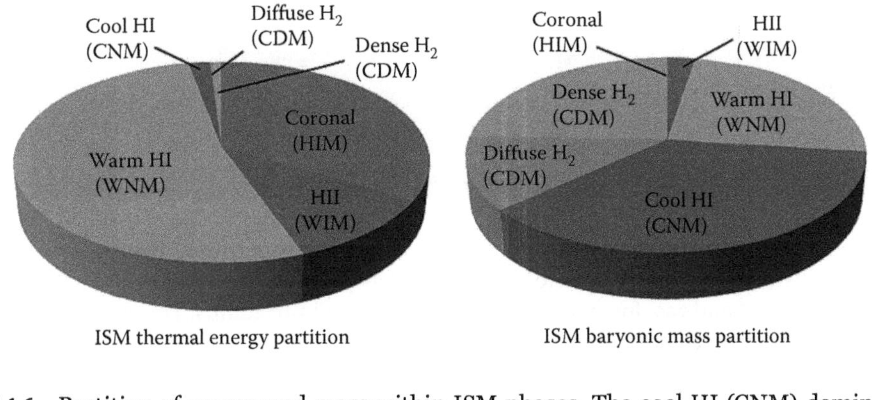

FIGURE 1.6 Partition of energy and mass within ISM phases. The cool HI (CNM) dominates the mass, while warm HI (WNM) dominates the energy.

1.3 LIFECYCLE OF THE ISM

Even after half a century of study, key questions about the evolution of the ISM in the Milky Way and external galaxies remain.

- ∞ How and where are interstellar clouds made, and how long do they live?
- ∞ Under what conditions and at what rate do clouds form stars?
- ∞ How do stars return enriched material back to the galaxy?
- ∞ How do these processes sculpt the evolution of galaxies?

These questions can be presented in the form of a lifecycle for the ISM, in which each of the phases described above plays a role (see Figure 1.7). At the top of the cycle, overlapping spherical supernova shockwaves interfere constructively to produce local over-densities of warm, neutral, and atomic hydrogen clouds (WNM). The WNM clouds are converted to cool HI clouds (CNM) through a thermal instability and phase transition. These clouds are most often observed through the collisionally excited, hyperfine, spin-flip transition of hydrogen at 1420 MHz ($\lambda \sim 21$ cm). The clouds also radiate away energy by emission in the ionized carbon fine structure line ([CII]). At densities >10^5 cm^{-3}, gas thermal energy can also be

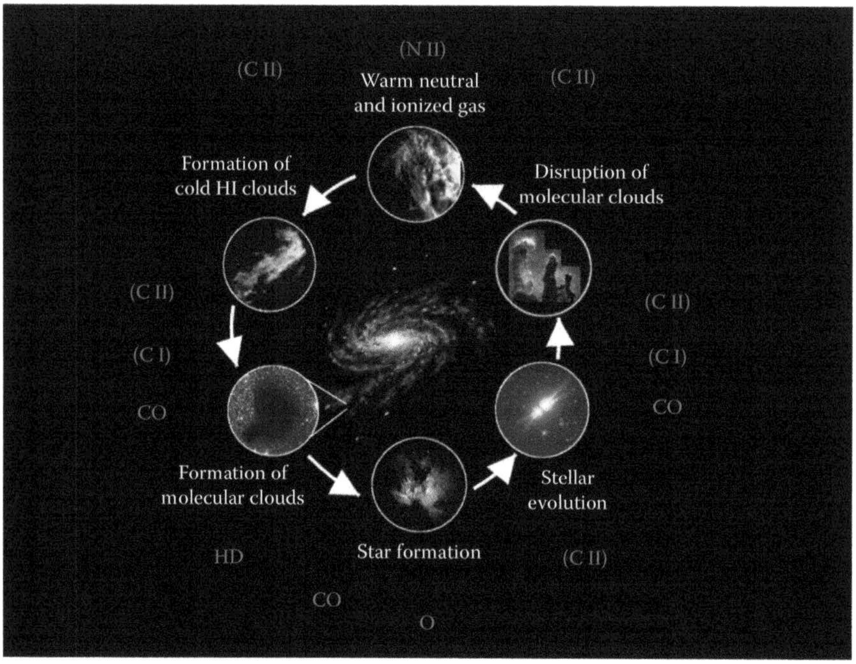

FIGURE 1.7 Lifecycle of the ISM. Starting at the top, overlapping supernovae shock waves sweep up material and constructively interfere to form warm ionized/neutral atomic clouds. The clouds cool by radiation from dust and gas, form molecules, and fragment into clumps. The clumps continue to radiatively cool, and then collapse under their own weight to form stars (bottom of cycle). Through stellar radiation, winds, and shockwaves, the parent cloud dissipates, returning raw materials back to the diffuse ISM. (Adapted from Kulesa, C., 2011, *IEEE Trans. Terahertz Sci. Technol.*, 1(1), 232. With permission.)

efficiently radiated away by dust grains. This radiation causes the kinetic temperature (the temperature due to the microscopic interactions of gas particles) to decrease proportionally. When the number density of hydrogen molecules within a cloud becomes high enough, and the gas kinetic temperature low enough, gravity overwhelms the expansion effects of thermal energy, and a cloud contracts under its own gravity. This situation was first described by Jeans (1904). The mass of a cloud beyond which it will collapse due to self-gravity is called the Jeans mass, M_J. An expression for M_J can be derived from the virial theorem.

The virial theorem states that, for a system of particles to be gravitationally bound, it must possess at least twice as much gravitational potential energy, U, as it does kinetic energy, K, such that

$$\begin{aligned} 0 &= 2K + U \\ K &= -\frac{1}{2}U \end{aligned} \tag{1.3}$$

If we assume the kinetic energy is dominated by thermal motions within the gas, then

$$K = N\left(\frac{3}{2}kT\right), \quad \text{for } N = \frac{M_J}{m} \tag{1.4}$$

where

N = number of particles in the gas
M_J = Jeans mass (g)
m = mean mass per particle (gm)
k = Boltzmann's constant (1.38×10^{-16} ergs K^{-1})
T = gas temperature (K)

The gravitational potential energy of the cloud is given by

$$U = -\frac{3}{5}\frac{GM_J^2}{R} \tag{1.5}$$

where

G = gravitational constant (6.674×10^{-8} $cm^3\ g^{-1}\ s^{-2}$)
R = radius of cloud core (cm)

Assuming the cloud is spherical, with a uniform density, its radius is related to its mass and density through the relation,

$$R = \left(\frac{3M_J}{4\pi\rho}\right)^{1/3} \tag{1.6}$$

where

ρ = gas mass density (g cm^{-3}).

Substituting the above expressions into Equation 1.3 and solving for M_J we find

$$M_J = \left(\frac{5kT}{Gm}\right)^{3/2}\left(\frac{3}{4\pi\rho}\right)^{1/2} \tag{1.7}$$

The expression for the sound speed, c_S, in a gas is

$$c_s = \left(\frac{\gamma kT}{m}\right)^{1/2} \tag{1.8}$$

where

c_S = sound speed in the gas (cm/s)
γ = adiabatic index (1.4 for diatomic and 1.7 for monatomic molecules)

Rearranging Equation 1.7 and substituting in the physical constants, we then have

$$M_J = 5.24\frac{c_S^3}{G^{3/2}\rho^{1/2}}\,\text{gm} \tag{1.9}$$

Gas clouds are composed largely of hydrogen. We can then rewrite Equation 1.9 in terms of hydrogen gas number density, n_{H} (cm^{-3}) as:

$$M_J \approx 2.4\times 10^{23}c_S^3 n_H^{-1/2}\,\text{gm} \tag{1.10}$$

Cloud contraction will continue as long as the cloud can efficiently radiate away its heat, that is, thermal energy. As the cloud becomes denser and denser, hydrogen atoms are able to provide some degree of "self-shielding," blocking the passage of FUV photons into the cloud's interior. With the FUV photons blocked, molecular hydrogen can form through an exothermic reaction on dust grains (Hollenbach and Salpeter, 1971). Heat released through the reaction is able to "pop" the newly minted H_2 molecules from the grain surfaces, and the cloud's interior transitions from atomic to molecular, advancing us to the next step in the ISM cycle (CDM). As the gas column density of hydrogen, *NH*, (i.e., the number of hydrogen atoms per unit area along a line of sight) continues to increase, so does the dust column density, N_d. Dust column density is often expressed in terms of magnitudes of visual extinction, A_V. Along each line of sight with one A_V of dust extinction, there is a corresponding gas column density of

$$N_H \approx 1.79\times 10^{21}\,\frac{\text{H atoms}}{\text{cm}^2} \tag{1.11}$$

Dust grains are very efficient absorbers of FUV photons. Once an $A_V \geq 1$ is achieved, more and more molecules are able to form. Models of the gas/dust chemistry and energetics were used by Hollenbach and Tielens (1997) to predict the intensity of atomic and molecular probes within and between ISM phases (see Figure 1.8). The probes include emission (or absorption) lines of carbon, oxygen, nitrogen, and carbon monoxide, *all of which fall within the THz frequency regime.*

We can derive a characteristic size scale for collapse within a cloud by substituting Equation 1.7 back into Equation 1.6. Doing so, we find,

$$R_J \propto \left(\frac{T}{\rho}\right)^{1/2} \tag{1.12}$$

The size scale over which a cloud becomes self-gravitating, that is, the Jeans length (λ_J), is ≈2 R_J. Equation 1.12 tells us that as long as a collapsing cloud can radiate away its heat, the increase in density will cause the size scale, over which collapse can occur, to

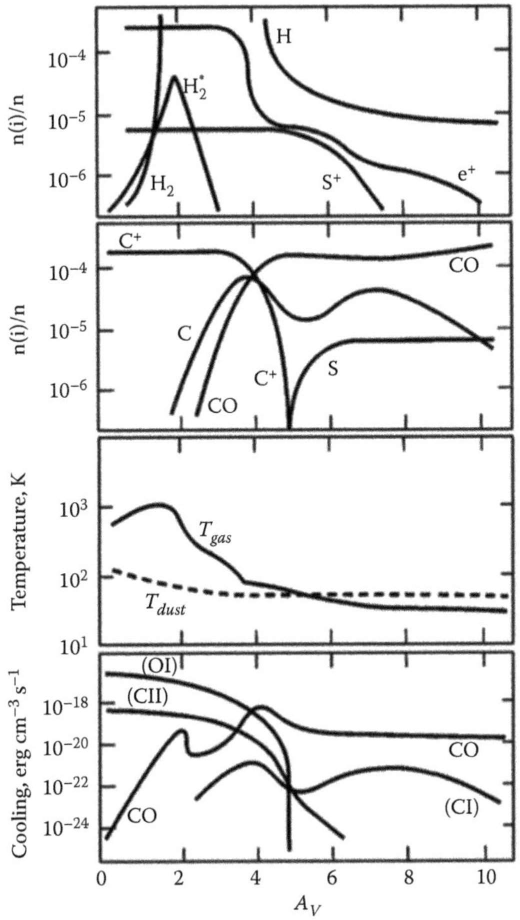

FIGURE 1.8 PDR structure in Orion. Calculated structure of the PDR in Orion for density of $n_{H2} = 2.3 \times 10^5$ cm^{-3}, $G_o = 10^5$ as a function of visual extinction, A_v, into the PDR (Tielens and Hollenbach, 1985b). The ionizing source is to the left. Top two panels: Abundances relative to total hydrogen. Third panel: Gas and dust temperatures. Bottom panel: Cooling in the various gas lines. (Reprinted from Hollenbach, D. J. and Tielens, A. G. G. M., 1995, *The Physics and Chemistry of Interstellar Molecular Clouds*, G. Winnewisser and G. Pelz (ed.), Springer, Berlin, 164. With permission.)

continually drop, making an extended cloud begin to fragment (Spitzer, 1978). The fragmentation process will cause the cloud to break into denser clumps, and these clumps to break into even denser cloud cores. The prestellar cores can have densities of >10^4 H_2 cm^{-3} and, depending on prevailing cloud conditions (e.g., ambient magnetic fields and rotational support), may or may not be stable against further collapse. If a core is unstable, then collapse may continue from the inside out (Shu, 1977), until the core becomes optically thick to its own thermal radiation. At this point, the core begins to heat up, until its thermal energy achieves equilibrium with gravitational energy, and the initial infall is halted, with central molecular hydrogen densities of $n_{H2} \approx 10^{10}$ cm^{-3}. What was a prestellar core is now a protostar. A protostar is a young star in the process of accreting the bulk of the mass it will have on the main sequence. Warmer cloud cores require more gravity to overcome thermal energy, so they tend to form high-mass stars. Lower temperature cores tend to form lower-mass stars. At this point, we are at the bottom of the ISM lifecycle of Figure 1.7.

During the star formation process, rotational momentum is conserved, and the protostar spins up as it contracts. Magnetic fields, partially frozen into the gas through interaction with ions, may produce some degree of "magnetic braking," and serve to slow the rotation. One important byproduct of rapid core rotation is the creation of a disk (much like the formation of a pizza crust by spinning a ball of dough). Once formed, the young protostar will continue to accrete material from its parent cloud through the disk. The infall process can be observed at THz frequencies through the distorted line profiles of optically thick emission lines observed along lines of sight to the core (see Chapter 2). The resulting disk accretion shock will cause the release of copious amounts of FUV/EUV photons that will ionize the surrounding ISM. In an effort to transfer rotational angular momentum form the disk back to the ISM, the protostar will form a bipolar outflow, consisting of ~100 km/s jets of gas that can extend for light years from each pole (Shu et al., 2000). These jets can be seen at both optical (see Figure 1.9) and radio wavelengths. At THz frequencies the jets manifest themselves as wing emission in molecular and/or atomic lines (e.g., carbon monoxide, CO, and neutral atomic carbon, [CI]) that trace cloud material swept up by the jets (see Snell et al., 1980—for a single protostar; Walker et al., 1990—for a protobinary system; Walker et al., 1993—for [CI]). Later (~10^5 years), as accretion subsides, so will the bipolar outflow. What was once the accretion disk will cool and become a protoplanetary disk. Disk evolution can best be observed at THz frequencies using the high angular resolution provided by an interferometer (e.g., ALMA—the Atacama Large Millimeter Array), or from a single dish telescope, through the use of high excitation lines (Walker et al., 1994; Williams and Cieza, 2011). The star will continue to pump FUV and EUV photons into the ISM as a byproduct of nuclear fusion. The amount of ionizing flux produced by a star increases exponentially with its mass. Not surprisingly, it is the massive O and B stars that are responsible for the majority of the ionizing and FUV flux observed in the ISM. The EUV flux may cause the formation of a "blister HII" region around the young star (e.g., the Orion in Figure 1.3). Within the centers of stars, the raw ISM material is enriched with heavy elements through nuclear fusion. High-mass stars go supernova, giving back most of their mass to the ISM, with further enrichment occurring in the maelstrom of the supernova process itself. The combination of fragmentation, UV flux, and stellar winds (both from within and without) will serve to disrupt the parent cloud back into the WNM and WIM components, completing the ISM lifecycle.

How long the cycle takes to complete is subject to environmental conditions and debate, but is estimated to be between 5 and 10 million years (Ballesteros Paredes et al., 1999;

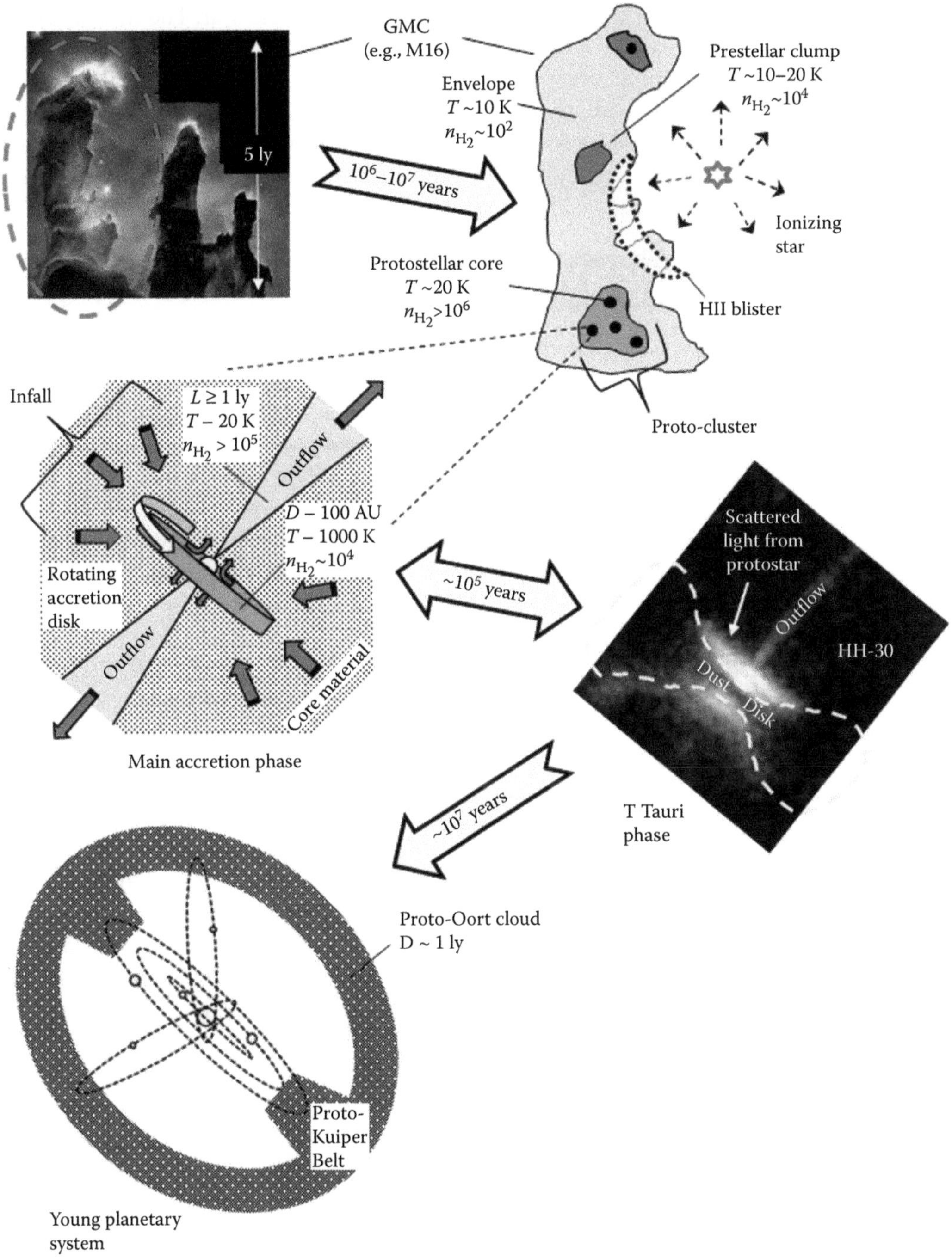

FIGURE 1.9 Protostellar evolution. In just ~10^7 years a young planetary system can form from the hard vacuum of interstellar space, providing a potential island for life in a vast cosmic sea. (Image credits: M16: NASA, Jeff Hester, and Paul Scowen; HH-30: NASA Hubble. With permission.)

Mordecai-Mark Mac Low, 2004), that is, the lifetime of a high-mass star. The time it takes for a prestellar core to evolve to a young star with a protoplanetary disk is only ~10^5 years. The time it takes for the protoplanetary disk to cool and evolve into a planetary system is estimated to be ~10^7 years. All in all, the time it takes for a planetary system to form from the WIM/WNM can be as little as ~0.1% of the lifetime of a solar mass star. The key milestones in protostellar evolution are depicted in Figure 1.9.

1.4 PROBING THE LIFECYCLE OF THE ISM

Driven by nucleosynthesis in stars, the three most abundant elements in the Universe after hydrogen and helium are oxygen, carbon, and nitrogen. With respect to hydrogen, their relative abundances are 6×10^{-4}, 3×10^{-4}, and 1×10^{-4}, respectively (Stacey, 2011). In diffuse gas, about half the oxygen and carbon are lost from the gas phase in grains. Neutral, ionized, and molecular manifestations of these elements play a central role in the ISM and have THz transitions that provide a means of exploring the underlying chemistry and physics of the ISM. Key spectroscopic probes associated with each ISM phase are shown in Figures 1.4 and 1.6. Their THz transitions and frequencies are listed in Table 1.2. Since these transitions occur at THz frequencies, the photons being emitted or absorbed are relatively large (~100–1000 μm) compared to the size of a typical interstellar dust grain (~0.1 μm), and, therefore, do not suffer the dust extinction effects common to UV, visible, and even IR studies. Some transitions in Table 1.2 are from neutral molecules (e.g., CO). Many are associated with fine-structure lines of atoms or ions.

1.4.1 ORIGIN OF FINE-STRUCTURE LINES

When one thinks of atomic emission or absorption lines, what usually comes to mind are electrons moving from one energy shell to another, about an atomic nucleus. Such motions involve significant changes of energy, ΔE, which results in the emission or absorption of a photon at optical or ultraviolet wavelengths. The gross structure of these energy levels is described by the principle quantum number, n, which indicates the distance of the energy level from the nucleus (i.e., the size of the shell). In addition to orbital distance, the electron orbital angular momentum, $\mathbf{l}_i$, and spin angular momentum, $\mathbf{s}_i$, are also quantized. Summing over all electrons within the atom, we have total spin and angular momentum quantum numbers, $\mathbf{S} \equiv \Sigma \mathbf{s}_i$ and $\mathbf{L} \equiv \Sigma \mathbf{l}_i$. The interaction of **S** and **L** split the energy shell of a given **L** into subshells. Transitions between these subshells are relatively low-energy and are the source of THz fine-structure lines. These transitions are denoted by $^{2S+1}\mathrm{P}_J$, where the superscript $2S + 1$ is equivalent to the total number of levels into which the spin term is split, P denotes the total orbital angular momentum (e.g., S for $\mathbf{L} = 0$, P for $\mathbf{L} = 1$, and D for $\mathbf{L} = 2$..), and **J** denotes the magnitude of the vector sum of **L** and **S**: $\mathbf{J} \equiv \mathbf{L} + \mathbf{S}$ (see Figure 1.10). Furthermore, each **J** level has a degeneracy given by $g_J = 2J + 1$. All the fine-structure lines in Table 1.2 are either ^{2}P or ^{3}P, which indicates that the associated atom or ion has either 1, 2, 4, or 5 p electrons in its ground state configuration (Stacey, 2011). Since these transitions do not result in a change of electronic configuration, they are classified as "forbidden lines," meaning they radiate through less efficient magnetic dipole transitions, not through electric

TABLE 1.2 Bright THz Spectral Lines

Species ⇒ Line[a]	Trans.	E.P. (K)[b]	ν (THz)	λ (μm)	A (s^{-1})	n_{H_2} (cm^{-3})[c]
O^0 ⇒ [OI]	$^3P_1 \rightarrow {}^3P_2$	228	4.745	63.18	9.0×10^{-5}	4.7×10^5
	$^3P_0 \rightarrow {}^3P_1$	329	2.060	145.53	1.7×10^{-5}	9.4×10^4
O^{++} ⇒ [OIII]	$^3P_2 \rightarrow {}^3P_1$	440	5.786	51.82	9.8×10^{-5}	3.6×10^3[d]
	$^3P_1 \rightarrow {}^3P_0$	163	3.393	88.36	2.6×10^{-5}	510[d]
C^+ ⇒ [CII]	$^2P_{3/2} \rightarrow {}^2P_{1/2}$	91	1.901	157.74	2.1×10^{-6}	2.8×10^3 50[d]
N^+ ⇒ [NII]	$^3P_2 \rightarrow {}^3P_1$	188	2.459	121.90	7.5×10^{-6}	310[d]
	$^3P_1 \rightarrow {}^3P_0$	70	1.461	205.18	2.1×10^{-6}	48[d]
N^{++} ⇒ [NIII]	$^2P_{3/2} \rightarrow {}^2P_{1/2}$	251	5.230	57.32	4.8×10^{-5}	2.1×10^3[d]
C^0 ⇒ [CI]	$^3P_2 \rightarrow {}^3P_1$	63	0.893	370.42	2.7×10^{-7}	1.2×10^3
	$^3P_1 \rightarrow {}^3P_0$	24	492.2	609.14	7.9×10^{-8}	4.7×10^2
^{12}CO	$J = 13 \rightarrow 12$	503	1.497	200.23	2.2×10^{-4}	2.5×10^6
	$J = 11 \rightarrow 10$	365	1.267	236.60	1.3×10^{-4}	1.4×10^6
	$J = 9 \rightarrow 8$	249	1.037	289.12	7.3×10^{-5}	8.4×10^5
	$J = 7 \rightarrow 6$	155	0.8067	371.65	3.4×10^{-5}	3.9×10^5
	$J = 6 \rightarrow 5$	116	0.6915	433.56	2.1×10^{-5}	2.6×10^5
	$J = 4 \rightarrow 3$	55	0.4610	650.76	6.12×10^{-6}	6.12×10^{-4}
	$J = 3 \rightarrow 2$	33	0.3458	867.55	2.5×10^{-6}	2.5×10^{-4}

Source: Data from Stacey, G. J., 2011, *IEEE Trans. Terahertz Sci Technol*, 1(1), 241. For a more extensive listing, see Appendix 1.

[a] Species refers to the type of gas, with O^0 indicating neutral, and O^+ and O^{++} indicating singly and doubly ionized states. The designations for spectroscopic lines use roman numerals to indicate the degree of ionization, with 'I' being neutral, 'II' being single ionized, and 'III' being doubly ionized. Brackets '[ ]' around a line name means it is forbidden to having electric dipole radiation.

[b] Excitation potential = equivalent thermal energy above ground state; for CO, $T_{min} \approx \dfrac{\nu h(J+1)}{2k}$.

[c] Density at which spontaneous radiative de-excitation and collisional de-excitation are equal; for CO, $n_{crit} \approx \dfrac{A_{UL}}{10^{-10}}$ for $T_{H_2} \sim 100\,\mathrm{K}$.

[d] Collision partners are electrons, otherwise H and H_2.

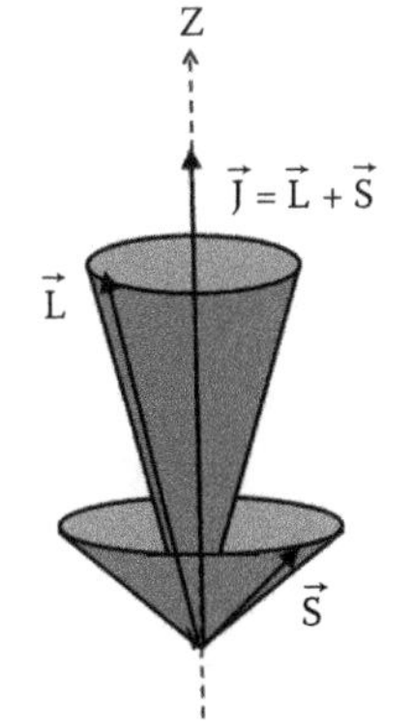

FIGURE 1.10 Atomic fine-structure lines originate through transitions between low-energy subshells generated through the quantization of the total angular momentum, **J**, due to interactions between orbital, **L**, and spin, **S**, electron angular momentum vectors.

TABLE 1.3 Elemental Properties

Element	Z	Abundance[a]	I → II[b]	II → III	III → IV
H	1	1	13.5984		
He	2	9.55×10^{-2}	24.5874	54.416	
C	6	2.95×10^{-4}	11.2603	24.383	47.888
N	7	7.41×10^{-5}	14.5341	29.601	47.449
O	8	5.37×10^{-4}	13.6181	35.121	54.936

[a] Protosolar relative abundance from Asplund et al. (2009).
[b] Ionization potentials (eV) from Ralchenko et al. (2010).

dipole transitions. Compared to the higher probability of an electric dipole transition, magnetic dipole transitions are metastable, with transition times measured in days.

The origin of the PDR emission line regions of Figures 1.4, 1.7, and 1.8 can be traced back to the ionization potentials of hydrogen, carbon, nitrogen, and oxygen (see Table 1.3). Hydrogen, being the most abundant element by far, permeates the ISM and acts as a "sink" for EUV/soft x-ray photons with energies ≥ its ionization potential, 13.6 eV. Elements in higher ionization states, such as singly ionized oxygen [OII], singly ionized nitrogen [NII], doubly ionized oxygen [OIII], and doubly ionized nitrogen [NIII], can only be found in regions ever closer to the ionizing star, where there is a sufficient number of high-energy photons to ionize the surrounding hydrogen, with enough left over for the oxygen and nitrogen.

1.4.2 IMPORTANCE OF [CII]

Amongst the most abundant elements, only carbon has an ionizing potential less than that of hydrogen. This means C^+ and its emission line [CII] can potentially be found in *both* ionized regions close to a star, where the visual extinction is $<5\,A_V$, and faraway in the otherwise neutral phases of the ISM. The [CII] emission line is the dominant coolant in diffuse neutral atomic gas (CNM phase), as well as molecular cloud surfaces in the atomic and molecular (H_2) layers before CO forms. In these regions, the heating is dominated by the photoelectric ejection of electrons from dust grains and molecules by FUV photons. If the density and temperature are sufficiently high ($n_{H2} > 10^4$, $T > 90$ K), as can be found in a dense PDR, the neutral atomic oxygen [OI] line can also contribute to gas cooling. The excitation of [CII] is not by direct collisions with the primary photoelectrons, but by the hydrogen and helium that are heated by thermalization of the photoelectrons. Also, the [CII] transition dominates in the CNM, where densities are only ~50 cm^{-3}, because there is no other atomic or molecular transition (except for neutral atomic hydrogen) that can be excited at such low densities. *The ubiquity and relatively easy excitation of the [CII] line makes it the most important spectroscopic probe of the ISM. [CII] is the diagnostic thread on which all phases of the ISM is strung* (see Figure 1.7). This is true not just for our Milky Way, but for all galaxies through cosmic time. *Indeed, the [CII] line can be the brightest single emission line from a galaxy as a whole, accounting for between 0.1% and 1% of the total far-infrared luminosity of the system* (Stacey et al., 1991). This means our Milky Way radiates ~70 million solar luminosities in this one line, while in distant hyper-luminous galaxies with $L_{Far\text{-}IR} > 10^{13} L_\odot$, ~10 billion solar luminosities ($L_\odot$) are emitted in this one line (Stacey et al., 2010)! In such hot, dense regions, the [OI] 63 μm and [OIII] lines may rival the [CII] line in intensity, while the other fine-structure lines in Table 1.2 may be only 5–10 times weaker (Brauher et al., 2008; Stacey, 2011). PDR gas is primarily cooled through the collisional (thermal) excitation of C+ and O, and subsequent radiation of the energy in [CII] and [OI] lines. Being so strong and optically thin, fine-structure lines radiate energy freely into interstellar space and play an important role in cooling the ISM throughout the Universe, providing gravity a helping hand in its eternal tug-of-war with thermal energy. Appearing in all phases of the ISM can make interpreting the origin of [CII] emission in large-scale maps difficult. Fortunately, the [NII] 205 μm line has similar excitation requirements as [CII] (see Table 1.2), meaning that where conditions are ripe for [NII] emission, [CII] will emit as well. However, due to its significantly higher ionization potential, [NII] emission will only occur in strongly ionized regions. Therefore, by comparing the [CII] and [NII] maps of a region, it is possible to determine if the [CII] is arising from ionized or neutral gas. Large-scale maps of [CII] and [NII] emission within the Milky Way were first made by the cosmic background explorer (COBE) satellite using the far-infrared absolute spectrophotometer (FIRAS) instrument (see Figure 1.11) (Bennett et al., 1994). COBE was optimized for measuring cosmic background radiation. With its 7° field of view and 5% spectral resolution (~1000 km/s), it lacked the angular and velocity resolution to discern the origin of the fine structure emission lines.

As discussed earlier, the Milky Way is rotating. Along any given line of sight, multiple gas clouds may be observed. The observed velocity of a cloud is a function of its distance from the galactic center. With sufficient velocity resolution, it is possible to estimate the location of a [CII] emitting region within the galaxy from its line velocity. Sensitive, heterodyne instruments are needed to provide the required high spectral resolution needed to separate the [CII] emitting clouds in velocity space. A sample [CII] spectrum taken along one line of sight through the Milky Way with the heterodyne instrument for the far-infrared (HIFI) (see Chapter 6) aboard the 3.5 m *Herschel space observatory* is shown in Figure 1.12. As part of the

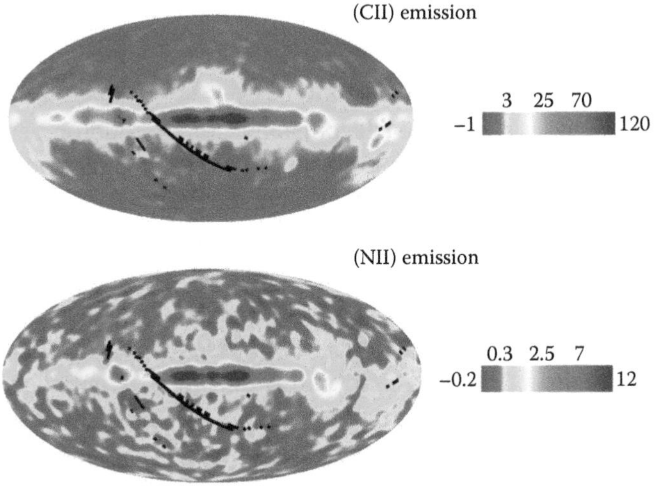

FIGURE 1.11 COBE FIRAS images of Milky Way [CII] 158 μm emission (top) and [NII] 205 μm emission (bottom). Images were made at 7° angular and ~1000 km/s velocity resolution. Black swaths indicate unobserved regions of sky. Color bars indicate emission intensity in units of nW m^{-2} sr^{-1}. (From Fixsen, D. J., Bennett, C. L., and Mather, J. C., 1999, *Ap. J.*, 526, 207. With permission.)

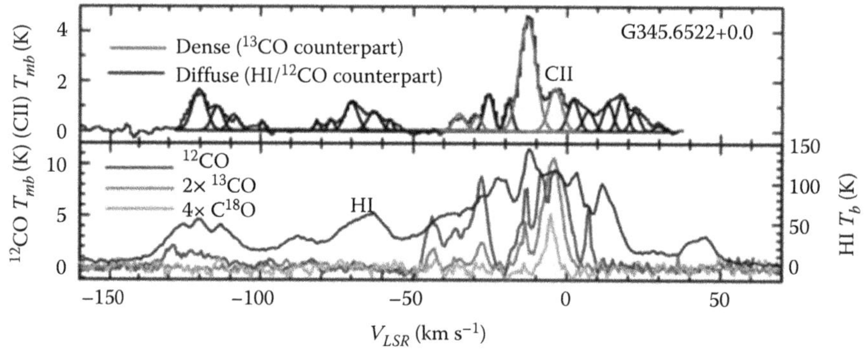

FIGURE 1.12 *Herschel* [CII] spectrum (top, continuous) along line of sight toward galactic coordinates $l = 345.65°$ and $b = 0°$. Multiple velocity components can be seen corresponding to different locations in the galactic plane. Bottom spectrum shows complementary spectra from CO and HI. (From Langer, W. et al., 2010, *Astron. Astrophys.*, 521, L17. With permission.)

Galactic Observations of Terahertz (GOT) C+ study, *Herschel* was able to observe ~500 such lines of sight in a volume weighted sampling of the Milky Way at ~12″ angular and ~0.1 km/s velocity resolution (Pineda et al., 2013). From combining these observations with those of CO (see below) and HI, the study suggests that in the galactic plane ~47% of the observed [CII] is associated with PDRs, ~28% comes from "CO-dark" clouds, ~21% emerges from atomic gas, and ~4% originates in ionized gas. The study indicates that most of the atomic gas within the solar radius is in the CNM, while the atomic gas in the outer galaxy is in the WNM. The [CII] study also shows that the CNM accounts for ~43% of all atomic gas in the Milky Way.

1.4.3 DIFFUSE GAS DENSITY

The ratio of two ionized fine-structure lines within the same species, for example, the 122 and 205 μm lines of [NII], can be used to estimate the gas density within highly ionized regions. As explained by Stacey (2011), the two lines have different Einstein spontaneous emission rates (*A*'s), which leads them to have different critical densities, $n_{crit}^{122\mu m}$ and $n_{crit}^{205\mu m}$. When the gas density, $n < n_{crit}$, each collision leads to a photon being emitted from each line, so their ratio remains constant. As n increases, first the 205 μm and then the 122 μm line becomes thermalized, causing a big swing in their line ratio between gas densities of 20 and 2000 cm^{-3}. Beyond 2000 cm^{-3}, both transitions are thermalized, and the line ratio becomes constant again (see Figure 1.13). In the transition region between $n_{crit}^{205\mu m}$ and $n_{crit}^{122\mu m}$, the [NII] 122/205 μm line ratio can be used as a sensitive density indicator.

[NII] emission can be found throughout the galaxy (see Figure 1.11), but is localized to active star formation regions. To determine volume densities, n, outside these regions we can use a combination of [CII] and HI line measurements. The observed [CII] intensity, $I_{[CII]}$, is related to the column density, N_C^+, (number of particles per cm^2) of C$^+$ atoms along a line of sight through the galaxy by (Goldsmith et al., 2012)

$$I_{[CII]} = N_{C^+}\left[3.05 \times 10^{15}\left(1 + 0.5\left(1 + \frac{A_{ul}}{R_{ul}n}\right)e^{91.21/T_{kin}}\right)\right]^{-1} \tag{1.13}$$

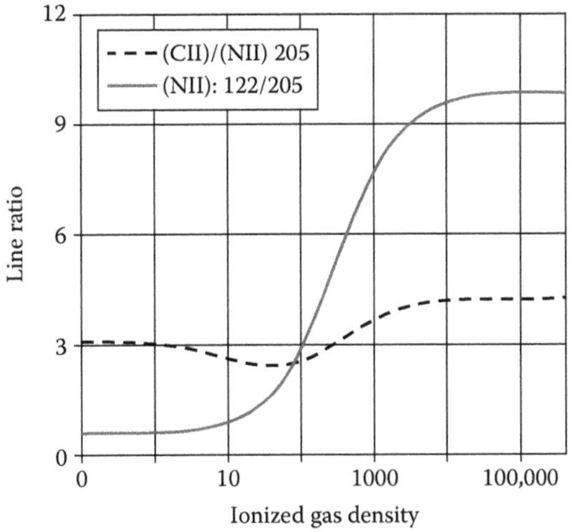

FIGURE 1.13 Fine structure line ratios as density and phase probes. The [NII] line only arises in ionized regions. A low [CII] 158 μm/[NII] 205 μm line ratio indicates [CII] emission arising in ionized gas. The ratio is relatively insensitive to density variations. The [NII] 122 μm/[NII] 205 μm can be used as an effective density tracer between 20 and 2000 cm^{-3}. (From Stacey, G. J., 2011, *IEEE Trans. Terahertz Sci. Technol.*, 1(1), 241. With permission.)

where

$I_{[CII]}$ = integrated intensity of [CII] line (K km s^{-1})
N_{C}^{+} = column density of [CII] ([CII] atoms/cm^2)
n = volume density of collision partner, be it electrons, H, or H_2 (cm^{-3})
A_{ul}= 2.3 × 10^{-6} s^{-1} = spontaneous emission coefficient for [CII] 158 μm line
R_{ul} = collisional de-excitation rate coefficient (cm^3 s^{-1})
T_{kin} = cloud kinetic temperature (K)

For collisions with electrons, the collisional de-excitation rate is given by

$$R_{ul} = 8.7 \times 10^{-8}\left(\frac{T_e}{2000}\right)^{-0.37}, \quad \text{for } 100\text{K} \le T_e \le 20{,}000\text{K} \tag{1.14}$$

where

T_e = electron temperature (K).

For collisions with hydrogen atoms, the collisional de-excitation rate is given by

$$R_{ul} = 7.6 \times 10^{-10}\left(\frac{T_{kin}}{100}\right)^{0.14}, \quad \text{for } 20\text{K} \le T_{kin} \le 2000\text{K} \tag{1.15}$$

The value of R_{ul} can be seen to be only weakly dependent on cloud temperature.

For a hydrogen cloud temperature of T_{kin}= 100 K (approximately the excitation potential of [CII]), substitution yields, $R_{ul} = 7.6 \times 10^{-10}$ cm³ s⁻¹. Adopting this value and solving for n we find

$$n = 1.15 \times 10^{19} \left[\frac{N_{C^+}}{I_{[CII]}} - 6.83 \times 10^{15} \right]^{-1} \text{cm}^{-3} \tag{1.16}$$

If we can derive an independent estimate of N_C^+, we can use Equation 1.16, together with an observed value of $I_{[CII]}$, to obtain n. N_C^+ can be estimated from the column density of HI. The term HI refers to the 1420 MHz (21 cm) emission line associated with the spin-flip transition of neutral atomic hydrogen that is observable throughout the Milky Way. For T_{kin}= 100 K gas, the hydrogen column density, N_H, can be derived from the observed integrated intensity of HI, I_{HI}, using the expression (Pineda et al., 2013),

$$N_H = 2.37 \times 10^{18} I_{HI} \tag{1.17}$$

where

I_{HI} = integrated intensity of HI line (K km s⁻¹)
N_H = column density of hydrogen, H, (H atoms cm⁻²)

To derive an estimate of N_C^+ from N_H, we use the abundance relation (Pineda et al., 2013),

$$\frac{[C]}{[H]} = 5.5 \times 10^{-4} 10^{-0.07 R_{Gal}} \tag{1.18}$$

where

R_{Gal} = distance from the galactic center (kpc).
Combining Equations 1.18 and 1.17 we have,

$$N_{C^+} = 1.3 \times 10^{15-0.07 R_{Gal}} I_{HI} \text{ cm}^{-2} \tag{1.19}$$

An expression for the gas density n can now be obtained by substituting Equation 1.19 into Equation 1.16.

$$n = 1.15 \times 10^{19} \left[1.3 \times 10^{15-0.07 R_{Gal}} \left(\frac{I_{HI}}{I_{CII}} \right) - 6.83 \times 10^{15} \right]^{-1} \text{cm}^{-3} \tag{1.20}$$

The above expression applies to the cool HI (CNM) clouds. *Herschel* observations (Langer et al., 2014) indicate that a significant amount of [CII] emission arises in diffuse H_2 (CNM) clouds, not detected in HI. Therefore, the value of n derived using Equation 1.20 should be considered as an upper limit to the true gas density along a line of sight through the galaxy.

1.4.4 STAR FORMATION RATE AND INFRARED (IR) LUMINOSITY

Stars spend most of their lifetimes on the main sequence, defined to be the location on the Hertzsprung–Russell diagram (i.e., luminosity vs. surface temperature plot), where they are burning hydrogen into helium within their central cores. The greater the mass, M, of a star, the greater will be the temperature within its core, and the greater will be its luminosity, L, and UV flux. For most stars, the relationship between M and L is exponential (Salaris et al., 2005).

$$
\begin{aligned}
L &\approx 0.23\left(\frac{M}{M_\odot}\right)^{2.3}, \quad (M < 0.43M_\odot) \\
L &\approx \left(\frac{M}{M_\odot}\right)^{4}, \quad (0.43M_\odot < M < 2M_\odot) \\
L &\approx 1.5\left(\frac{M}{M_\odot}\right)^{3.5}, \quad (2M_\odot < M < 20M_\odot)
\end{aligned}
\tag{1.21}
$$

where

L = stellar luminosity in solar luminosities ($L_\odot$), and M = stellar mass in solar masses ($M_\odot$).

EXAMPLE 1.1

For the brightest [CII] emission peak of Figure 1.12, what is the density, n, of collision partners responsible for the gas excitation? Assume a gas kinetic temperature of 100 K and a distance from the galactic center of, R_{kpc}, of 5 kpc.

We can find n by using Equation 1.20, but first we must determine the integrated intensities of the [CII] and HI lines. Since their emission peaks seems to be well-fitted by a Gaussian, we can relate the line emission peak, T_{pk}, and full-width-at-half-maximum line-width, ΔV_{FWHM}, to its integrated intensity, I, (i.e., the area under the line) using the relation,

$$I = \frac{1}{2}\left[\frac{\pi}{1n(2)}\right]^{1/2} T_{pk}\Delta V_{FWHM} \approx 1.11 \times T_{pk}\Delta V_{FWHM}$$

The half-width velocity of the [CII] line in Figure 1.12 appears to be ~7 km/s. The peak in [CII] is ~4.6 K and in HI is ~140 K. Assuming a Gaussian fit to the underlying emission line, the corresponding integrated intensities are:

$$
\begin{aligned}
I_{\mathrm{HI}} &\sim 1.1 \times T_{\mathrm{HI}}^{pk}\Delta V_{FWHM} = 1.1 \times (140\mathrm{K})(7\ \mathrm{km/s}) = 1078\ \mathrm{K\ km/s} \\
I_{[\mathrm{CII}]} &\sim 1.1 \times T_{[\mathrm{CII}]}^{pk}\Delta V_{FWHM} = 1.1 \times (4.6\mathrm{K})(7\ \mathrm{km/s}) = 35.4\ \mathrm{K\ km/s}.
\end{aligned}
$$

Substituting into Equation 1.20, we find,

$$n = 1.15 \times 10^{19}\left[1.38 \times 10^{15-0.07(5\mathrm{kpc})}\left(\frac{1078\ \mathrm{K\ km/s}}{35.4\ \mathrm{K\ km/s}}\right) - 6.83 \times 10^{15}\right]^{-1} \mathrm{cm}^{-3} = 963\,\mathrm{cm}^{-3}$$

The hotter a star burns, the shorter will be its lifetime on the main sequence, τ_{MS},

$$\tau_{MS} \approx 10^{10}\left(\frac{M}{M_{\odot}}\right)^{-2.5} \tag{1.22}$$

where

τ_{MS} = stellar lifetime on the main sequence (years).

Massive stars have the greatest luminosities, are the largest contributors to the galactic UV flux, and have the shortest lifetimes. Their presence is indicative of recent star-forming activity and can be traced by the PDRs and atomic fine structure line emission they generate. In particular, the [NII] 205 μm line, with an ionization potential just higher than hydrogen, has been demonstrated to be an excellent probe of the star formation rate (SFR) and infrared dust luminosity (L_{IR}) (Zhao et al., 2013),

$$\log \mathrm{SFR} = (-5.31 \pm 0.32) + (0.95 \pm 0.05) \log L_{[\mathrm{NII}]} \tag{1.23}$$

$$\log L_{\mathrm{IR}} = (4.51 \pm 0.32) + (0.95 \pm 0.05) \log L_{[\mathrm{NII}]} \tag{1.24}$$

where

SFR = star formation rate ($M_{\odot}\ \mathrm{yr}^{-1}$)

$L_{[\mathrm{NII}]}$ = luminosity of [NII] line ($L_{\odot}$)

L_{IR} = luminosity of dust in IR ($L_{\odot}$)

EXAMPLE 1.2

If the infrared luminosity, L_{IR}, of a dust core is 1000 L_{ε}, what is the expected luminosity of the [NII] line?

Start by solving Equation 1.24 for $L_{[\mathrm{NII}]}$, assuming a zero error.

$$\log L_{\mathrm{IR}} = 4.51 + 0.95 \log L_{[\mathrm{NII}]}$$

$$L_{[\mathrm{NII}]} = \mathrm{antilog}\left[\frac{\log L_{\mathrm{IR}} - 4.51}{0.95}\right]$$

Substitution then yields,

$$L_{[\mathrm{NII}]} = 10^{(-1.59)}$$
$$= 0.026 L_{\odot}$$

The [NII] 205 μm line is less contaminated from the emission of older stars and less subject to instrument parameters than luminosity estimates made in the infrared. Therefore, when available, $L_{[\mathrm{NII}]}$ may be a more accurate indicator of the star formation rate (SFR) than the more conventional L_{IR}-derived estimates.

1.4.5 SPECTRAL CLASSIFICATION OF IONIZING STAR

Since many of the fine-structure lines exist due to the presence of energetic UV photons, they can be used to measure the hardness of the ambient interstellar radiation field. The [NIII] 57 μm/NIII] 205 μm line ratio can be used to probe the effective surface temperature, T_{eff}, and spectral type of ionizing stars between B2V and O7.5 V (see Figure 1.14). For the hottest stars, the [OII]/[NII] line ratios are an even more sensitive tracer of spectral type, due to the greater ionization of the [OII] line (Stacey, 2011). Multiple [OII]/[NII] line ratios can be used to further constrain the spectral type of the ionizing star for a wide range of electron densities, n_e (see Figure 1.15).

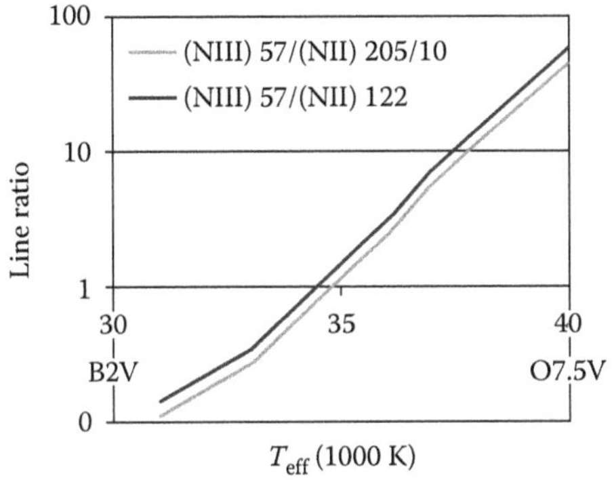

FIGURE 1.14 [NII] line ratios probe stellar type. N^+ is produced by ionizing UV photons. [NII] line ratios can be used to probe the effective surface temperature, T_{eff}, of a star, from which its spectral type can be ascertained. (From Stacey, G. J., 2011, *IEEE Trans. Terahertz Sci. Technol.*, 1(1), 241. With permission.)

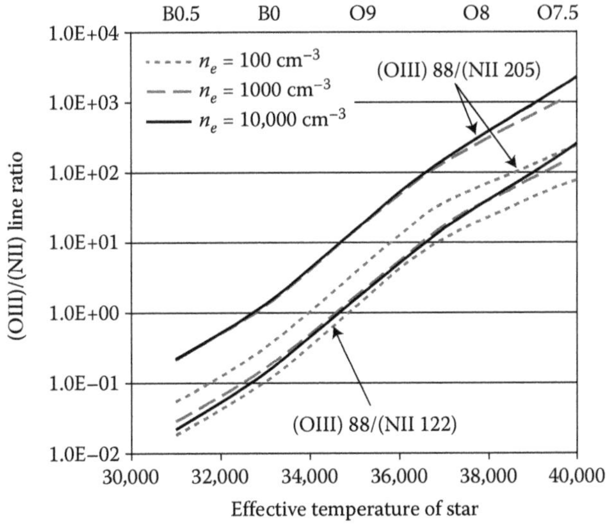

FIGURE 1.15 [OIII]/[NII] line ratios probing stellar type. O^{++} requires a harder UV field to form than N^+. The resulting [OIII] to [NII] line ratios can be used to probe T_{eff}, and is relatively insensitive to electron density, n_e. (From Stacey, G. J., 2011, *IEEE Trans. Terahertz Sci. Technol.*, 1(1), 241. With permission.)

1.4.6 THz MOLECULAR LINES

1.4.6.1 ROTATIONAL TRANSITIONS

The THz portion of the electromagnetic spectrum is the host to a wide variety of molecular rotational transitions that can be observed in emission and/or absorption. Rotational transitions arise when a molecule has an unbalanced charge distribution, that is, one side of the molecule is more positive (or negative) than the other. The degree of the charge imbalance is often expressed in units of Debye, D, where, $1\ D = 10^{-10}$esu $\approx 3.335 \times 10^{-30}$ C m. The dipole moment, p_o, of diatomic molecules ranges from 0 to 11 D. An unequal charge distribution produces a minute "molecular voltage," V_m, across the molecule. Referring to Figure 1.16, we have

$$V_m \approx \frac{1}{4\pi\varepsilon_o}\frac{p_o}{d^2} \tag{1.25}$$

where

p_o = dipole moment (C m)
d = inter-nuclear distance of molecule (m)
ε_o = permittivity of free space (8.85×10^{-12} C^2 N^{-1} m^{-2})

EXAMPLE 1.3

What is the effective voltage, V_m, between atoms in a CO molecule?

The dipole moment of a CO molecule is $p_o = 0.122D$, where 1 Debye (D) = 3.335×10^{-36} C m. The intermolecular distance is $d = 112.8 \times 10^{-12}$ m. Substitution into Equation 1.25 yields,

$$V_m^{CO} = 0.288\ \mu\text{V}$$

When the molecule undergoes a collision (with H_2, for example), it will begin to rotate. Analogous to the case of individual atoms, where electrons are only permitted to orbit a nucleus with specific angular momentum, and, thus, at specific distances/energies, molecules can only rotate with specific angular momenta, and, thus, rotational energies, E_J. In the case of diatomic molecules (see Figure 1.16),

$$E_J = \frac{h^2}{8\pi^2 I}J(J+1) \tag{1.26}$$

where

$I = \mu_m r_m^{\ 2}$ = moment of inertia (g cm^2)
$\mu_m = m_1 m_2/m_1 + m_2$ = reduced mass (g)
r_m = inter-nuclear spacing of the nuclei of the molecule (cm)
m_1 = mass of molecule 1 (g)

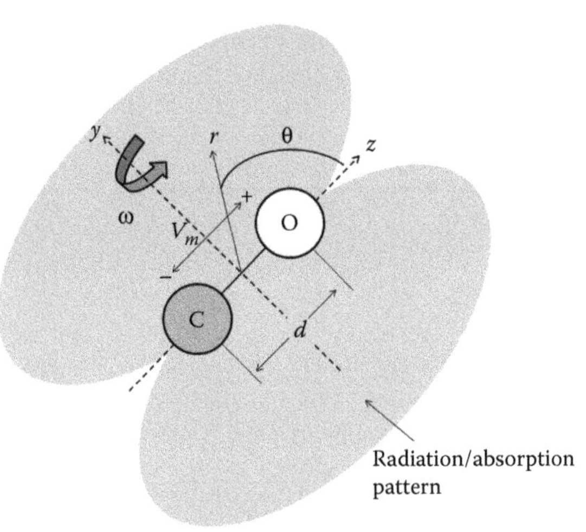

FIGURE 1.16 Many diatomic molecules, like CO, have an uneven charge distribution that produces a nonzero dipole moment (or voltage) across its length. When viewed edge-on, the polarity of the rotating molecule appears to move (i.e., oscillate) back and forth, producing dipole radiation, analogous to what happens when voltages are flipped back and forth across a dipole antenna when it is attached to a radio transmitter.

m_2 = mass of molecule 2 (g)
J = rotational quantum number
h = Planck's constant = 6.626×10^{-27} cm^2 g s^{-1}

Which J state the molecule finds itself in after a collision, depends to first order on the impact velocity, which is a function of the gas temperature. As the molecule rotates, an edge-on observer would see a time varying molecular charge, $q(t)$, moving back and forth, such that (Griffiths, 1999),

$$q(t) = q_o \cos(\omega t) \tag{1.27}$$

where
$p_o = q_o d$ = maximum value of the dipole moment, p, (C)
$\omega = 2\pi f$ = angular frequency (rad s^{-1})
t = time (s)

The molecule behaves as an oscillating electric dipole or, more figuratively, like a "molecular battery" rotating end-over-end with an angular frequency, ω.

$$p(t) = p_o \cos(\omega t)\hat{z} \tag{1.28}$$

where $\hat{z}$ = unit vector in z direction.

In the parlance of antenna theory, the situation is analogous to a radio transmitter applying a time-varying voltage at the feed point of a dipole antenna, forcing charge to swing first one way, and then another, across the antenna. The energy radiated by a classic oscillating electric dipole is determined by the Poynting vector,

$$\vec{S} = \frac{1}{\mu_o}(\vec{E} \times \vec{B}) = \frac{\mu_o}{c}\left\{\frac{p_o\omega^2}{4\pi}\left(\frac{\sin\theta}{r}\right)\cos\left[\omega\left(t - \frac{r}{c}\right)\right]\right\}^2 \hat{r} \tag{1.29}$$

where

$\vec{S}$ = radiated flux (watts m^{-2})
$\vec{E}$ = electric field intensity (V m^{-1})
$\vec{B}$ = magnetic flux density (tesla = weber m^{-2} = wb m^{-2})
μ_o = permeability of vacuum ($4\pi \times 10^{-7}$ N A^{-2})
c = speed of light (3×10^8 m s^{-1})
r = distance from neutral point of charge distribution to observer (m)
$\hat{r}$ = unit vector in r direction
θ = angle between molecule's major axis and r (rad)
t = time (s)

The observed intensity, $\vec{I}_\nu$, is found by averaging the above relation in time over a complete rotation (i.e., cycle).

$$\vec{I}_\nu = \langle\vec{S}\rangle = \left(\frac{\mu_o p_o^2\omega^4}{32\pi^2 c}\right)\frac{\sin^2\theta}{r^2}\hat{r} \tag{1.30}$$

The value of $\vec{I}_\nu$ is maximum broadside to the dipole (where the maximum charge separation is observed), and zero off the ends of the dipole (where no charge separation is observed). This makes the radiation pattern of electric dipoles, in general, look like a doughnut with its hole aligned with the molecular axis (see Figure 1.16).

The total power radiated by the dipole is found by integrating $\vec{I}_\nu$ over the surface of a sphere of radius r.

$$\langle P\rangle = \int \vec{I}_\nu \cdot d\vec{a} = \frac{\mu_o p_o^2\omega^4}{12\pi c} \tag{1.31}$$

where

$\langle P\rangle$ = radiated power (W)
$d\vec{a}$ = infinitesimal surface area of sphere (m^2)

From the above equation, we can infer that the larger the dipole moment of a molecule, the more effective an emitter (or absorber) it will be.

In classical electromagnetics, an oscillating dipole will radiate continually as long as the energy is resupplied, say by a transmitter. However, in the case of molecules, quantum mechanics comes into play, with the emission and absorption of photons being subject to transition probabilities (i.e., the Einstein *A*'s and *B*'s—see Section 2.2). When a molecule does lose rotational energy, it will jump from an upper-energy state, E_U, to a lower-energy state, E_L, and emit a photon with energy (Townes and Schawlow, 1975),

$$\Delta E_{UL} = E_U - E_L \tag{1.32}$$

and frequency, ν_{UL},

$$\nu_{UL} = \frac{\Delta E_{UL}}{h} = \frac{h}{8\pi^2 I}\left[U(U+1) - L(L+1)\right] \tag{1.33}$$

where

$U = J_{\text{upper}}$
$L = J_{\text{lower}}$
I = moment of inertia (g cm^2)
h = Planck's constant (6.63×10^{-34} J s)

Only $\Delta J \pm 1$ transitions are allowed, such that $U = L + 1$. This gives us

$$\nu_{ul} = 2B(L+1) \tag{1.34}$$

where

$$B = \frac{h}{8\pi^2 I} = \text{rotational constant of molecule (Hz)}$$

In case a molecule absorbs a photon of energy, ΔE_{UL}, it will "spin-up" and produce an absorption line at the frequency given by Equation 1.33.

The most important diatomic molecule in the THz regime is ^{12}CO (and, to a lesser degree, its isotopes), which can produce an emission line "ladder" in active star-forming regions in both galactic and extragalactic sources (e.g., Figures 1.18 through 1.20). As we will see in Chapter 2, by taking the ratio of these line intensities, we can probe both the temperature and densities of molecular clouds, while their shapes and Doppler shifted frequencies can be used to investigate cloud dynamics.

1.4.6.2 H_2O LINES

For life on Earth, water is essential to the formation and destruction of the peptide bonds which link monomers to polymers, allowing the creation of complex organic molecules. Water is also of great astrophysical importance, serving as a gas coolant in molecular clouds, thereby affecting the balance between thermal and gravitational energy within dense cloud cores. In this role, water helps to regulate the star formation process. Water also plays an essential role in astrochemistry, particularly when it relates to reactions involving oxygen—the third most abundant element in the Universe. It is also the most common type of ice mantle covering the surface of dust grains. The presence of a water–ice mantle is believed to aid in the dust coagulation process that may ultimately lead to planet formation (Bergin et al., 2008). It is this "freezing-out" of water vapor onto dust grains that makes its presence and abundance in interstellar clouds highly variable. The H_2O to H_2 abundance ratio can vary from ~10^{-4} in warm regions (T_{dust}> 100 K), where significant amounts of water remain in the gas phase, to ~10^{-8} in denser, lower-temperature regions, where freeze-out onto dust grains takes place (Emprechtinger et al., 2013). Besides being seen toward molecular clouds, water vapor emission has also been observed in the envelopes of evolved stars (Neufeld et al., 2011), emanating from comets (Neufeld et al., 2000) and, more recently, asteroids (Küppers et al., 2014).

Water is an asymmetric top, triatomic molecule composed of one oxygen and two hydrogen atoms. Other examples of asymmetric rotators observed in the ISM are SO_2, H_2S, methanol (CH_3OH), ethanol (C_2H_5OH), and formaldehyde (H_2CO), among many. Water has three principal rotation axes, each with its own rotational constant. The interaction between rotation axes causes each rotational quantum number J to be split into $2J + 1$ levels. These levels are identified by using two additional labels, K_{-1} and K_1, where $K_{-1} = 0$, ... J and $K_1 = J - K_{-1}$ or $K_1 = J - K_{-1} + 1$ (except when $K_{-1} = 0$, in which case $K_1 = J - K_{-1}$). A particular rotational energy level is designated using the nomenclature ($J_{K_{-1}}$, K_1) (Kwok, 2006). This molecular structure leads to water having many rotational energy levels, between which there are transitions in the THz range. The nuclear spins of the two hydrogen atoms can be parallel or antiparallel. Water molecules with parallel and antiparallel spins are designated as para- and ortho-water species, respectively. Figure 1.17 is an energy level diagram of water showing lower energy transitions that are observable at THz frequencies (Goldsmith, 2009). Observations of THz lines of water by the HIFI instrument at the *Herschel Space Observatory* reveal an H_2O ortho/para abundance ratio of ~3 (Emprechtinger et al., 2013).

Because of its relatively low excitation requirements and frequency, the ground state, 1_{10}−1_{01} transition of ortho-water at 557 GHz has been a major target of study for the submillimeter wave astronomy satellite (*SWAS*), *Odin*, and *Herschel*/HIFI space missions. The transition is excited principally through collisions with molecular hydrogen, H_2. If we assume each collisional excitation from the 1_{01}−1_{10} energy level is followed by radiative decay, then we say the transition is "effectively thin." Under this assumption, the average column density of ortho-H_2O column density within a telescope's beam can be estimated from the observed 557 GHz line emission, using the expression (Bergin et al., 2003),

$$\begin{aligned} N(\text{o-H}_2\text{O}) &= \frac{4\pi}{hc^3}\frac{2\nu_{ul}^2 k}{C_{ul} n_{H_2}} e^{\frac{h\nu_{ul}}{kT_g}} \int T_{mb} dV \\ &= \frac{4\pi}{hc^3}\frac{2\nu_{ul}^2 k}{C_{ul} n_{H_2}} e^{\frac{h\nu_{ul}}{kT_g}} (1.11 \times T_{mb}^{pk} \Delta V_{FWHM}) \end{aligned} \quad (1.35)$$

where

$N(o\text{-}H_2O)$ = column density of ortho-water (cm^{-2})

T_{mb} = main beam corrected line temperature (see Chapter 8) (K)

T_{mb}^{pk} = peak main beam temperature (K)

ν_{ul} = line frequency (557 GHz)

T_g = gas temperature (K)

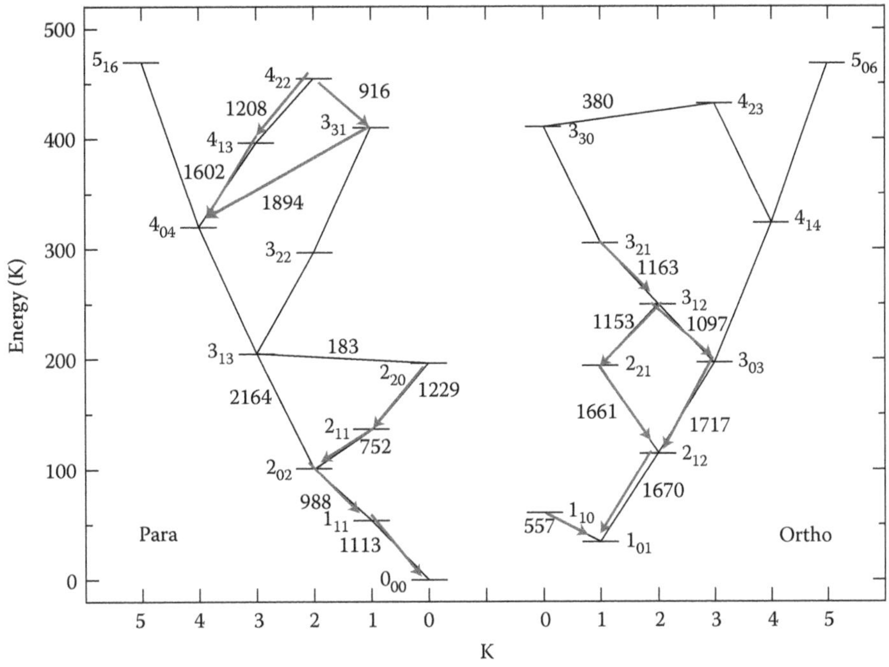

FIGURE 1.17 Para- and ortho-water energy level diagram. Transition frequencies are indicated in GHz. (After Goldsmith, P. F., 2009, *Astronomy in the Submillimeter and Far Infrared Domains with the Herschel Space Observatory*, EAS Publications Series, 34, 89–105. With permission.) Ortho-H_2O is ~3 times as abundant as para-H_2O in the ISM. The *y*-axis indicates the energy to excite a transition (in units of Kelvin) above the ground state. For collisional excitation, the energy associated with the upper level of a transition can be used as an estimate of the gas temperature required to populate that level, thereby allowing the transition to occur. Due to its low frequency and excitation requirements, the ground-state 1_{10}–1_{01} transition of ortho-water at 557 GHz is an important probe of water in the ISM. (Courtesy of P. F. Goldsmith.)

C_{ul} = collision rate between o-H_2O and H_2 at temperature T_{ex}
n_{H_2} = number density of H_2 (cm^{-3})
ΔV_{FWHM} = full-width at half-maximum line width ($cm\ s^{-1}$)

Here we assume all the H_2 is in its ground rotational state, that is, $j = 0$ (Neufeld et al., 2000). Values of C_{ul} can be found in Phillips et al. (1996) for many H_2O transitions at temperatures ranging from 20 to 140 K. For convenience, the values of C_{ul} for the 557 GHz transition are provided in Table 1.4.

TABLE 1.4 Excitation Rates of Ortho-H_2O in Collisions with H_2 ($j = 0$)

	Temperature (K)						
Transition	20	40	60	80	100	120	140
$1_{10} - 1_{01}$	1.17×10^{-11}	1.60×10^{-11}	1.89×10^{-11}	2.16×10^{-11}	2.44×10^{-11}	2.72×10^{-11}	3.03×10^{-11}

1.5 THz SPECTRAL ENERGY DISTRIBUTIONS (SEDs)

Along any given line of sight (LOS) through a galaxy there will likely be both gas and dust. The observed spectral energy distribution (SED, i.e., a plot of observed flux versus frequency—see Chapter 3) along the LOS will be a superposition of diluted, thermal black-body radiation from the dust, and, if the physical conditions are appropriate, line emission from whichever atoms and molecules are present in sufficient abundance. The largest number of THz emission features will occur along the lines of sight toward the star-forming regions in the galactic disk and nucleus.

Figure 1.18 is a model SED toward a low-mass protostar (e.g., IRAS 16293) in the galactic disk. Here, a ~32 K dust core with a bolometric luminosity of ~27 $L_\odot$ provides the continuum atop which molecular and atomic fine-structure lines can be observed. The bolometric luminosity is proportional to the area under the SED (see Equation 3.15). In a low-mass protostar, the luminosity is dominated by the accretion shock and is orders of magnitude higher than it will be when the star is burning on the main sequence. In the case of IRAS 16293, there are 100s of lower-level molecular emission lines that occur in this spectral regime, with CO (due to its high relative abundance and easy excitation) being the brightest (see e.g., van Dishoeck et al., 1995). At frequencies lower than ~200 GHz, free–free

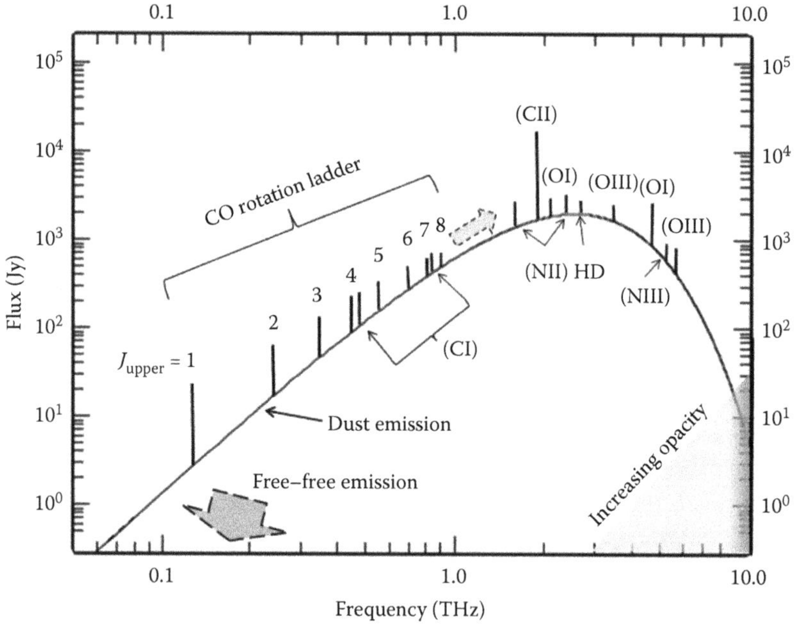

FIGURE 1.18 Schematic THz SED of a low-mass protostar. The SED is a composite of the dust continuum and line emission from the protostellar dust–gas core. The primary source of energy is from the protostar's accretion shock. Both molecular and atomic fine-structure lines are present. Free–free emission becomes important at low frequencies and dust opacity at high frequencies. The dust continuum takes the form of a diluted (i.e., gray) black body with an effective temperature, $T \approx 32$ K, an emissivity, $\beta \approx 1.3$, and source size, $\Omega_s \approx 2.67 \times 10^{-9}$ rad^2.

emission from the compact HII region surrounding the protostar will begin to contribute noticeably to the SED. On the high-frequency end, the photon wavelengths become comparable to the size of dust grains and the protostellar core becomes optically thick to its own radiation, that is, a true black body.

A schematic THz SED toward the central 2 pc of the galactic center is shown in Figure 1.19. The data come from Goicoechea et al. (2013), Genzel et al. (1985), and references therein. Toward the galactic center, we see the superposition of emission from externally/shock-heated clouds, HII regions, PDRs, and the central black hole. The bolometric luminosity is ~$3 \times 10^6 L_\varepsilon$ (assuming a distance of 10 kpc), with an effective dust temperature of ~55 K. Many of the molecular and fine structure emission lines listed in Table 1.1 are present. The total luminosity within the atomic lines integrated over the source is $(3-6) \times 10^4$ L_ε (Genzel and Townes, 1987). As is the case toward the nuclei of starburst galaxies, the [CII] and [OI] transitions serve as important cooling lines, accounting for ~0.5% of the total bolometric luminosity. Cooling through CO emission lines also plays an important role in cooling the gas, particularly in regions where $T \leq 300$ K (Genzel et al., 1985). Due to the exceptionally high UV flux, free–free emission becomes significant below ~300 GHz.

The appearance of the galactic center SED is not unique. Many such SEDs have been observed. For example, Figure 1.20 is the rest-frame THz SED of the merging starburst

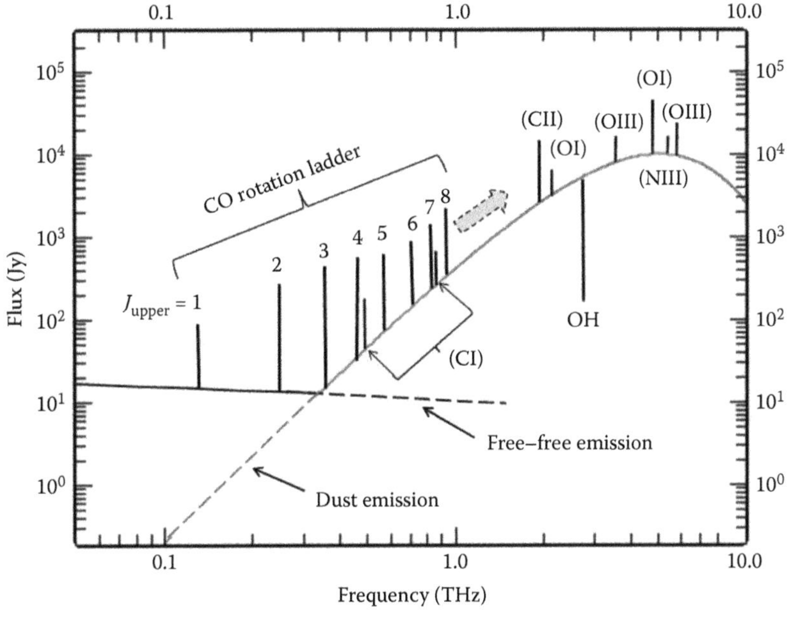

FIGURE 1.19 Schematic THz SED of the galactic center. The SED is a composite of the dust continuum and line emission from the central ~4 pc (~40″) of the Milky Way, containing externally/shock heated clouds, HII regions, PDRs, and the central black hole. Many bright THz atomic and molecular emission lines are present, with a combined radiating power of ~$10^7 L_\varepsilon$. Free–free emission becomes significant below ~300 GHz. Here, the dust continuum takes the form of a diluted (i.e., gray) black body with an effective temperature, $T \approx 55$ K, an emissivity, $\beta \approx 1.5$, and source size, $\Omega_s \approx 2.2 \times 10^{-9}$ rad^2.

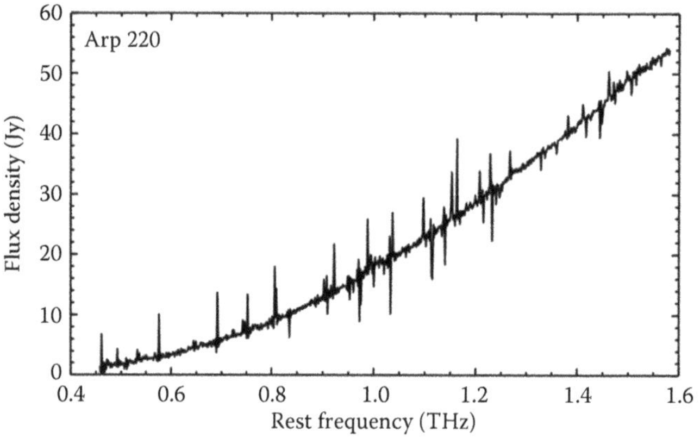

FIGURE 1.20 Rest-frame far infrared and submillimeter spectrum of the merging starburst galaxy Arp 220 ($z = 0.0181$) obtained with the SPIRE instrument on the *Herschel space observatory.* Notable features apparent in the spectrum include: warm dust continuum emission rising to high frequency, rotational emission lines of CO molecules ranging from $J = 4 \rightarrow 3$ at 461 GHz to $J = 13 \rightarrow 12$ at 1.497 THz in intervals of 115 GHz, an array of H2O emission lines, and absorption lines of HCN, OH+, and H_2O+. (From Rangwala, N. et al., 2011, *Ap. J.*, 743, 94. With permission; Figure and caption courtesy of J. Glenn.)

galaxy Arp 220 as seen through the spectral and photometric imaging receiver (SPIRE) instrument on the *Herschel Space Observatory.* The CO line ladder from $J = 4 \rightarrow 3$ to $J = 13 \rightarrow 12$ and lines of H_2O, HCN, OH+, and H_2O+ are clearly visible (Rangwala et al., 2011). Indeed, by performing large-scale spectroscopic and dust continuum surveys of the Milky Way, a high fidelity template for the properties and behavior of our own ISM can be created, and serve as a Rosetta Stone for interpreting observations of the more distant Universe.

CONCLUSION

In this chapter we have learned the ISM has a lifecycle, transitioning from one phase to another—from atomic to molecular, and then back again, over millions of years. It is while in the molecular phase that a small fraction of the ISM gives birth to stars and planets. Within the centers of stars the raw ISM material is enriched with heavy elements through nuclear fusion. High mass stars go supernova, giving back most of their mass to the ISM, with further enrichment occurring through the supernova process itself. The cycle then begins again. Over the lifetime of the Milky Way a parcel of gas could have gone through ~100s or 1000s of such cycles. Through each phase of the lifecycle, THz lines of atoms and molecules can be used to probe physical conditions. The [CII] line can be present in each phase. Therefore, it is particularly well suited to this task. Rotational transitions of molecules such as CO and H_2O play a significant role in determining the energy balance within dense molecular clouds. In the next chapter, we will develop the analytical tools needed to derive physical conditions from line emission.

PROBLEMS

1. If you want to observe the [CII] line at 1.9 THz, what would be the atmospheric transmission from an observatory at an altitude of 4.2 km, 14 km, and 32 km?

2. What is the approximate spatial extent of the Orion Nebula on the sky?

3. What is the Jeans mass of a 1 ly diameter cloud with a mean temperature of 15 K, molecular hydrogen density of 1×10^4 cm^{-3}, and a velocity dispersion of 1 km/s?

4. What is the hydrogen column density toward a protostar with an $A_V = 20$?

5. If you were on a starship passing through a molecular cloud along your flight trajectory, what would be the order of atomic and molecular species you would encounter along the way?

6. Which atomic and/or molecular transitions are well-suited for probing interstellar gas in regions with gas densities and temperatures of $\geq 1 \times 10^4$ cm^{-3} and $\geq$100 K?

7. What percentage of a 1 $M_\odot$ star's life is spent in the formation process?

8. What is the expected intensity of a [CII] line being emitted from an interstellar cloud 4 kpc from the galactic center with an HI integrated intensity of 100 K km/s?

9. What is the expected luminosity in an [NII] line from the vicinity of a dust core with an IR luminosity of 1000 $L_\odot$?

10. Compute the frequency of the ^{12}CO $J = 12 \rightarrow 11$ rotational transition.

11. Looking toward a star forming region, you observe a 557 GHz water emission line. The line has an integrated intensity of 1.89 K km s^{-1}. From dust observations you estimate the temperature and n_{H_2} of the core to be ~40 K and 10^5 cm^{-3}. What is the column density of H_2O along the line of sight through the region? What would you estimate the corresponding H_2 column density to be?

REFERENCES

Asplund, M., Grevesse, N., Sauval, A., and Scott, P., 2009, The chemical composition of the sun, *ARAA*, 47, 481.

Ballesteros-Paredes, J., Hartmann, L., and Vazquez-Semadeni, E., 1999, Turbulent flow-driven molecular cloud formation: A solution to the post-TTauri problem? *Ap. J.*, 527, 285.

Bally, J., 2008, Overview of the orion complex, in *Handbook of Star Forming Regions Vol. I* ASP Conference Series, Bo Reipurth.

Bennett, C. L., Fixsen, D. J., Hinshaw, G. et al., 1994, Morphology of the interstellar cooling lines detected by COBE, *Ap. J.*, 434, 587.

Bergin, E., Blake, G., Goldsmith, P., Harris, A., Melnick, G., and Zimuidzinas, J., 2008, Design reference mission case study, *Stratospheric Observatory for Infrared Astronomy Science Steering Committee.* https://www.sofia.usra.edu/Science/science_cases/Bergin2008.pdf

Bergin, E. A., Kaufman, M. J., Melnick, G. J., Snell, R. L., and Howe, J. E., 2003, A survey of 557 GHz water vapor emission in the NGC 1333 molecular cloud, *Ap. J.*, 582, 830.

Brauher, J., Dale, D., and Helou, G., 2008, A compendium of far-infrared line and continuum emission for 227 galaxies observed by the infrared space observatory, *Astrophys. J. Suppl.*, 178, 280.

Burrus, C., 1963, Formed-point-contact varactor diodes utilizing a thin epitaxial gallium arsenide layer, *Proc. IEEE (Correspondence)*, 51, 1777–1778.

Day, P., LeDuc, H., Mazin, B., Vayonakis, A., Zimuidzinas, J., 2003, A broadband superconducting detector suitable for use in large arrays, *Nature*, 425, 817.

Deul, E. R. and van der Hulst, J. M., 1987, A survey of the neutral atomic hydrogen in M33, *A&AS*, 67, 509.

Dolan, G. J., Phillips, T. G., and Woody, D. P., 1979, Low-noise 115 GHz mixing in superconducting oxide barrier tunnel junctions, *Appl. Phys. Lett.*, 34, 347.

Draine, B. T., 2011, *Physics of the Interstellar Medium*, Princeton University Press, Princeton, NJ.

Emprechtinger, M., Lis, D. C., Rolffs, R., Schilke, P., Monje, R. R., Comito, C., Ceccarelli, C., Neufeld, D. A., and van der Tak, F. F. S., 2013, The abundance, ortho/para ratio, and deuteration of water in the high-mass star-forming region NGC 6334 I, *Ap. J.*, 765, 61.

Engargiola, G., Plambeck, R. L., Rosolowsky, E., and Blitz, L., 2003, Giant molecular clouds in M33. I. BIMA all-disk survey, *Ap. JS.*, 149, 343.

Fixsen, D. J., Bennett, C. L., and Mather, J. C., 1999, COBE far infrared absolute spectrophotometer observations of galactic lines, *Ap. J.*, 526, 207.

Genzel, R. and Townes, C. H., 1987, Physical conditions, dynamics, and mass distribution in the center of the galaxy, *ARA&A*, 25, 377.

Genzel, R., Watson, D. M., Crawford, M. K., and Townes, C. H., 1985, The neutral-gas disk around the galactic center, *Ap. J.*, 297, 766.

Gershenzon E., Gol'tzman G., Gogidze I., Gusev Y., Elant'ev A., Karasik B., and Semenov A., 1990, Millimeter and submillimeter range mixer based on electronic heating of superconducting films in the resistive state, *Sov. Phys. Supercond.*, 3, 1582.

Goicoechea, J., Etxaluze, M., Cernicharo, J. et al., 2013, Herschel far-infrared spectroscopy of the galactic center. Hot molecular gas: Shocks versus radiation near sgr A, *Ap. J.*, 769, L13.

Goldsmith, P. F., 2009, The astrophysics of spectroscopic studies with the herschel apace observatory, *Astronomy in the Submillimeter and Far Infrared Domains with the Herschel Space Observatory*, L. Pagani and M. Gerin (eds.), EAS Publications Series, 34, 89–105.

Goldsmith, P. F., Langer, W. D., Pineda, J. L., and Velusamy, T., 2012, Collisional excitation of the [CII] fine structure transition in interstellar clouds, *Ap. JS.*, 203, 13.

Gol'tsman, G. N., Karasik, B. S., Okunev, O. V. et al., 1995, NbN hot electron superconducting mixers for 100 GHz operation, *IEEE Trans. Appl. Supercond.*, 5, 3065.

Griffiths, D., 1999, *Introduction to Electrodynamics*, Prentice-Hall, Upper Saddle River, NJ.

Hollenbach, D. J. and Salpeter, E. E., 1971, Surface recombination of hydrogen molecules, *Ap. J.*, 163, 155.

Hollenbach, D. J. and Tielens, A. G. G. M., 1997, Dense photodissociation regions (PDRs), *Annu. Rev. Astron. Astrophys.*, 35, 179–215.

Irwin, K. D., Nam, S. W., Cabrera, B., Chugg, B., Park, G. S., Welty, R. P., and Martinis, J. M., 1995, A self-biasing cryogenic particle detector utilizing electrothermal feedback and a SQUID readout, *IEEE Trans. Appl. Supercond.*, 5(2 pt.3), 2690.

Jeans, J., 1904, *The Dynamical Theory of Gases*, Cambridge University Press, London.

Jansen, D. J., Chapter 2: Excitation of Molecules in Dense Clouds, http:home.strw.leidenuniv.nl/~jansen/research/ch2.pdf.

Kulesa, C., 2011, Terahertz spectroscopy for astronomy: From comets to cosmology, *IEEE Trans. Terahertz Sci. Technol.*, 1(1), 232.

Küppers, M., O'Rourke, L., Bockelée-Moran, D. et al., 2014, Localized sources of water vapour on the dwarf planet (1) Ceres, *Nature*, 505, 525.

Kwok, S., 2006, *Physics and Chemistry of the Interstellar Medium*, University Science Books, Mill Valley, USA, p. 225.

Langer, W., Velusamy, T., Goldsmith, P. F., Li, D., Pineda, J., and Yorke, H., 2010, C+ detection of warm dark gas in diffuse clouds, *Astron. Astrophys.*, 521, L17.

Langer W. D., Velusamy, T., Pineda, J. L., Willacy, K., and Goldsmith P. F., 2014, A Herschel [C II] galactic plane survey. II. CO-dark H_2 in clouds, *A&A*, 561, A122.
Low, F. J., 1961, Low-temperature germanium bolometer, *J. Opt. Soc. Am.*, 51(11) 1300.
Mac Low, M. M. and Klessen, R. S., 2004, The control of star formation by supersonic turbulence, *Rev. Mod. Phys.*, 76, 125–194.
Melnick, G. J., 1988, On the road to the large deployable reflector (LDR): The utility of balloon-borne platforms for far-infrared and submillimeter spectroscopy, *Int. J. Infrared Milli. Waves*, 9(9) 781.
Neufeld, D. A., González-Alfonso, E., Melnick, G. et al., 2011, The widespread occurrence of water vapor in the circumstellar envelopes of carbon-rich asymptotic giant branch stars: First results from a survey with herschel/HIFI, *Ap. J.*, 727, L29.
Neufeld, D. A., Stauffer, J. R., Bergin, E. A. et al., 2000, Submillimeter wave astronomy satellite observations of water vapor toward comet C/1999 H1 (Lee), *Ap. J.*, 539, L151.
Peñarrubia J., Ma Y.-Z., Walker M., and McConnachie A., 2014, A dynamical model of the local cosmic expansion. *Monthly Notices of the Royal Astronomical Society* 433(3), 2204–2222.
Phillips, T. G. and Jefferts, K. B., 1973, A low temperature bolometer heterodyne for millimeter wave astronomy, *Rev. Sci. Instrum.*, 44, 1009.
Phillips, T. R., Maluendes, S., and Green, S., 1996, Collisional excitation of H_2O by H_2 molecules, *Ap. JS.*, 107, 467.
Pineda, J. L., Langer, W. D., Velusamy, T., and Goldsmith, P. F., 2013, A Herschel [C II] galactic plane survey. I. The global distribution of ISM gas components, *A&A*, 554, 103.
Ralchenko, Y., Kramida, A., Reader, J., and NIST ASD Team, 2010, NIST Atomic Spectra Database (ver. 4.0); NIST Physical Measurement Laboratory, http://www.nist.gov/pml/data/asd.cfm.
Rangwala, N., Maloney, P., Glenn, J. et al., 2011, Observations of Arp 220 using *Herschel*-SPIRE: An unprecedented view of the molecular gas in an extreme star formation environment, *Ap. J.*, 743, 94.
Richards, P. L., Shen, T. M., Harris, R. E., and Lloyd, F. L., 1979, Quasiparticle heterodyne mixing in SIS tunnel junctions, *Appl. Phys. Lett.*, 34, 345.
Salaris, M. and Cassisi, S., 2005, *Evolution of Stars and Stellar Populations*, Wiley, New York.
Shu, F. H., 1977, Self-similar collapse of isothermal spheres and star formation, *Ap. J.*, 214, 488.
Shu, F. H., Najita, J., Shang, H., and Li, Z. Y., 2000, X-Winds theory and observations, in V. Mannings, A. P. Boss, and S. S. Russell (eds.), *Protostars and Planets IV* (University of Arizona Press, Tucson, 2000), pp. 789ff.
Snell, R. L., Loren, R. B., and Plambeck, R. L., 1980, Observations of CO in L1551—Evidence for stellar wind driven shocks, *Ap. J.*, 239L, 17.
Spitzer, L., 1978, *Physical Processes in the Interstellar Medium*, WILEY-VCH Verlag GmbH & Co. KGaA.
Stacey, G. J., 2011, THz low resolution spectroscopy for astronomy, *IEEE Trans. Terahertz Sci. Technol.*, 1(1), 241.
Stacey, G. J., Geis, N., Genzel, R., Lugten, J. B., Poglitsch, A., Sternberg, A., and Townes, C. H., 1991, The 158 micron C II line—A measure of global star formation activity in galaxies, *Ap. J.*, 373, 423.
Stacey, G. J., Hailey-Dunsheath, S., Ferkinhoff, C., Nikola, T., Parshley, S. C., Benford, D. J., Staguhn, J. G., and Fiolet, N., 2010, A 158 μm [C II] Line Survey of Galaxies at z ~ 1–2: An indicator of star formation in the early universe, *Ap. J.*, 724, 957.
Townes, C. and Schawlow, A., 1975, *Microwave Spectroscopy*, Dover Publications, New York, p. 4.
van Dishoeck, E. F., Blake, G. A., Jansen, D. J., and Groesbeck, T. D., 1995, organic and deuterated species toward IRAS 16293–2422, *Ap. J.*, 447, 760.
Walker, C. K., Adams, F. C., and Lada, C. J., 1990, 1.3 millimeter continuum observations of cold molecular cloud cores, *Ap. J.*, 349, 515.
Walker, C. K., Carlstrom, J. E., and Bieging, J. H., 1993, The IRAS 16293–2422 cloud core—A study of a young binary system, *Ap. J.*, 402, 655.
Walker, C. K., Lada, C. J., Young, E. T., and Margulis, M., 1988, An unusual outflow around IRAS 16293–2422, *Ap. J.*, 332, 335.
Walker, C. K., Narayanan, G., and Boss, A., 1994, Spectroscopic signatures of infall in young protostellar systems, *Ap. J.*, 431, 767.

Weinreb, S., 1961, Digital radiometer, *Proc. IEEE*, 49(6), 1099.
Weinreb, S., 1980, *Electronics Division Internal Report*, No. 202, NRAO.
Williams, J. P. and Cieza, L. A., 2011, Protoplanetary disks and their evolution, *ARAA*, 49, 67–117.
Wilson, R. W., Jefferts, K. B., and Penzias, A., 1970, Carbon monoxide in the orion nebula, *Ap. J.*, 161, L43.
Wolfire, M. G., Hollenbach, D., and McKee, C. F., 2010, The dark molecular gas, *Ap. J.*, 716, 1191.
Wolfire, M. G., McKee, C. F., Hollenbach, D., and Tielens, A. G. G. M., 2003, Neutral atomic phases of the interstellar medium in the galaxy, *Ap. J.*, 87, 278.
Young, D. and Irvin, J., 1965, Millimeter frequency conversion using Au-n-type GaAs Schottky barrier epitaxial diodes with a novel contacting technique, *Proc. IEEE (Correspondence)*, 53, 213G2131.
Zhao, Y., Lu, N., Xu, C. K. et al., 2013, A Herschel survey of the [N II] 205 μm line in local luminous infrared galaxies: The [N II] 205 μm emission as a star formation rate indicator, *Ap. J.*, 765, L13.

2

THz RADIATIVE TRANSFER BASICS AND LINE RADIATION

PROLOGUE

Beyond the boundaries of the solar system, the only physical evidence we have about the nature of the Universe is discerned from the light we receive from it. Radiative transfer is the mathematical framework by which we interpret this light and derive the physical conditions in the distant Universe. In this chapter, the basics of radiative transfer will be discussed, and expressions provided for determining gas temperatures, optical depths, column densities, masses, and velocity fields in astrophysical sources. These same analytical tools can also be used in a wide variety of remote-sensing applications.

2.1 EQUATION OF RADIATIVE TRANSFER

Imagine a cloud of gas and dust out in space (see Figure 2.1). The atoms, molecules, and dust grains that make up the cloud are absorbing and emitting photons in some stochastic way. Once we have successfully captured these photons with our telescopes, and detected them with our instruments, how do we interpret their meaning? The most important equation for achieving this goal is the equation of radiative transfer (ERT),

$$\frac{dI_\nu}{ds} = j_\nu - k_\nu I_\nu \tag{2.1}$$

where

I_ν = intensity (erg/s/cm^2/rad^2/Hz)

j_ν = emission coefficient (erg/s/cm^3/rad^2/Hz)

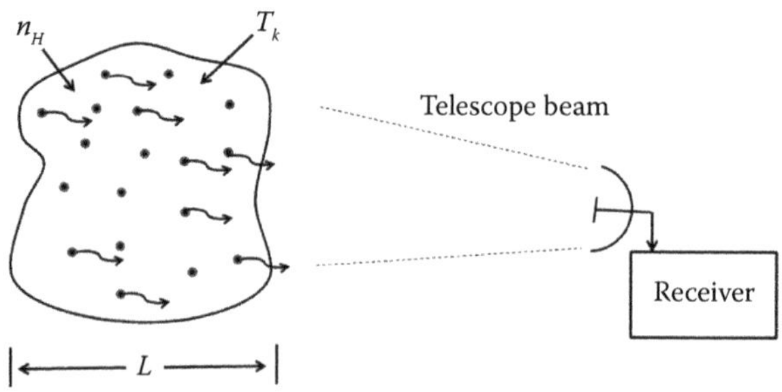

FIGURE 2.1 Interstellar gas cloud emitting and absorbing photons. A small subset of the emitted photons will intercept the telescope's beam pattern.

k_ν = absorption coefficient (cm^{-1})
s = path-length traveled by emitted photon (cm)

The subscript ν indicates the parameter for a single frequency, that is, monochromatic.

Once emitted from an atom, molecule, or dust grain, a photon will travel along a path, *ds*. Let us assume our line of sight (LOS) is aligned with the *z* axis of a Cartesian coordinate system, as illustrated in Figure 2.2. As observers, we are sensitive to the presence or absence of photons traveling back along our LOS, that is, traveling on the component of *ds* projected onto *z*.

$$\begin{aligned} dz &= ds\cos\phi = -ds\cos\theta \\ ds &= -\frac{dz}{\cos\theta} \end{aligned} \tag{2.2}$$

Rewriting Equation 2.1 in terms of *dz*, we find

$$\begin{aligned} -\cos\theta\frac{dI_\nu}{dz} &= j_\nu - k_\nu I_\nu \\ \cos\theta\frac{dI_\nu}{dz} &= k_\nu I_\nu - j_\nu \end{aligned} \tag{2.3}$$

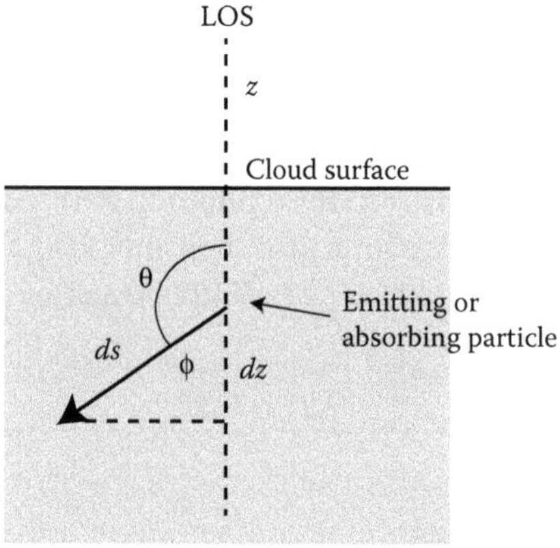

FIGURE 2.2 Geometry of observing along a line of sight (LOS) into a cloud.

Since we are projecting back along our LOS, $\theta = 0$ and Equation 2.3 reduces to

$$\frac{dI_\nu}{dz} = k_\nu I_\nu - j_\nu \quad (2.4)$$

We can now rewrite the equation in a much more convenient form:

$$\frac{dI_\nu}{kdz} = I_\nu - \frac{j_\nu}{k_\nu} \quad (2.5)$$

$$\frac{dI_\nu}{d\tau_\nu} = I_\nu - S_\nu \quad (2.6)$$

where

$$S_\nu = \frac{j_\nu}{k_\nu} = \text{source function} \quad (2.7)$$

and

$$\tau_\nu = \int_0^L k_\nu dz = \text{optical depth} \quad (2.8)$$

S_ν can be broken up into scattering and nonscattering parts. In a scattering process, a photon maintains its identity through the absorption and emission process—its direction and frequency may change, but a one-to-one correspondence is maintained. Examples of elastic scattering of photons include Mie scattering (by particles about the same size or larger than the photon), and Rayleigh scattering (by particles much smaller than the size of the photon). Inelastic scattering processes such as Raman, Compton, and inverse Compton scattering, also exist. In THz astronomy, where the photon's wavelength is relatively long (e.g., ~10–1000 μm), and the interstellar medium is relatively low-density (e.g., ~100–10^6 molecular hydrogen molecules per cm^3 and low ion fraction), scattering processes, with a few notable exceptions, can be largely ignored. In a nonscattering process, the identity of the photon is, at least temporarily, lost through absorption by an atom, molecule, or dust grain, and there is only a statistical relationship between absorbed and emitted photons. Nonscattering processes also include the important case of atoms, molecules, or dust grains acting as "fundamental" sources of photons, where thermal/kinetic energy is transformed into light. For these reasons, we will focus our attention on nonscattering processes.

2.2 SOLUTION TO THE EQUATION OF RADIATIVE TRANSFER UNDER LOCAL THERMODYNAMIC EQUILIBRIUM (LTE)

The equation of radiative transfer is a first order differential equation, which can be solved using the integration factor method. Let us imagine we observe a line of sight (LOS) through a cloud, as shown in Figure 2.3. At the cloud's boundary closest to us, the optical depth, $\tau'_\nu = 0$. As we proceed through the cloud, the value of τ builds up until we penetrate the far side of the cloud. At this point, the value of τ has increased to τ''_ν.

Multiplying Equation 2.6 by $e^{-\tau}$ we find

$$\frac{dI_\nu}{d\tau_\nu}(e^{-\tau_\nu}) - I_\nu e^{-\tau_\nu} = -S_\nu e^{-\tau_\nu}$$

Using the product rule, we find the two terms on the left side of the equation are equivalent to $d/d\tau_\nu(I_\nu e^{-\tau_\nu})$. Substitution then yields

$$\Big|_{\tau'}^{\tau''} I_\nu e^{-\tau_\nu} = -\int_{\tau'_\nu}^{\tau''_\nu} S_\nu e^{-\tau_\nu} d\tau \tag{2.9}$$

Integrating the source function, S_ν, and optical depth, τ_ν, along the LOS, we find the observed intensity coming from the front surface of the cloud ($\tau = \tau_\nu = 0$) to be

$$\begin{aligned} I_\nu(\tau'_\nu) &= I_\nu(\tau''_\nu)e^{-\tau''_\nu} + \int_{\tau'_\nu}^{\tau''_\nu} S_\nu e^{-\tau_\nu} d\tau \\ &= I_\nu(\tau''_\nu)e^{-\tau''_\nu} + [-S_\nu e^{-\tau''_\nu} + S_\nu e^{-\tau'_\nu}] \\ &= I_\nu(\tau''_\nu)e^{-\tau''_\nu} + S_\nu(1 - e^{-\tau''_\nu}) \end{aligned} \tag{2.10}$$

where $I_\nu(\tau''_\nu)$ = intensity hitting the cloud from behind, at the $\tau_\nu = \tau''_\nu$ surface, for example, background radiation. As the background radiation propagates through the cloud, it is attenuated by a factor, $e^{-\tau''_\nu}$. The term $S_\nu(1 - e^{-\tau''_\nu})$ describes radiation originating from within the cloud itself.

A standard observing mode in astronomy is to measure the intensity along a LOS to the object of interest (e.g., an interstellar cloud), I_ν^{ON}, and then along a LOS just off the

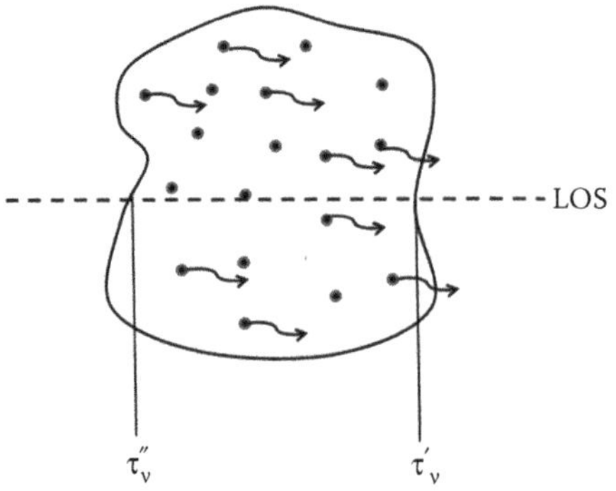

FIGURE 2.3 Observing along a line of sight (LOS) through a cloud. τ'_ν is the optical depth at the cloud's front surface and τ''_ν is the optical depth accumulated through the cloud.

object, I_ν^{OFF}. The intensities are then subtracted from each other in the receiver system (see Figure 2.4). This is done to subtract what is common between the two measurements (e.g., cosmic background radiation, atmospheric noise, and instrument noise), and leave what we are trying to observe, the intensity from the object itself, ΔI_ν,

$$\Delta I_\nu = I_\nu^{ON} - I_\nu^{OFF} \tag{2.11}$$

and

$$= [S_\nu - I_\nu](1 - e^{-\tau_\nu''}) \tag{2.12}$$

where

I_ν^{ON} = intensity from the object + background (erg/s/cm²/rad²/Hz)
I_ν^{OFF} = intensity from background (erg/s/cm²/rad²/Hz)
τ_ν'' = optical depth of the object along LOS at a given frequency (neper = Np)
ΔI_ν = background subtracted intensity of object (erg/s/cm²/rad²/s/Hz)

To understand how emission, j_ν, and absorption, k_ν, coefficients are related to the atoms and molecules in a cloud, we must introduce the Einstein coefficients. Imagine a two-level system (e.g., atom or molecule) as shown in Figure 2.5, then,

A_{ul} = spontaneous emission coefficient from level u to l, the probability per unit time that an atom or molecule in the upper state will emit a photon, and transition to the lower state (s^{-1}).

B_{ul} = induced emission coefficient from level u to l, the probability per unit time per unit intensity at a given frequency that an atom or molecule in the upper state will emit a photon and transition to the lower state (s^{-1} erg^{-1} cm^2 rad^2).

B_{lu} = absorption coefficient from level l to u; induced emission from level u to l; the probability per unit time per unit intensity at a given frequency that an atom or molecule in the lower state will absorb a photon and transition to the upper state (s^{-1} erg^{-1} cm^2 rad^2).

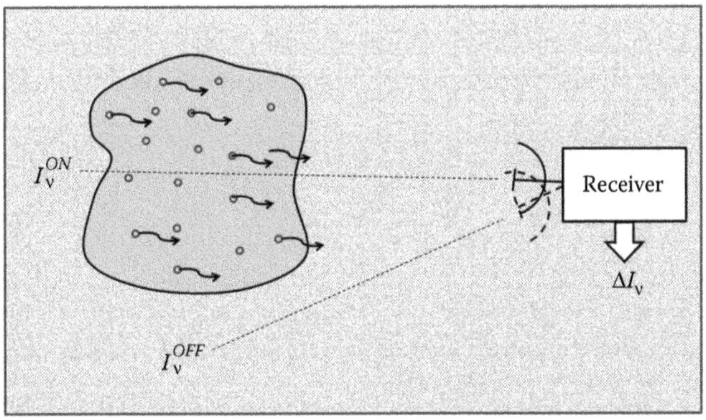

FIGURE 2.4 Differencing intensities observed toward ON and OFF positions to derive the intensity from the cloud itself, ΔI_ν.

FIGURE 2.5 Two-level energy diagram.

These rates govern how fast the transitions will occur. They are atomic constants that depend only on the atom, molecule, and levels in question. They do not depend on the environment in which the emitting particles find themselves.

We can construct emission and absorption coefficients from the Einstein coefficients,

$$j_\nu = \frac{h\nu}{4\pi} n_u A_{ul} \tag{2.13}$$

and

$$k_\nu = \frac{h\nu}{4\pi}(n_l B_{lu} - n_u B_{ul}) \tag{2.14}$$

where

n_l = number of atoms or molecules/cm³ in level 1
n_u = number of atoms or molecules/cm³ in level 2

In *continuum* observations: n_l and n_u are the number of electron–ion pairs per unit relative kinetic energy per unit volume. They are proportional to the electron and ion density.

In *spectral line* observations, n_l and n_u are level populations associated with the atom or molecule. A line profile function, φ_ν, must be used to keep the Einstein coefficients independent of ambient conditions (see below).

Now, we can define a source function, S_ν, in terms of Einstein coefficients:

$$S_\nu = \frac{j_\nu}{k_\nu} = \frac{n_u A_{ul}}{n_l B_{lu} - n_u B_{ul}} = \frac{A_{ul}/B_{ul}}{(n_l B_{lu}/n_u B_{ul}) - 1} \tag{2.15}$$

If we assume the gas is in thermal dynamic equilibrium (TE), then the population ratio between the two energy levels is given by the Boltzmann excitation relation,

$$\frac{n_l}{n_u} = \frac{g_l}{g_u} e^{h\nu/kT_{ex}} \tag{2.16}$$

where

T_{ex} = gas excitation temperature (K)
g_n = number of quantum mechanical states corresponding to a particular energy
$g_l = 2l + 1$ = statistical weight of lower state
$g_u = 2u + 1$ = statistical weight of upper state
l = quantum number associated with upper state
u = quantum number associated with lower state

Regardless of whether the gas is in TE, the following relationships between the Einstein coefficients are valid:

$$\frac{A_{ul}}{B_{ul}} = \frac{2h\nu^3}{c^2} \tag{2.17}$$

and

$$\frac{B_{lu}}{B_{ul}} = \frac{g_u}{g_l} \tag{2.18}$$

Substituting these expressions into Equation 2.15, we find

$$S_\nu = \frac{(2h\nu^3/c^2)}{g_u n_l / g_l n_u - 1} \Leftrightarrow \frac{2h\nu^3}{c^2}\frac{1}{e^{h\nu/kT_{ex}} - 1} \tag{2.19}$$

In other words,

$$S_\nu = B_\nu \Leftarrow \text{Planck Blackbody Law} \tag{2.20}$$

We see the nonscattering part of the source function is equivalent to the Planck function for the case of TE. Outside of TE, Equation 2.19 does not strictly hold. True TE is something that is rarely found in nature. However, in regions of relatively high density (e.g., within molecular clouds), it is often appropriate to define a region of local thermodynamic equilibrium (LTE) within which collisions between particles are frequent enough, so that they take on a Maxwellian velocity distribution. In such instances, collisional excitation processes dominate over radiative excitation processes, and the region can be characterized by a single kinetic temperature, T_k, where $T_{ex} \approx T_k$. In low density regions, where radiative processes dominate, the region can be characterized by a single radiative temperature, T_{RAD}, where $T_{ex} \approx T_{RAD}$. In true TE, the conditions are such that $T_{ex} = T_k = T_{RAD}$. There are numerous approaches for estimating source functions under non-LTE conditions, when radiative excitations and de-excitations become important in determining level populations. These will be introduced later in the chapter. For now, we will adopt the simplifying assumption of LTE, and use it to derive a number of important properties of molecular clouds.

For a cloud in LTE, we can substitute Equation 2.20 into Equation 2.12 to yield the very useful relation

$$\Delta I_\nu = [B_\nu(T_{ex}) - B_\nu(T_{BG})](1 - e^{-\tau_\nu}) \tag{2.21}$$

where

$B_\nu(T_{ex})$ = Planck function set to the excitation temperature and observational frequency of the cloud

$B_\nu(T_{BG})$ = Planck function set to the background temperature and observational frequency of the cloud

$\tau_\nu = \int_0^L k_\nu \, ds$ = optical depth through the cloud at the observational frequency

k_ν = absorption coefficient at observing frequency (cm^{-1})

ds = differential path-length along LOS (cm)
L = total path-length through cloud (cm)

The last term in Equation 2.21, $(1 - e^{-\tau_\nu})$, has the form of a power loss (see Equation 5.49), and indicates the fraction of the particles along the LOS radiating with the intensity, $(B_\nu(T_{ex}) - B_\nu(T_{BG}))$.

For situations where $h = kT$, often encountered in astrophysical sources below ~100 GHz, one can make use of the Rayleigh–Jeans approximation to Planck's Law (Equation 2.22) to further simplify Equation 2.21.

$$B_\nu(T) = \frac{2\nu^2 kT}{c^2} \propto T \tag{2.22}$$

$$\Delta T = [T_{ex} - T_{BG}](1 - e^{-\tau_\nu}) \tag{2.23}$$

where
c = speed of light
k = Boltzmann's constant
ΔT = kinetic temperature of the cloud (K)
T_{ex} = excitation temperature of gas (K)
T_{BG} = temperature of background (K)
τ_ν = optical depth at observational frequency

At THz frequencies, the full Planck function should be used in deriving temperatures.

We can gain insight into how radiative transfer works, by examining Equation 2.21 at its limits. As illustrated in Figure 2.6, a cloud will be seen in emission ($\Delta I_\nu > 0$) when $B_\nu(T_{ex}) > B_\nu(T_{BG})$, and in absorption ($\Delta I_\nu < 0$) when $B_\nu(T_{ex}) < B_\nu(T_{BG})$. At the limits of τ_ν, we find

$$\Delta I_\nu = [B_\nu(T_{ex}) - B_\nu(T_{BG})] \quad \text{for } \tau_\nu \gg 1 \tag{2.24}$$

and, by making use of the Taylor series for e^x,

$$\Delta I_\nu = [B_\nu(T_{ex}) - B_\nu(T_{BG})]\tau_\nu \quad \text{for } \tau_\nu \ll 1 \tag{2.25}$$

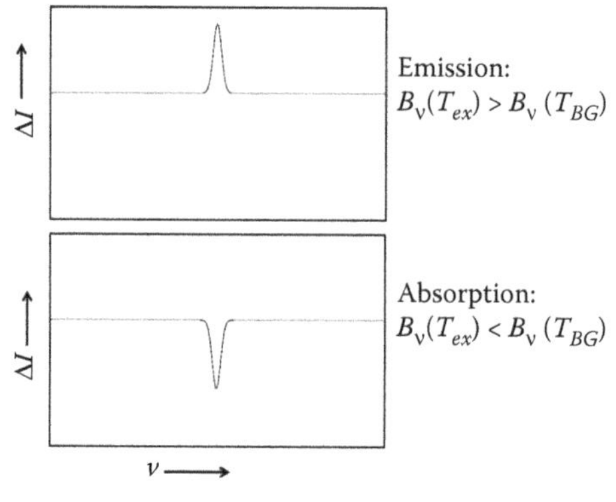

FIGURE 2.6 Criteria for generation of emission and absorption lines.

Equation 2.24 tells us optically thick line (e.g., ^{12}CO) or dust emission can be used to determine the temperature of a cloud. Equation 2.25 tells us the intensity of optically thin lines (e.g., ^{13}CO or $C^{18}O$) or dust emission is directly proportional to the cloud's optical depth. From knowing the cloud's optical depth, we will soon learn it is possible to determine the gas (or dust) column density, and, ultimately, the cloud's mass.

Gas particles (e.g., molecules or atoms) within a cloud will either emit or absorb photons, depending on their level populations. The interaction of photons between particles depends, to a large extent, on the velocity fields within the cloud. For one gas particle to interact with the photons from another, the two particles must be in a comoving frame, that is, moving at the same velocity. If they do not match in velocity space, the associated Doppler shift between particles means they will not match in frequency space either. The particles are effectively invisible to each other. The velocity match does not need to be exact; microturbulence within the cloud broadens the velocity range over which incoming photons can be received. Mathematically, these effects are introduced into our radiative transfer calculations through the line profile function, ϕ_ν.

The line profile function gives the probability that a photon will appear at a given frequency, ν.

$$\phi_\nu d\nu = -p(v)dv = p\left(-\frac{c\Delta\nu}{v}\right)\frac{cd\nu}{\nu}$$

Here, $p(v)dv$ is the probability that an atom has a LOS velocity between v and $v + dv$. The emergent photon from an atom or molecule will be Doppler shifted by an amount determined by the value of v. If the velocity distribution results from elastic collisions between gas particles participating in thermal motions, then $p(v)dv$ is described by a one-component Maxwellian distribution, which takes the form of a normalized Gaussian function.

$$p(v)dv = \frac{1}{\Delta v_D\sqrt{\pi}}e^{-(v/\Delta v_D)^2}dv$$

For the case of simple thermal broadening Δv_D is given by the gas sound speed,

$$\Delta v_D = \left(\frac{2kT}{m}\right)^{1/2} \tag{2.26}$$

where

T = temperature of gas (K)

m = mass of atom or molecule (typically, $H_2 = 2.33$ amu $= 3.8 \times 10^{-24}$ g)

If a microturbulent velocity, v_{turb}, is included, then

$$\Delta v_D = \left(\frac{2kT}{m} + v_{turb}^2\right)^{1/2} \tag{2.27}$$

Macroturbulent motions (i.e., systematic motions with velocities > Δv_D) within a cloud do not always change ϕ_v, but cause it to be centered about a frequency different than ν_o.

$$\phi_\nu d\nu = (\Delta\nu_D)^{-1}\phi_v dv = \frac{1}{\Delta\nu_D\sqrt{\pi}}e^{-u^2}du \tag{2.28}$$

where

$$u = \begin{cases} \dfrac{\nu-\nu_o}{\Delta\nu_D}, \text{ thermal and/or microturbulent} \\ \dfrac{\nu-\nu_o}{\Delta\nu_D} - \dfrac{v(r)(\nu_o/c)\mu}{\Delta\nu_D}, \text{ macroturbulent and microturbulent/thermal} \end{cases}$$

$\mu = \cos\theta$ = projection of the velocity vector back along LOS toward observer
$\Delta\nu_D = \Delta v_D(\nu_o/c)$
$v(r)$ = radial velocity taken to be positive toward the observer

Since ϕ_ν is a probability,

$$\int_0^\infty \phi_\nu d\nu = 1 \tag{2.29}$$

and we can insert it directly into our expressions for j_ν and k_ν (Equations 2.13 and 2.14).

$$j_\nu = \frac{h\nu}{4\pi}n_u A_{ul}\phi_\nu \tag{2.30}$$

$$k_\nu = \frac{h\nu}{4\pi}(n_l B_{lu} - n_u B_{ul})\phi_\nu \tag{2.31}$$

The equation for differential optical depth then becomes

$$d\tau_\nu = k_\nu dz = \frac{h\nu}{4\pi}\phi_\nu(n_l B_{lu} - n_u B_{ul})dz \tag{2.32}$$

2.3 RADIATIVE TRANSFER OF ROTATIONAL TRANSITIONS OF LINEAR MOLECULES IN LTE

Under LTE conditions, it is the ambient temperature, optical depth, and velocity fields within a cloud, which determine the appearance of spectral lines. Through careful study of the appearance of these lines, it is often possible to determine the mass, structure, and dynamics of objects. At THz frequencies, rotational transitions of diatomic molecules such

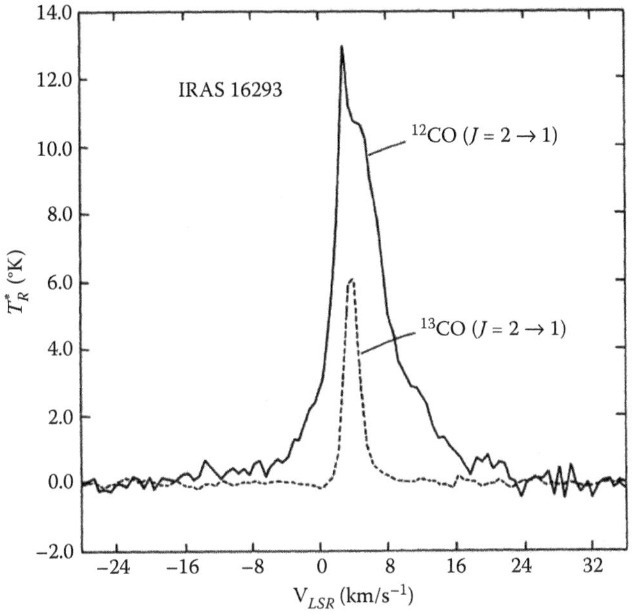

FIGURE 2.7 ^{12}CO and ^{13}CO $J = 2 \rightarrow 1$ spectra taken toward the proto-binary system IRAS 16293. (After Walker, C. K. et al., 1988, *Ap. J.*, 332, 335. With permission.)

as CO, CS, CN, HCO+, and others are often used to determine the physical conditions within a molecular cloud. Due to their abundance and relatively low excitation threshold, carbon monoxide (^{12}CO) and its isotopes (e.g., ^{13}CO and C^{18}O) are particularly useful. A plot of the ^{12}CO and ^{13}CO $J = 2 \rightarrow 1$ spectra, taken toward the protobinary source IRAS 16293, is shown in Figure 2.7 (Walker et al., 1988). We will discuss the origin of emission line profiles within protostars in Chapter 4. For now, let us use these CO lines to discuss the methodology of how to use the LTE form of the equation of transfer to derive cloud properties. We begin by determining the line's optical depth, then its excitation temperature, followed by its column density. From the line's column density, we can estimate the total gas column density, and mass along the LOS.

2.3.1 DETERMINING GAS OPTICAL DEPTH

By starting with Equation 2.32, and making substitutions for B_{lu} using Equation 2.18, N_u using Equation 2.16, and B_{ul} using Equation 2.17, the following LTE expression for optical depth in a molecular line can be derived:

$$d\tau_\nu = \frac{c^2}{8\pi\nu_{ul}^2}\frac{g_u}{g_l}A_{ul}n_l(1 - e^{-(h\nu_{ul}/kT_{ex})})\phi_\nu dz \qquad (2.33)$$

where

$d\tau_\nu$ = differential optical depth along LOS (nepers)
ν_{ul} = line center frequency (Hz)
$g_l = 2l + 1$ = statistical weight of lower state, l

$g_u = 2u + 1$ = statistical weight of upper state, u

A_{ul} = spontaneous emission coefficient from u to l (s^{-1}) = power radiated by the molecule (or atom) divided by its energy

$$= \frac{\langle P_{ul} \rangle}{h\nu_{ul}} = \left(\frac{64\pi^4 \nu_0^3}{3hc^3} \right) |\mu_{ul}|^2$$

$|\mu_{ul}|^2 = (\mu^2 (l+1)/(2l+1))$

μ = dipole moment of molecule (Debye)

n_l = number density of molecules in l (cm^{-3})

T_{ex} = excitation temperature (K)

dz = differential path-length along LOS (cm)

ϕ_ν = line profile function

The value of the optical depth, τ_ν, itself can often be determined directly from observations

In radio astronomy, intensities are often expressed in terms of an effective radiation temperature, $J_\nu(T) = T_R$, with the units of K.

$$J_\nu(T) = \frac{c^2 \Delta I_\nu}{2\nu_{ul}^2 k} = \frac{(h\nu_{ul}/k)}{(e^{(h\nu/kT)} - 1)} \tag{2.34}$$

The optical depth of a molecular transition can be derived from the ratio of the observed radiation temperatures, R_1, of two isotopes. Let us take the example of ^{12}CO, and its rarer isotope ^{13}CO.

$$R_1 = \frac{T_R^{12}}{T_R^{13}} = \frac{(J_\nu(T_{ex}^{12}) - J_\nu(T_{BG}))\,(1 - e^{-\tau_\nu^{12}})}{(J_\nu(T_{ex}^{13}) - J_\nu(T_{BG}))\,(1 - e^{-\tau_\nu^{13}})} \tag{2.35}$$

The optical depth in the ^{12}CO line is related to the optical depth in the ^{13}CO line by

$$\tau_\nu^{12} = X_{13}^{12} \tau_\nu^{13} \tag{2.36}$$

where X_{13}^{12} = the abundance ratio between ^{12}CO and ^{13}CO. In this analysis, let us assume the isotopes are intermixed and share the same physical conditions, that is, $T_{ex}^{12} = T_{ex}^{13}$ and are influenced by the same velocity fields. For a given transition ($J = 1 \rightarrow 0$, $J = 2 \rightarrow 1$, $J = 3 \rightarrow 2$, etc.), the frequency of the two isotopes are close enough together that we can make the approximation

$$J_\nu(T_{ex}^{12}) \approx J_\nu(T_{ex}^{13}) \tag{2.37}$$

The background temperature (typically taken to be that of the cosmic background, $T_{BG} = 2.735$ K) will also be the same for both transitions. Substitution of Equations 2.36 and 2.37 into Equation 2.35 yields

$$R_1 = \frac{(1 - e^{-X_{13}^{12}\tau_\nu^{13}})}{(1 - e^{-\tau_\nu^{13}})} \tag{2.38}$$

from which we can derive an observed value for τ_ν^{13}, and, through Equation 2.36, τ_ν^{12}.

For values of $\tau_\nu^{13} \ll 1$, Equation 2.36 reduces to

$$R_1 \approx \frac{\tau_\nu^{12}}{\tau_\nu^{13}} \approx X_{13}^{12} \tag{2.39}$$

Equation 2.39 tells us that if the ^{13}CO line is optically thin, then the value of R_1 should be the same as the abundance ratio of the two species. Observations show the value of X_{13}^{12} changes with the galactic radius, with values ranging from ~24 in the galactic center to ~70 in the outer galaxy (Langer and Penzias, 1990). An average value of $X_{13}^{12} \approx 40$ is commonly assumed for the galactic disk. In many sources, we find

$$3 \leq R_1 \leq 10$$

Such low values of R_1 indicate that ^{12}CO emission is, in many cases, optically thick and will serve as an effective thermometer. A good rule of thumb is that ^{12}CO will be optically thick in clouds where the visual extinction, A_V, is ≥4.

2.3.2 DERIVATION OF GAS EXCITATION TEMPERATURE

By substituting T_R for $J_\nu(T)$ in Equation 2.34, and solving for T_{ex}, we obtain

$$T_{ex} = \left[\frac{k}{h\nu_{ul}} \ln\left[\frac{h\nu_{ul}}{k}\left[\frac{T_R}{f_a(1-e^{-\tau_\nu})} + J_\nu(T_{BG})\right]^{-1} + 1\right]\right]^{-1} \tag{2.40}$$

where we have included a beam (or area) filling factor, f_a. This factor describes what fractional area of the beam is filled by the cloud. For arcminute sized beams in nearby

> EXAMPLE 2.1
>
> What is the optical depth, τ_ν^{13}, of the ^{13}CO $J = 2 \rightarrow 1$ line at line center (4.0 km/s) in Figure 2.6?
>
> We can estimate the optical depth, τ_ν^{13}, using Equation 2.38, if we know the ratio, R_1, between the ^{12}CO to ^{13}CO line emission at the velocity of interest, and their abundance ratio, X_{13}^{12}.
>
> From examination of Figure 2.6, we find at line center (where $V_{LSR} = 4.0$ km/s) $^{12}T_R^* \approx 13$K and $^{13}T_R^* \approx 6$K. This yields a value of $R_1 \approx 2.2$. The object (IRAS 16293) is associated with the galactic disk, so let's assume $X_{13}^{12} \approx 40$. Since $R_1 = X_{13}^{12}$, we know right away that the ^{13}CO line emission is getting optically thick. Substituting R_1 and X_{13}^{12} into Equation 2.38, and solving iteratively for τ_ν^{13}, we find,
>
> $$\tau_\nu^{13} \approx 0.6$$

galactic clouds at velocities close to line center, the filling factor is likely to be close to one. In situations where the clouds are unresolved clumps in the beam, the filling factor is not well known. Here it is best to use line ratios to derive T_{ex}. In the case of ^{12}CO, two commonly observed lines are the $J = 3 \rightarrow 2$ and $J = 1 \rightarrow 0$ transitions. Their line ratios are given by

$$R_2 = \frac{T_R^{32}}{T_R^{10}} = \frac{f_a^{32}\,(J_\nu(T_{ex}^{32}) - J_\nu(T_{BG}))\,(1 - e^{-\tau_\nu^{32}})}{f_a^{10}\,(J_\nu(T_{ex}^{10}) - J_\nu(T_{BG}))\,(1 - e^{-\tau_\nu^{10}})} \tag{2.41}$$

where

T_R^{32} = observed ^{12}CO $J = 3 \rightarrow 2$ line temperature
T_R^{10} = observed ^{12}CO $J = 1 \rightarrow 0$ line temperature
f_a^{32} = areal filling factor of ^{12}CO $J = 3 \rightarrow 2$ emission
f_a^{10} = areal filling factor of ^{12}CO $J = 1 \rightarrow 0$ emission

If the $J = 3 \rightarrow 2$ and $J = 1 \rightarrow 0$ observations are made with the same beam size, and we assume the emission arises from the same volume, then the filling factors cancel out. In the optically thick limit, Equation 2.41 becomes

$$R_2 = \frac{T_R^{32}}{T_R^{10}} = \frac{(J_\nu(T_{ex}^{32}) - J_\nu(T_{BG}))}{(J_\nu(T_{ex}^{10}) - J_\nu(T_{BG}))} \quad \text{for } \tau \gg 1 \tag{2.42}$$

and in the optically thin limit,

$$R_2 = \frac{T_R^{32}}{T_R^{10}} = \frac{(J_\nu(T_{ex}^{32}) - J_\nu(T_{BG}))\,\tau_{32}}{(J_\nu(T_{ex}^{10}) - J_\nu(T_{BG}))\,\tau_{10}} \quad \text{for } \tau \ll 1 \tag{2.43}$$

where

$$\frac{\tau_{32}}{\tau_{10}} = \frac{\nu_{32}\,|\mu_{32}^2|\,n_2\,[1 - e^{-(h\nu_{32}/kT_{ex})}]\,\phi_{32}}{\nu_{10}\,|\mu_{10}^2|\,n_0\,[1 - e^{-(h\nu_{10}/kT_{ex})}]\,\phi_{10}} \tag{2.44}$$

If we assume the lines have the same shape, then the line profile functions will also cancel out. To derive the ratio of the level populations, we must first determine the ratio of the ground state populations of each to the total number density of molecules, N_{tot}, along the LOS. For a linear molecule in LTE, with $kT_{ex} = hB$, this is given by

$$\frac{n_l}{n_{tot}} = (2l + 1)\exp\left(-\frac{l\,h\nu_{ul}}{2\,kT_{ex}}\right)\left[\frac{hB}{kT_{ex}}\right] \tag{2.45}$$

where

B = rotational constant of molecule (Hz)
$= \nu_{ul}/2(l + 1)$

Once we know (n_2/n_{tot}), and (n_0/n_{tot}), we take their ratio to get a value for (n_2/n_0) to substitute into Equation 2.44.

Figures 2.8 and 2.9 show the behavior of Equations 2.46 and 2.47 for a range of ^{12}CO emission lines and excitation temperatures commonly observed at THz frequencies. Examination of the figures reveals that the ^{12}CO $J=2\rightarrow 1/J=1\rightarrow 0$ line ratio rapidly becomes insensitive to changes in T_{ex} for $T_{ex} > 15$ K. Higher frequency transitions requiring greater collisional/thermal energies for excitation are better probes of T_{ex} in warmer gas, with optically thin line ratios being the most sensitive.

EXAMPLE 2.2

Based on the spectra of Figure 2.6, what would you estimate the physical (kinetic) temperature, T_k, of the IRAS 16293 cloud core to be? Assume the source areal beam filling factor, f_a, is one, and the main beam efficiency, η_{mb}, is 0.6.

Under the assumption of LTE, the gas kinetic temperature, T_k, and excitation temperature, T_{ex}, are the same, so we can use Equation 2.40 to estimate T_k. If we choose to use the very optically thick ^{12}CO line, then the expression simplifies to

$$T_{ex} = \left[\frac{k}{h\nu_{ul}}\ln\left[\frac{h\nu_{ul}}{k}\left[\frac{T_R}{f_a}+J_\nu(T_{BG})\right]^{-1}+1\right]\right]^{-1}$$

For this purpose, we take T_R to simply be the peak line temperature divided by η_{mb},

$$T_R \approx \frac{T_R^*}{\eta_{mb}} \approx \frac{13\text{ K}}{0.6} \approx 22\text{ K}$$

T_{BG} is the cosmic background temperature, 2.7 K. ν_{ul} is the frequency of the line being used. For the ^{12}CO $J = 2 \rightarrow 1$, this is 230.538 GHz.

Using Equation 2.34, we get

$$J_\nu(2.7\text{ K}) = 0.19\text{ K}$$

Substituting into Equation 2.40, we find

$$\begin{aligned} T_k \approx T_{ex} &= \left[\frac{1}{11.07}\ln\left[11.07\left[\frac{22}{1.0}+0.19\right]^{-1}+1\right]\right]^{-1} \\ &= \left[\frac{1}{11.07}\ln(1.5)\right]^{-1} \\ &= 27.3\text{ K} \end{aligned}$$

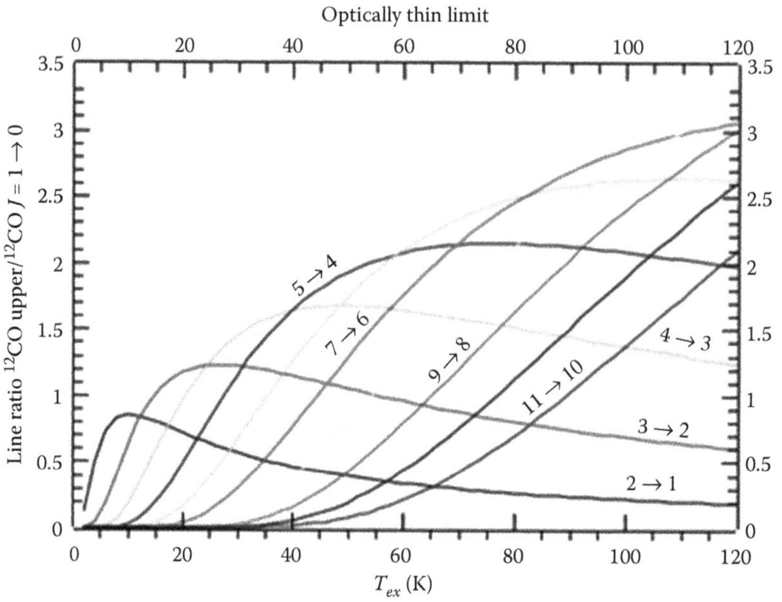

FIGURE 2.8 Relationship between the ratios of ^{12}CO line intensities in higher lying transitions to the ^{12}CO $J = 1 \rightarrow 0$ line in the optically thick limit.

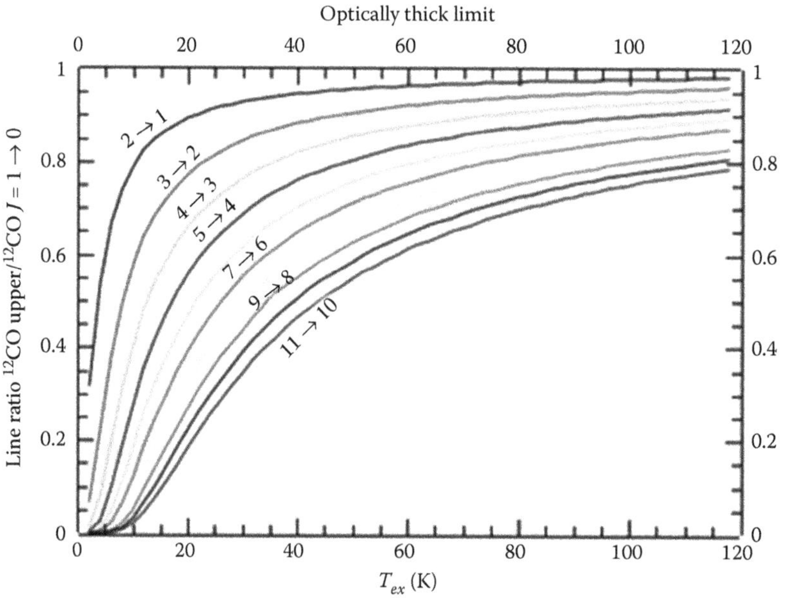

FIGURE 2.9 Relationship between the ratios of ^{12}CO line intensities in higher lying transitions to the ^{12}CO $J = 1 \rightarrow 0$ line in the optically thin limit.

2.3.3 DERIVATION OF GAS COLUMN DENSITY WITH OPTICAL DEPTH

Column density, N_{thin}^{ν}, is the number of particles (be they atoms, molecules, or dust grains) per unit area projected along the line of sight (see Figure 2.10). An accurate measurement of the column density requires that the measurement be made using an optically thin tracer. Since our tracers have a frequency dependence, so does our column density measurement.

$$N_{thin}^{\nu} = n_{tot}L = \left(\frac{n_l}{n_{tot}}\right)^{-1} n_l L \tag{2.46}$$

where

N_{thin}^{ν} = total column density of particles along LOS (particles/cm²)
n_{tot} = average number density of particles along LOS (particles/cm³)
L = path length through cloud
n_l = average number density of particles in lower energy state along LOS (particles/cm³)
N_l^{ν} = total column density of particles in lower energy state along LOS (particles/cm²)

Our expression for τ_ν (Equation 2.33) can be rearranged such that

$$n_l\,dz = d\tau_\nu \left[\frac{c^2}{8\pi\nu_{ul}^2}\frac{g_u}{g_l}A_{ul}(1 - e^{-(h\nu_{ul}/kT_{ex})})\phi_\nu\right]^{-1} \tag{2.47}$$

Integrating along our LOS, we have:

$$n_l \int_0^L dz = \left[\frac{c^2}{8\pi\nu_{ul}^2}\frac{g_u}{g_l}A_{ul}(1 - e^{-(h\nu_{ul}/kT_{ex})})\phi_\nu\right]^{-1} \int_0^L d\tau_\nu \tag{2.48}$$

and

$$n_l L = \tau_\nu \left[\frac{c^2}{8\pi\nu_{ul}^2}\frac{g_u}{g_l}A_{ul}(1 - e^{-(h\nu_{ul}/kT_{ex})})\phi_\nu\right]^{-1} \tag{2.49}$$

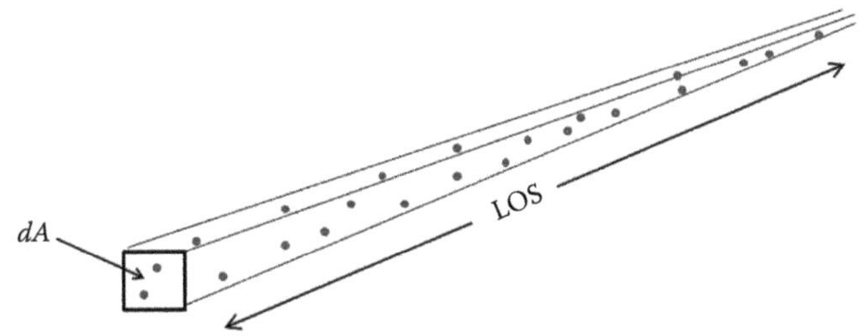

FIGURE 2.10 Column density, N_{thin}^{ν}, is the number of particles per unit area, dA, projected along the LOS.

Substitution of Equations 2.49 and 2.45 into Equation 2.46, then yields

$$N_{thin}^{\nu} = \tau_\nu T_{ex} \left[\frac{c^2 g_u}{8\pi\nu_{ul}^2} \left[\frac{hB}{k} \right] A_{ul} (1 - e^{-(h\nu_{ul}/kT_{ex})}) \right]^{-1} \exp\left(\frac{l\ h\nu_{ul}}{2\ kT_{ex}} \right) (\phi_\nu)^{-1} \tag{2.50}$$

where

N_{thin}^{ν} = total column density of particles along LOS (particles/cm^2)
B = rotational constant of molecule (Hz)
T_{ex} = excitation temperature (K)
φ_ν = line profile function
ν_{ul} = line center frequency (Hz)
$g_u = 2u + 1$ = statistical weight of upper state, u
A_{ul} = spontaneous emission coefficient from u to l (s^{-1})

Equation 2.50 tells us that the column density measured through line emission or absorption is directly proportional to the observed optical depth and, to first order, the excitation temperature.

In many instances, particularly when the lines are weak, τ, T_{ex}, and N_{thin}^{ν} are derived using line integrated intensities, that is, averaged over the frequency range, $\Delta\nu$, or the velocity width, ΔV, of the lines, where

$$\Delta\nu = \Delta V \frac{\nu_{ul}}{c} \tag{2.51}$$

Since we are averaging over the full line profile,

$$\begin{aligned} \phi_\nu \Delta\nu &\approx 1 \\ \Delta\nu &\approx \frac{1}{\phi_\nu} \end{aligned} \tag{2.52}$$

Substitution of Equations 2.52 and 2.51 into Equation 2.50 yields

$$N_{thin} = \tau_\nu T_{ex} \left[\frac{c^3 g_u}{8\pi\nu_{ul}^3} \left[\frac{hB}{k} \right] A_{ul} (1 - e^{-(h\nu_{ul}/kT_{ex})}) \right]^{-1} \exp\left(\frac{l\ h\nu_{ul}}{2\ kT_{ex}} \right) \Delta V \tag{2.53}$$

2.3.4 DERIVATION OF GAS COLUMN DENSITY IN THE OPTICALLY THIN LIMIT

Oftentimes, one does not have the luxury of observing both an optically thick and thin isotope. However, one can still estimate a lower limit to the gas column density in the optically thin limit. In the optically thin limit, where $\tau \ll 1$, background radiation passes through a cloud largely unimpeded, such that when you observe ON and OFF the cloud and subtract, the background term in Equation 2.12 cancels out, leaving

$$\Delta T_{obs} = J_{\nu}(T_{ex})(1 - e^{-\tau_{\nu}}) \tag{2.54}$$

where, as before,

$$J_{\nu}(T_{ex}) = \frac{(h\nu/k)}{e^{(h\nu/kT_{ex})} - 1}$$

Substitution yields

$$\Delta T_{obs} = \frac{h\nu/k}{e^{(h\nu/kT_{ex})} - 1}(1 - e^{-\tau}) \tag{2.55}$$

Let us rewrite this expression with an additional term, which will aid in simplification later (Goldsmith and Langer, 1999).

$$\Delta T_{obs} = \frac{h\nu/k}{e^{(h\nu/kT_{ex})} - 1}\frac{(1 - e^{-\tau})}{\tau}\tau \tag{2.56}$$

EXAMPLE 2.3

Assuming LTE conditions, what is the column density, N_{ν}^{thin}, of the ^{13}CO $J = 2 \rightarrow 1$ line at 4.0 km/s in Figure 2.6?

We can estimate N_{ν}^{thin} by plugging into Equation 2.53. Let's make a list of the input parameters we need:

$\tau_{\nu} = 0.8$ (from Example 2.1)
$T_{ex} = 27.3$ K (from Example 2.2)
$\nu_{ul} = 220.399$ GHz (Appendix 1)
$g_u = 2u + 1 = 2(2) + 1 = 5$

$$B = \frac{\nu_{ul}}{2(J+1)} = \frac{220.399 \times 10^9\,\text{Hz}}{2(1+1)} = 55.1 \times 10^6\,\text{Hz};$$

$h = 6.626 \times 10^{-27}$ erg s
$k = 1.38 \times 10$ erg K^{-1}

Two things are left to find: A_{ul} and ΔV. If A_{ul} is not readily available from a table, we can calculate it using the molecule's electric dipole moment, μ, and the following relations:

$$A_{ul} = \left(\frac{64\pi^4\nu_{ul}^3}{3hc^3}\right)|\mu_{ul}|^2$$

where

$$|\mu_{ul}|^2 = \frac{\mu^2(l+1)}{2l+1}$$

For the CO molecule, $\mu = 0.112$ Debye, and for the $J = 2 \rightarrow 1$ transition, $l = 1$. When substituting into the above expression, one must first put μ in units of statC × cm, by using the conversion factor, 1 Debye = 1D1 × 10^{-18} statC × cm. Substitution yields, $A_{21} = 1 \times 10^{-6}\ \mathrm{s}^{-1}$.

ΔV is the velocity width of the central line peak, here one spectrometer channel. Inspection of the figure gives

$$\Delta V \approx 0.35/\mathrm{s} = 0.35 \times 10^5\ \mathrm{cm/s}.$$

Careful substitution into Equation 2.53 then yields

$$N_{thin} \approx 2.16 \times 10^{18}\ \mathrm{cm}^{-2}$$

N_{thin} can be converted to N_{H_2} by multiplying by the appropriate H_2 to ^{13}CO abundance ratio, $X_{thin}^{H_2} \approx 7.5 \times 10^5$ (Frerking et al., 1982). We then find,

$$N_{H_2} \approx N_{thin} \times X_{thin}^{H_2} \approx 1.6 \times 10^{24}\ \mathrm{cm}^{-2}$$

IRAS 16293 is a heavily embedded protostellar cloud core. Values of N_{H_2} along the lines of sight to most toward molecular clouds are typically two orders of magnitude *less* than toward IRAS 16293.

The integrated optical depth along the line of sight through the cloud can be written as

$$\tau = \frac{h}{\Delta V} N_u B_{ul} \left(e^{(h\nu/kT_{ex})} - 1\right) \tag{2.57}$$

where

ΔV = full-width-half-maximum line width (km/s)
N_u = line of sight column density in upper energy state
B_{ul} = induced emission coefficient from level u to l (radiation density/ unit frequency)
Substitution of Equation 2.57 into Equation 2.56 yields

$$\Delta T_{obs} = \frac{h^2\nu}{k\Delta V} N_u B_{ul} \left(\frac{1-e^{-\tau}}{\tau}\right) \tag{2.58}$$

Here, for unit consistency, we relate B_{ul} to A_{ul} through the relation (Hilborn 1982),

$$B_{ul} = \frac{c^3}{8\pi h \nu^3} A_{ul} \tag{2.59}$$

Let us define an integrated intensity, II, equal to the area under all or part of the spectral line being considered, such that,

$$II = \int \Delta T_{obs} dV \tag{2.60}$$

Rewriting Equation 2.58 with Equations 2.59 and 2.60,

$$II = \frac{c^3 h}{8\pi k \nu^2} A_{ul} N_u \left(\frac{1 - e^{-\tau}}{\tau} \right) \tag{2.61}$$

In the optically thin limit, $\tau = 1$, the term in parentheses $\rightarrow 1$. Under this circumstance, Equation 2.61 can be rearranged to yield

$$N_u = \frac{8\pi k \nu^2}{c^3 h A_{ul}} II \tag{2.62}$$

From the Boltzmann relation, we know the column density of gas in the lower energy level is, then,

$$N_l = N_u \frac{g_l}{g_u} e^{(h\nu/kT_{ex})} \tag{2.63}$$

This is all good, but what we really want to know is the total column density, N_{thin}, along our LOS. By substituting Equations 2.63 and 2.62 into Equation 2.16, and rearranging Equation 2.45, we can derive an expression for N_{thin}.

$$N_{thin} = \frac{8\pi k^2 \nu^2}{g_u c^3 h^2 A_{ul} B} e^{(h\nu/kT_{ex})(1-(l/2))} T_{ex} \times II \tag{2.64}$$

where

$g_u = 2u + 1 =$ statistical weight of upper state, u
$B = (\nu/2(l + 1)) =$ rotation constant of molecule (Hz)
$T_{ex} =$ excitation temperature of gas (K)
$l =$ lower rotational level

Plugging in the constants, Equation 2.64 simplifies to

$$N_{thin} \approx 4.4 \times 10^{-5} \frac{\nu^2}{g_u A_{ul} B} e^{(h\nu/kT_{ex})(1-(l/2))} T_{ex} \times II \tag{2.65}$$

where

II = integrated intensity (K km/s)

$\approx 1.1 \times T_{pk} \times \Delta V_{FWHM}$, when integrated over full Gaussian line profile

T_{pk} = calibrated peak temperature of line corrected for antenna–source coupling efficiency (K)

ΔV_{FWHM} = full-width-half-maximum (FWHM) of Gaussian line profile (km/s)

2.3.5 ESTIMATING GAS DENSITY AND MASS

Now that we are able to determine the gas column density along our LOS through Equation 2.53 or Equation 2.65, we can combine this knowledge with a map of the cloud to determine the cloud's characteristic particle density and mass. Let L be the characteristic width (or size scale) of the cloud on the sky, then the cloud's characteristic H_2 density, $\tilde{n}_{H_2}$, is given by

$$\tilde{n}_{H_2} = \frac{N_{H_2}}{L} \tag{2.66}$$

where

$\tilde{n}_{H_2}$ = cloud's characteristic density (cm^{-3})

N_{H_2} = column density of H_2 (cm^{-2})

$= X_{thin}^{H_2} N_{thin}$

$X_{thin}^{H_2}$ = H_2 to optically thin tracer abundance ratio

L = cloud characteristic size scale (cm)

Assuming the cloud is spherical, its H_2 mass can then be found from

$$M_{H_2} = n_{H_2} \times \left(\frac{4}{3}\right)\pi\left(\frac{L}{2}\right)^3 \times m_{H_2} \tag{2.67}$$

where

m_{H_2} = 2.33 amu = mass of H_2 molecule

amu = 1.64×10^{-24} g

2.4 NON-LTE APPROACH

Under non-LTE conditions, level populations are effected both by collisions and the ambient radiation field in which the atom or molecule finds itself. By using the Boltzmann distribution (Equation 2.26), we enforce the condition of TE (or LTE), where collisions alone set the level populations. If this assumption is lifted, we can use statistical equilibrium to describe the flow of energy into and out of an energy level. For our two-level atom or molecule (Figure 2.5) we can write (Mihalas, 1970)

$$n_l(B_{lu}\int \phi_\nu J_\nu d\nu + C_{lu}) = n_u(A_{ul} + B_{ul}\int \phi_\nu J_\nu d\nu + C_{ul}) \tag{2.68}$$

where

$$J_\nu = \frac{1}{4\pi}\int I_\nu d\omega = \frac{1}{4\pi}\int_0^{2\pi}\int_0^{\pi} I_\nu(\theta,\phi)\sin\theta d\theta d\phi \quad (2.69)$$

J_ν = mean intensity = I_ν integrated over all directions
C_{ul} = rate of collisional de-excitation from u to l
C_{lu} = rate of collisional excitation from l to u; = $(n_u/n_l)C_{ul}$ for thermal occupation numbers
θ = polar angle in spherical coordinates
ϕ = azimuthal angle in spherical coordinates

Equation 2.68 can be rewritten to provide a new expression for n_u/n_l:

$$\frac{n_u}{n_l} = \frac{B_{lu}\int\phi_\nu J_\nu d\nu + C_{lu}}{A_{ul} + B_{ul}\int\phi_\nu J_\nu d\nu + C_{ul}} \quad (2.70)$$

Making use of the Einstein relations (Equations 2.17 and 2.18), substitution of Equation 2.70 into Equation 2.15 then yields a more general expression for the source function that includes both radiative and collisional processes.

$$S_l = \frac{\int\phi_\nu J_\nu d\nu + \varepsilon' B_\nu}{1+\varepsilon'} \equiv (1-\varepsilon)\bar{J}_\nu + \varepsilon B_\nu \quad (2.71)$$

where

$$\varepsilon' = \frac{C_{ul}(1 - e^{-(h\nu/kT)})}{A_{ul}}$$
$$\varepsilon \equiv \frac{\varepsilon'}{1+\varepsilon'} \quad (2.72)$$

As discussed by Mihalas (1970), in the above source function expression, Equation 2.71, $\bar{J}_\nu$ represents a noncoherent scattering term, and $\varepsilon' B_\nu$ a thermal source term. The thermal source term describes photons that are contributed to the radiation field by atoms or molecules that were collisionally excited. ε' is a sink term representing photons that are taken from the radiation field by collisional de-excitation. $\bar{J}_\nu$ represents the net result of the photon source and sink terms over the region. When $\varepsilon' \gg 1$, then the collision rate, C_{ul}, exceeds A_{ul}, in which case $S_l \rightarrow B_\nu(T)$ and the gas is in LTE.

The collisional excitation rate, C_{lu}, is the product of the particle density in level l, and the rate constant, γ_{lu} (Green and Thaddeus, 1976),

$$C_{lu} = n_{lu}\gamma_{lu} \quad (2.73)$$

where

$\gamma_{lu} = \langle\sigma_{lu}\nu\rangle$ = collision rate constant (cm³ s⁻¹)

σ_{lu} = collision cross section = πr^2 (for $J = 1 \rightarrow 0$; $r_{CO} = 1.13 \times 10^{-8}$ cm, $\sigma_{CO} = 4 \times 10^{-16}$ cm^2)
$\langle v \rangle$ = relative velocity of particles (cm s^{-1})
= for a Maxwell–Boltzmann distribution $\approx$ (8 kT/nm)$^{1/2}$

A useful metric for determining if a particular transition can be considered to be in LTE or non-LTE is the critical density, n_{crit}, which is the density above which collisional processes dominate over radiative processes.

$$n_{crit} = \frac{A_{ul}}{\gamma_{ul}} \tag{2.74}$$

Collision rates for transitions of CO, CS, OCS, and HC_3N for temperature ranges observed in molecular clouds, can be found in Green and Chapman (1978).

With collision rates, Einstein coefficients, and some initial assumptions of the ambient radiation field, we should have what we need to solve the statistical equilibrium equation (Equation 2.68) for our level populations. Once the level populations are known, Equations 2.46 through 2.55 can be used to estimate cloud properties. A difficulty in achieving this goal arises from the interdependency of the level populations, and the local radiation field. The existence of the interdependency requires both simplifying assumptions to be made, and an iterative approach to obtain a solution. Over the past 50+ years, several numerical approaches have been developed to handle this problem. The most commonly used methods adopt an escape probability formalism, such that,

$$\bar{J}_\nu = S_\nu(1 - \beta) + \beta B_\nu(T_{BG}) \tag{2.75}$$

where

$\bar{J}_\nu$ = average mean intensity in cloud
S_ν = source function
β = probability photons will escape the medium where they were created

The β depends only on the optical depth, τ, but can take on different forms, depending on assumptions about the cloud geometry and ambient velocity fields (van der Tak et al., 2007). For moderate-to-large velocity gradients in an expanding shell (e.g., in the expanding envelope around an evolved star), β takes the form (Mihalas, 1978; De Jong et al., 1980):

$$\beta_{LVG} = \frac{1}{\tau}\int_0^\tau e^{-\tau'}\,d\tau' = \frac{1 - e^{-\tau}}{\tau} \tag{2.75a}$$

In the case of a static, spherically symmetric medium (e.g., prestellar cores), β can be expressed as (Osterbrock and Ferland, 2006):

$$\beta_{sphere} = \frac{1.5}{\tau}\left[1 - \frac{2}{\tau^2} + \left(\frac{2}{\tau} + \frac{2}{\tau^2}\right)e^{-\tau}\right] \tag{2.76}$$

For simple plane parallel (i.e., slab) geometries (e.g., shocks), β takes the form (De Jong et al., 1975):

$$\beta = \frac{1 - e^{-3\tau}}{3\tau} \tag{2.77}$$

The public domain program RADEX (van der Tak et al., 2007) allows users to adopt any of these three escape probability formalisms for their calculations. If multiple lines have been observed, it is possible to constrain cloud column density, temperature, volume density, and molecular abundance ratios. An example of RADEX output is shown in Figure 2.11. Here, RADEX is used to plot column density versus volume density as a function of ^{12}CO $J = 3 \rightarrow 2/^{12}$ CO $J = 1 \rightarrow 0$ and ^{12}CO $J = 1 \rightarrow 0/^{13}$CO $J = 1 \rightarrow 0$ line ratios. For densities $>n_{crit}(\sim 10^4$ cm$^{-3})$, the lines become thermalized, rendering them insensitive to changes in volume density.

If the observational goal is to investigate microturbulent velocity fields (i.e., velocities less than the characteristic line width), over extended regions, the localized nature of an escape probability formalism reduces their effectiveness. Radiative transfer codes that work in the microturbulent regime include those that utilize either accelerated lambda iteration or Monte Carlo techniques (Bernes, 1979; Hogerheijde and van der Tak, 2000). In the Monte Carlo approach, all line photons are simulated by a number of model photons. The model photons are followed through the cloud, with their weight being modified

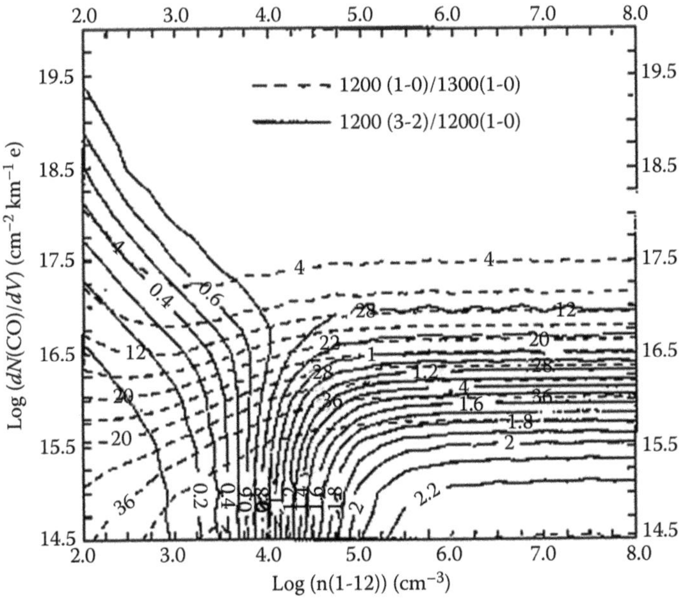

FIGURE 2.11 RADEX plot of ^{12}CO $J = 3 \rightarrow 2/^{12}$CO $J = 1 \rightarrow 0$ and ^{12}CO $J = 1 \rightarrow 0/^{13}$CO $J = 1 \rightarrow 0$ line ratios for a range of CO column densities per unit velocity (cm^{-2} km^{-1}) and volume densities (cm^{-3}) for a gas temperature of 20 K. (After Walker, C. E., 1991, A submillimeter–millimeterwave study of the molecular gas in the nuclear regions of three nearby starburst galaxies, PhD dissertation, University of Arizona.)

to account for absorptions and/or emissions that take place. The number of absorption events caused by the model photons is used to compute level populations. From the level populations, intensities are derived and compared to observations. The level populations can then be adjusted, and the processes repeated, until convergence with observations is achieved. Microturbulent codes are particularly useful in modeling regions of high optical depth ($\tau \geq 100$), where photon trapping can thermalize lines (van der Tak et al., 2007), for example, to model self-absorbed emission line profiles associated with the infall velocity fields of young protostellar objects (see Chapter 4 and Walker et al. 1986).

CONCLUSION

In this chapter, we have learned how to apply radiative transfer techniques to derive physical conditions in distant objects from the emission and absorption of atoms and molecules associated with them. In the next chapter, we will once more use radiative transfer, but for deriving physical conditions based upon thermal dust absorption and emission.

PROBLEMS

1. Prove the "On–Off" expression for observed intensity (Equation 2.21) starting with Equation 2.10.

2. How much does thermal broadening contribute to the width of a spectral line emitted from H_2 gas at 40 K?

3. If the gas in Problem 2 also has a turbulent velocity component of 1.3 km/s, what will be the observed line width?

4. You observe a $C^{32}S$ and $C^{34}S$ $J = 5 \rightarrow 4$ line emission toward a candidate protostellar core with a 12 m telescope. The peak temperatures of the lines are 4 and 0.4 K, respectively (e.g., Figure 4.5). If the abundance ratio between $C^{32}S$ and $C^{34}S$ is 22.6, what is the optical depth of the $C^{34}S$ gas at the peak of the line?

5. The rest frequencies of the $C^{32}S$ and $C^{34}S$ $J = 5 \rightarrow 4$ lines are 244.935686 GHz and 241.016176 GHz. The peak temperature of the $C^{34}S$ $J = 5 \rightarrow 4$ line is 0.4 K.

   a. From dust continuum observations (see Chapter 3) you estimate the physical temperature in a core to be ~20 K. Adopting this value for T_{ex}, a $C^{34}S$ optical depth of 0.05, and a line width of 0.3 km/s, what is the $C^{34}S$ column density along the observed line of sight (LOS)?

   b. Adopting an H_2 to $C^{32}S$ abundance ratio of 1.7×10^{-8}, what is the corresponding H_2 column density?

   c. What would you estimate the visual extinction, A_V, toward the core to be?

d. If the emission region is the size of 12 m telescope beam, what would you estimate the core mass to be if the object is at 160 pc? Express your answer in terms of solar masses ($M_\odot$).

6. If you assume the $C^{34}S$ emission in Problem 5 is optically thin, use Equation 2.65 to estimate its column density. How does this column density estimate compared to that of the $C^{34}S$ column density derived in Problem 5?

7. Using a 30 m telescope, you measure a CO $J = 1 \rightarrow 0$ integrated intensity of 30 K km/s. Along the same line of sight, you measure a CO $J = 3 \rightarrow 2$ integrated intensity of 21 K km/s with a 10 m telescope. Assuming LTE, what is the gas excitation temperature in the

   a. Optically thick limit?

   b. Optically thin limit?

8. Along a line of sight to a star forming region, you measure a CO $J = 1 \rightarrow 0$ and CO $J = 3 \rightarrow 2$ line integrated intensity of 25 K km/s, and 15 K km/s, respectively. You also measure a CO $J = 1 \rightarrow 0$ and $C^{13}O$ $J = 1 \rightarrow 0$ line ratio of 8. The FWHM line widths are 2 km/s. Using the RADEX plot of Figure 2.10, what would you estimate the CO column density and volume density to be?

REFERENCES

Bernes, C., 1979, A Monte Carlo approach to non-LTE radiative transfer problems, *A&A*, 73, 67.

de Jong, T., Boland, W., and Dalgarno, A., 1980, Hydrostatic models of molecular clouds, *A&A*, 91, 68.

de Jong, T., Dalgarno, A., and Chu, S.-I., 1975, Carbon monoxide in collapsing interstellar clouds, *Ap. J.*, 199, 69.

Frerking M. A., Langer W. D., and Wilson R. W., 1982, The relationship between carbon monoxide abundance and visual extinction in interstellar clouds, *Ap. J.*, 262, 590.

Goldsmith, P. F. and Langer, W. D., 1999, Population diagram analysis of molecular line emission, *Ap. J.*, 517, 209–225.

Green, S. and Chapman, S., 1978, Collisional excitation of interstellar molecules—Linear molecules CO, CS, OCS, and HC3N, *Ap. JS.*, 37, 169.

Green, S. and Thaddeus, P., 1976, Rotational excitation of CO by collisions with He, H, and H_2 under conditions in interstellar clouds, *Ap. J.*, 205, 766.

Hilborn, C., 1982, Einstein coefficients, cross sections, f values, dipole moments, and all that, *Am. J. Phys.*, 50, 982–986.

Hogerheijde, M. R. and van der Tak, F. F. S., 2000, An accelerated Monte Carlo method to solve two-dimensional radiative transfer and molecular excitation. With applications to axisymmetric models of star formation, *A&A*, 362, 697.

Langer, W. D. and Penzias, A. A., 1990, C-12/C-13 isotope ratio across the Galaxy from observations of C-13/O-18 in molecular clouds, *Ap. J.*, 357, L477.

Mihalas, D., 1970, *Stellar Atmospheres*, 1st edition, W. H. Freeman and Co., San Francisco.

Mihalas, D., 1978, *Stellar Atmospheres*, 2nd edition, W. H. Freeman and Co., San Francisco.

Osterbrock, D. E. and Ferland, G. J., 2006, *Astrophysics of Gaseous Nebulae and Active Galactic Nuclei*, University Science Books, Sausalito, CA.

van der Tak, F. F. S., Black, J. H., Schöier, F. L., Jansen, D. J., and van Dishoeck, E. F., 2007, A computer program for fast non-LTE analysis of interstellar line spectra. With diagnostic plots to interpret observed line intensity ratios, *A&A*, 468, 627.

Walker, C. E., 1991, A submillimeter–millimeterwave study of the molecular gas in the nuclear regions of three nearby starburst galaxies, PhD dissertation, University of Arizona.

Walker, C. K., Lada, C. J., Young, E. T., Maloney, P. R., and Wilking, B. A, 1986, Spectroscopic evidence for infall around an extraordinary IRAS source in Ophiuchus, *Ap. J.*, 309, L47.

Walker, C. K., Lada, C. J., Young, E. T., and Margulis, M., 1988, An unusual outflow around IRAS 16293-2422, *Ap. J.*, 332, 335.

3

THz CONTINUUM EMISSION

PROLOGUE

On a moonless night outside of cities, the clouds of stars associated with our Milky Way can be seen arching across the sky (see Figure 1.2). If it is an exceptionally dark site, one can begin to make out dark shapes among the dull white glow of stars. These dark objects are dust clouds blocking the light from the stars behind them. From how much the stars of a known brightness are dimmed, we can estimate how much dust is along our line of sight. Dust clouds not only obscure starlight, they also absorb and reradiate it at THz frequencies. In this chapter we will learn how the absorption and re-emission of starlight by dust can be used to derive ISM properties.

3.1 INTRODUCTION

In addition to cosmic background radiation, continuum emission can have contributions from synchrotron radiation produced by electrons spiraling around magnetic field lines, free–free emission produced by unbound electrons scattering off ions, or thermal dust emission produced by the reradiation of starlight by dust grains. Figure 3.1 is a plot displaying the relative intensities of these contributions as a function of frequency toward the starburst galaxy M82 (Klein et al., 1988; Carlstrom and Kronberg, 1991; Condon, 1992). At frequencies greater than ~300 GHz (i.e., $\lambda \geq 1$ mm), thermal dust emission is seen to dominate over other forms of continuum radiation. This behavior is due to the widespread existence of ~12–30 K dust (see Figure 3.2) and the wavelength dependence of the emission mechanisms. Dust grains are believed to form principally within the expanding atmospheres of evolved stars, where both the temperatures (1000–2000 K) and densities (~10^9 cm^{-3}) are suitable for the condensation and growth of grains. Dust is also observed to form in the ejecta of novae, and, perhaps, supernovae. Dust grains can grow or be shattered

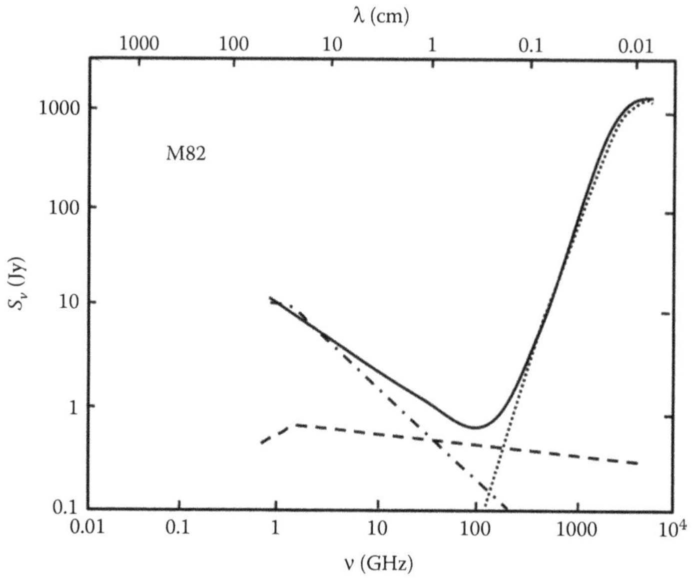

FIGURE 3.1 Radio/THz continuum spectrum of M82. Dotted line is for thermal dust, dashed line for free–free, and dot-dash line for synchrotron emission. Solid line is a fit to all observed flux values. (Data for fit from Condon, J., 1992, *Annu. Rev. Astron. Astrophys.*, 30, 575–611. With permission.)

FIGURE 3.2 Planck spacecraft map of the galactic plane shows an intricate web of cold (12–30 K) dust emission. (Image credit: Copyright ESA and the HFI Consortium, IRAS. With permission.)

through grain–grain collisions in clouds, can grow ice mantles, and be the site of molecule formation. Grains can be destroyed by supernova blast waves. Details concerning grain growth and destruction mechanisms are still largely unknown.

The extinction, A_ν, of starlight by dust is defined in terms of magnitudes of extinction,

$$\begin{aligned} A_\nu &= -2.5\log\left|\frac{S_\nu^o}{S_\nu^i}\right| \\ &= 1.086\tau_D \\ &= 1.086 N_D Q_{ext}(\pi a^2) \end{aligned} \tag{3.1}$$

where

S_ν^o = observed flux density (Jy)
S_ν^i = intrinsic flux density (Jy)
τ_D = dust optical depth
a = dust grain radius (cm)
N_D = dust column density (cm^{-2})
Q_ν^{ext} = emissivity of dust grain, that is, how much like a black body it appears to be; 1 ⇔ true black body, 0 ⇔ fully invisible

Q_ν^{ext} is due to a combination of scattering and absorption, such that

$$Q_\nu^{ext} = Q_\nu^{abs} + Q_\nu^{scat} \tag{3.2}$$

Expressed in terms of antenna theory, Q_ν^{ext} can be thought of as the aperture efficiency of a dust grain at a particular frequency (see Equation 5.4).

If the amount of extinction experienced by starlight is plotted as a function of wavelength, an extinction curve is obtained, whose features can provide clues to the composition and size distribution of dust grains (see Figure 3.3). Increased absorptions at 0.22, 10, and 18 μm are consistent with resonances due to the presence of graphite, silicon carbide, and silicon (Drain, 2003). Spectral features at 6.2, 7.7, 8.6, 11.3, and 12.7 μm suggest the presence of large numbers of polycyclic aromatic hydrocarbons (PAH) molecules. PAHs consist of carbon atoms organized into hexagonal, planar rings with hydrogen atoms attached at the edges. As much as 20% of the total far-infrared (i.e., THz) emission of a galaxy can present itself in PAH spectral features (Draine, 2011). From the slope of the extinction curve and width of the absorption features, dust grain size distributions can be estimated. Silicate material is believed to be a major component of interstellar dust grains. The corresponding interstellar absorption bands appear broad and smooth, suggesting that the silicate material is amorphous rather than crystalline. Draine and Lee (1984) show that dust grain models that include graphite and silicon are consistent with dust opacity measurements in the far-infrared. PAHs can be added to a graphite–silicate model either as a third component or as a small-particle extension of the graphitic material (Draine, 2011). Individual grain radii can extend from ≤0.005 (PAHs) to ~0.12 μm (silicate and carbonaceous). Under appropriate conditions, grains are capable of accumulating an ice mantle, as suggested by the presence of a 3.1 μm absorption feature in the interstellar extinction curve. In dense ($n_{H_2} \geq 10^6\ cm^{-3}$) protostellar cores, microscopic grains can coagulate to form fluffy aggregates upto macroscopic sizes (~0.1 mm), significantly

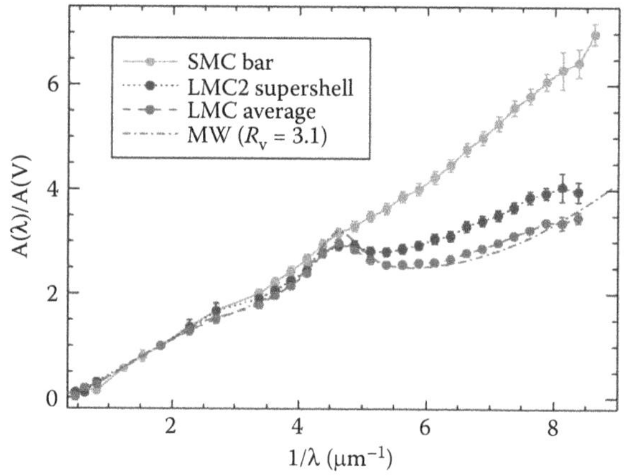

FIGURE 3.3 Interstellar extinction curves, as measured toward the Milky Way (MW), large magellanic cloud (LMC), and small magellanic clouds (SMC). Differences in features can be used to compare dust properties between objects. (Data published in Gordon, K. D. et al., 2003, *Ap. J.*, 594, 279. With permission.)

changing their effective opacity (Ossenkopf, 1993; Ossenkopf and Henning, 1994; Shirley et al., 2011; van de Marel et al., 2012).

Estimates of hydrogen column density can be made directly from extinction measurements using the following empirical relationship derived from x-ray observations (Predehl and Schmitt, 1995):

$$\frac{N_H}{A_v} \approx 1.79 \times 10^{21} \frac{\text{atoms}}{\text{cm}^2\,\text{mag}} \tag{3.3}$$

where

N_H = hydrogen column density (cm^{-2})

A_v = visual extinction (mag)

Dust grains account for only ~1% of the ISM (interstellar medium) by mass, but are responsible for absorbing ~30% of the Universe's starlight and reradiating it at THz frequencies. This is because dust is widespread, and has a relatively high emissivity (Q_V^{ext}) at the wavelengths where the brightest stars emit most of their energy, that is, in the UV/optical (Bernstein et al., 2002).

3.2 THz SPECTRAL ENERGY DISTRIBUTIONS

The wavelength of THz photons is large compared to the size of interstellar dust grains, making THz dust emission optically thin in all but the most extreme cases and a powerful probe of the ISM. The equation of transfer can be combined with what we know about interstellar dust grains to determine physical conditions within the ISM. For example, from a spectral energy distribution alone, a dust core's temperature, size, luminosity, and mass, can be estimated.

From Equation 2.5, we know

$$dI_\nu = k_\nu I_\nu dz - j_\nu dz \tag{3.4}$$

where

I_ν = intensity (erg/cm²/rad²/s/Hz)
k_ν = absorption coefficient (cm⁻¹)
j_ν = emission coefficient (erg/s/cm³/rad²/s/Hz)
z = path-length traveled by emitted photon (cm)

In the case of dust grains emitting as small black bodies, we have

$$k_\nu = N_d \pi a^2 Q_\nu^{abs} \tag{3.5}$$

and

$$j_\nu = k_\nu B_\nu(T_d) \tag{3.6}$$

At THz frequencies, the photons are much larger than a typical dust grain, so that scattering does not contribute significantly to the dust emissivity; $Q_\nu^{ext} \rightarrow Q_\nu^{abs}$. The dimensionless quantity Q_ν^{abs} is a function of the grain's radius, a, the imaginary part of the dielectric constant, ε_ν^d, and the frequency of observation, ν, (Hildebrand, 1983).

$$Q_\nu^{abs} = 8\pi a\left(\frac{\nu}{c}\right)\mathrm{Im}\left\{-\frac{\varepsilon_\nu^d - 1}{\varepsilon_\nu^d + 2}\right\} \tag{3.7}$$

Recalling expression Equation 2.8, we can now write an expression for dust optical depth, τ_d.

$$\begin{aligned}\tau_d &= k_\nu L = N_d \pi a^2 Q_\nu^{abs} \Delta z \\ &= \left(\frac{\nu}{\nu_o}\right)^\beta\end{aligned} \tag{3.8}$$

where

Δz = line of sight path-length through the dust cloud
ν_o = frequency at which the dust emission becomes optically thin (i.e., $\tau_\nu \approx 1$)
β = spectral index governing the grain emissivity along a line of sight

Scoville and Kwan (1976) have shown that in most sources, ν_o corresponds to the frequency ν_{max} of the peak in the spectral energy distribution. For a dust cloud at temperature, T_d, ν_{max} can be estimated using Wein's Displacement Law.

$$\nu_o \approx \nu_{max} \approx (0.103\ \mathrm{THz/K}) \cdot T_d \tag{3.9}$$

The solution to the equation of transfer expressed in terms of dust optical depth is, then,

$$I_\nu^d = B_\nu(T_d)(1 - e^{-\tau_d}) \tag{3.10}$$

where

I_ν^d = intensity from dust emission (erg/s/cm²/rad²/s/Hz)
$B_\nu(T_d)$ = Planck function at the temperature of dust grains, T_d, (K)
τ_d = optical depth of dust (nepers)

Astronomical sources subtend a solid angle Ω_s (rad²) on the sky. The source flux, F_ν, is simply the source intensity integrated over the associated solid angle.

$$F_\nu = \Omega_s I_\nu \tag{3.11}$$

Substitution yields

$$F_\nu = \Omega_s B_\nu(T_d)(1 - e^{-\tau_d}) \tag{3.12}$$

Since dust emission is largely optically thin, we can adopt the simplified solution to the equation of transfer (Equation 2.25), such that

$$\begin{aligned} F_\nu^d &= \Omega_s B_\nu(T_d)\tau_d \\ &= \Omega_s B_\nu(T_d)\left(\frac{\nu}{\nu_o}\right)^\beta \end{aligned} \tag{3.13}$$

A single temperature for dust grains in any environment is an approximation. In actuality, there is a range of temperatures present. Only in the most optically thick cases is a single temperature a good representation. As an example, Figure 3.4 shows a model spectral energy distribution, drawn using Equation 3.9, of a dust core associated with the protobinary system IRAS 16293-2422 (Walker et al., 1990). The dust core was assumed to have a temperature, $T_d = 32$ K, and a source size, $\Omega_s = 2.67 \times 10^{-9}$ rad². A convenient unit for flux often used in THz/radio astronomy is the Jansky (Jy), named after the radio astronomy pioneer Karl Jansky.

$$1\,\text{Jy} = 1 \times 10^{-26} \frac{\text{watts}}{\text{m}^2\,\text{Hz}} = 1 \times 10^{-23} \frac{\text{ergs}}{\text{s}\,\text{cm}^2\,\text{Hz}} \tag{3.14}$$

In Figure 3.4, three different values of β are drawn. A $\beta = 0$ corresponds to a truly opaque black body (where $Q_\nu^{abs} = 1$). Higher values of β result in lower values of F_ν and are said to yield a "diluted" black-body or grey-body distribution at frequencies where the dust becomes optically thin (i.e., $\nu < \nu_o$). Fits to the spectral energy distributions toward dust cores associated with star forming regions typically yield values of β between 1 and 2. For IRAS 16293-2422 model fits to observed fluxes yield a $\beta = 1.27$. The value of β can be used in combination with grain models to provide insight into the size and composition of dust grains.

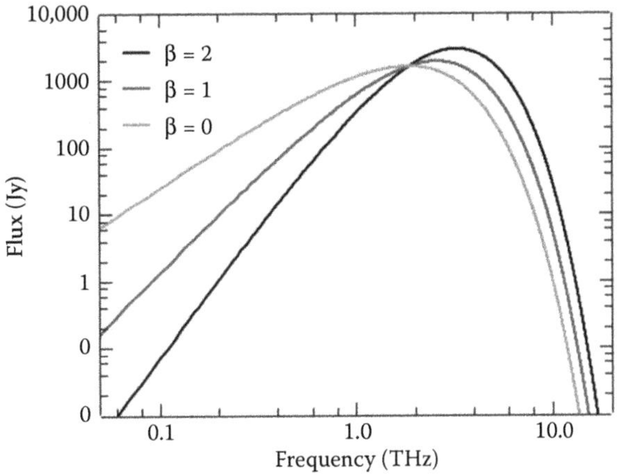

FIGURE 3.4 Model THz spectral energy distributions of dust emission toward a protostellar source. From this plot alone, the temperature, luminosity, mass, column density, and size of the dust core associated with the protostar can be estimated.

The luminosity of the underlying source heating a dust core is directly proportional to the area under the associated spectral energy distribution and can be determined by integrating Equation 3.9 over frequency and using the following relation:

$$L = 3.1 \times 10^{-10} D^2 \int F_\nu d\nu \tag{3.15}$$

where

L = luminosity in units of solar luminosity, ($1 L_\odot = 3.84 \times 10^{33}$ (ergs/s) $= 3.84 \times 10^{26}$ watts)
F_ν = flux (Jy)
D = distance in parsec (1 pc $= 3.09 \times 10^{18}$ cm)

3.3 DERIVING DUST OPTICAL DEPTH, COLUMN DENSITY, AND MASS

Once the black-body temperature is estimated from the spectral energy distribution, the dust optical depth, τ_d, can be estimated from a flux measurement at a given frequency, using Equation 3.13,

$$\tau_d = F_\nu [\Omega_m B_\nu(T)]^{-1} \tag{3.16}$$

where

F_ν = observed flux at ν
Ω_m = solid angle of the telescope beam used to observe the source

When $\tau_d \ll 1$, the dust mass can be found using the expression from Hildebrand (1983),

$$M_d = \left(\frac{4}{3}\right)\frac{F_\nu D^2}{Q_\nu B_\nu(T)} a\rho \tag{3.17}$$

where

$$Q_\nu = 7.5 \times 10^{-4}\left(\frac{\nu}{2.4\ \text{THz}}\right)^\beta \tag{3.18}$$

a = grain radius (~1×10^{-4} cm)
ρ = grain density (~3 g/cm³)
D = distance to the object (cm)

For a gas-to-dust ratio, f, (typically assumed to be ~100), the average gas mass (M_{H_2}), column density (N_{H_2}), and density (n_{H_2}) can also be determined from dust observations.

$$M_{H_2} = f \times M_d \tag{3.19}$$

$$N_{H_2} = \frac{\pi M_{H_2}}{4 m_{H_2} D^2 \Omega_m^2} \tag{3.20}$$

$$n_{H_2} = \frac{6 M_{H_2}}{\pi (D^3 \Omega_m^3)} \tag{3.21}$$

Here, we have assumed the dust emission arises from regions where the gas is essentially all in molecular form.

EXAMPLE 3.1

You wish to look for protostars in the Rho Ophiuchi molecular cloud (~160 pc away) using a 10 m radio telescope. Searching the IRAS database, you see there is an infrared source in the cloud with a spectral energy distribution (SED) that peaks at 3 THz (i.e., 100 microns). You use the telescope to map the object at 1.3 mm and find it to be about the size of your beam. The line of sight flux to the object is 6 Jy. You add your flux measurement to the SED and measure a spectral index longward of the peak of $\beta = 1.5$.

a. What is the average temperature of the dust core?
The average temperature of the dust core can be estimated using Wien's Displacement Law, Equation 3.9.

$$T = \frac{3\ \text{THz}}{0.103\ \text{THz/K}} \approx 29\ \text{K}$$

b. What is the mass of the protostellar dust core?
The dust core mass can be estimated from Equation 3.17. First determine the values of the various input parameters in compatible units.

$F_\nu = 6\ \text{Jy} = 6 \times 10^{-26}(\text{W/m}^2\ \text{Hz}) = 6 \times 10^{-23}(\text{erg/s cm}^2\ \text{Hz})$ = observed flux

$a \approx 0.1\ \mu\text{m} = 0.1 \times 10^{-4}\ \text{cm}$ = typical dust grain radius.

$\rho \approx 3(\text{g/cm}^{-3})$ = grain mass density.

$(M_g/M_d) \approx 100$ = gas-to-dust ratio.

$\nu = 230 \times 10^9\ \text{Hz}.$

$Q_\nu \approx 7.5 \times 10^{-4}\,(0.23\ \text{THz}/2.4\ \text{THz})^{1.5} = 2.2 \times 10^{-5}$ = dust emissivity (Equation 3.18).

$B_\nu(29\ K) = 3.87 \times 10^{-13}\ (\text{erg/cm}^2\ \text{sec Hz rad}^2)$ = Planck function at 230 GHz for dust core temperature (Equation 2.19).

$D = 160\ \text{pc} = 160(3.09 \times 10^{18}\ \text{cm}) = 4.9 \times 10^{20}\ \text{cm}$ = distance to source.

Substitution into Equation 3.17 yields

$$\begin{aligned} M_{H_2} &= \left(\frac{4}{3}\right)\frac{6 \times 10^{-23}(4.9 \times 10^{20})^2}{(2.2 \times 10^{-5})3.87 \times 10^{-13}}0.1 \times 10^{-4}(3)(100) \\ &= 6.8 \times 10^{33}\ \text{g} \\ &= 3.4\ M_\odot \end{aligned}$$

3.4 TEMPERATURE AND DENSITY DISTRIBUTIONS

As discussed by Westbrook et al. (1976), the distribution of optically thin dust emission at THz frequencies (i.e., the far-infrared) around a centrally heated dust core can serve as an indicator of the radial dependence of density, $\rho(r)$. If the dust temperature and observing frequency is such that $(h\nu/kT)$ is small and the Rayleigh–Jeans approximation can be used (e.g., >48 K at 1 THz), then the optically thin emission $E(r)$ varies with radius from the central heating source as

$$E(r) \propto \rho(r)T_d(r) \qquad (3.22)$$

where

$$\rho(r) \propto r^{-n}$$
$$T_d(r) \propto r^{-m}$$

Substitution yields

$$E(r) \propto r^{-k}, \quad \text{where } k = n + m \tag{3.23}$$

In general,

$$m = \begin{cases} 0.4 & \text{for } \beta = 1 \\ 0.33 & \text{for } \beta = 2 \end{cases}$$

From a knowledge of β, the temperature power law index, *m*, can be specified. Then from fits to the flux profile across the source, *E*(*r*), the density power law index, *n*, can be deduced. Knowledge of *n* can be used to constrain evolutionary models for a dust core. A fit to the observed 1.3 mm (i.e., 230 GHz) dust continuum flux profile, *E*(*r*), through the protostellar source, L1551 IRS5, is shown in Figure 3.5 (Walker et al., 1990). In performing fits, one should take care to account for the smoothing effect a telescope's beam has on the source distribution when observing unresolved sources (see Chapter 5). The observed flux profile for L1551 IRS5 is found to be consistent with the density profile $\rho(r) \propto r^{-1.5}$, expected in an infalling cloud core (Shu, 1977).

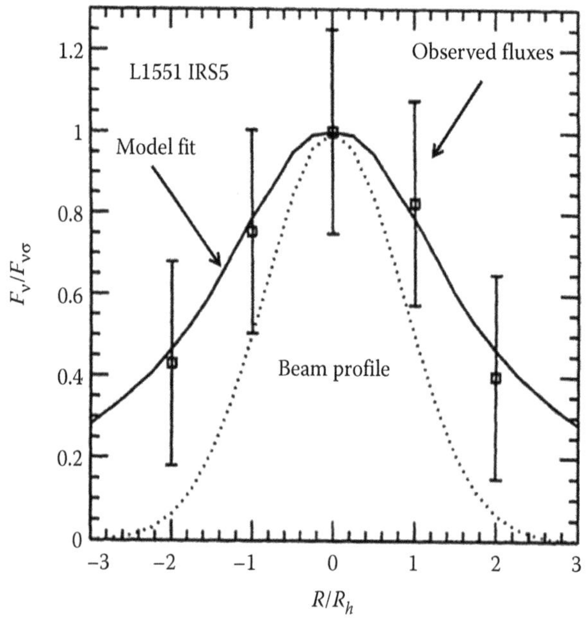

FIGURE 3.5 Observed and theoretical observed flux profiles through the protostellar source L1551 IRS5. Data points and error bars are from 1.3 mm dust continuum observations. The theoretical dust continuum distribution (solid line) was computed using the model of Adams, Lada, and Shu (1987). The theoretical flux profile was convolved with a 30″ Gaussian beam (dotted curve) to match observations. (After Walker, C. K. et al., 1990, *Ap. J.*, 349, 515. With permission.)

3.5 DUST ENERGY BALANCE IN CLOUDS

Dust grains are heated by starlight and reradiate this energy in the infrared. This situation is depicted in Figure 3.6, where a subset of the photons from a star of radius R_* is intercepted by a cloud at distance, R, composed of dust grains of radius a. The flux, F_ν^*, emanating from the star's surface, will be diluted by a factor $(R_*/R)^2$ before reaching the grains. The effective collecting (and transmitting) area of the grains is reduced from their physical area (πa^2) for grain emissivity, $Q_\nu^{abs} < 1$. Once absorbed, the photon's energy will go into heating the grain to a temperature (T_d) and be reradiated into space from the grain's surface ($4\pi a^2$) in the form of lower energy photons whose number and frequencies are dictated by the Planck function. Mathematically, the situation can be expressed in terms of the energy balance equation,

$$\underbrace{\int_0^\infty F_\nu^* \left(\frac{R_*}{R}\right)^2 \pi a^2 Q_\nu^{abs} d\nu}_{\text{Stellar photons}} = \underbrace{\int_0^\infty 4\pi a^2 Q_\nu^{abs} B_\nu(T) d\nu}_{\text{THz photons}} \tag{3.24}$$

where the left side of the equation describes the energy flow into a grain, and the right side, the energy flow out of a grain. Depending on the number and luminosity of the star or stars involved, the equilibrium value of T_d can range from ~10 to 1000 K.

3.6 DUST–GAS COUPLING

In most ISM environments, wherever there is dust, there will be gas. The dust is heated through the absorption of stellar photons. In dense regions, collisions with dust grains can

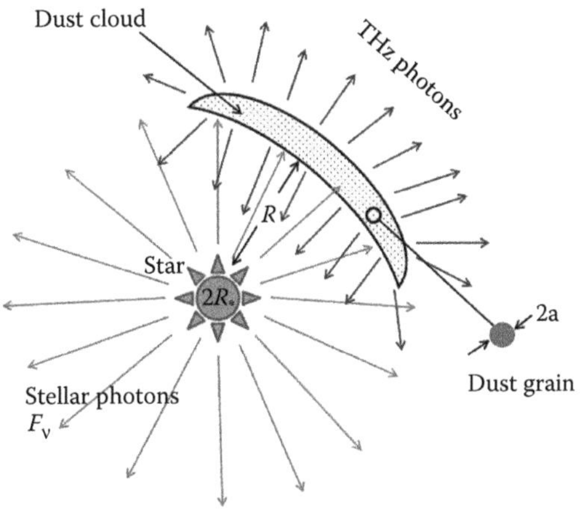

FIGURE 3.6 Energy balance in a dust cloud. The energy from stellar photons is intercepted and reprocessed into THz photons by dust grains.

transfer some of this thermal energy to the gas. For dust grains at a temperature, T_d, and gas at a temperature, T_g, the difference in thermal energy between them, ΔE_g^d, is

$$\Delta E_g^d = \frac{3}{2}k(T_d - T_g) \tag{3.25}$$

The dust-to-gas heating rate for a single gas particle, Γ_g^d, is the product of the dust–gas collision rate, C_g^d, and ΔE_g^d,

$$\Gamma_g^d = C_g^d \times \Delta E_g^d \tag{3.26}$$

where

$$C_g^d = n_d \sigma_d V_g^d$$

and

$$V_g^d = \left(\frac{8kT_g}{\pi m_{H_2}}\right)^{1/2}$$

Here, we take the collision velocity, V_g^d, to be the mean sound speed in the gas. Substitution yields

$$\Gamma_g^d = n_d(\pi a^2)\left(\frac{\gamma k T_g}{m_{H_2}}\right)^{1/2} \frac{3}{2}k(T_d - T_g) \tag{3.27}$$

where

Γ_g^d = heating rate of single a gas particle (ergs/s)
n_d = dust volume density (cm^{-3})
a = dust grain radius (cm)
m_{H_2} = mass of molecular hydrogen molecule (3.3×10^{-24} g)
γ = adiabatic index (= 1.6 for H_2 at 92 K)

The heating rate between all gas and dust particles, $\tilde{\Gamma}_g^d$, is, then,

$$\tilde{\Gamma}_g^d = n_d n_g(\pi a^2)\left(\frac{1.6k}{m_{H_2}}\right)^{1/2} \frac{3}{2}kT_g^{1/2}(T_d - T_g) \tag{3.28}$$

where $\tilde{\Gamma}_g^d$ is in units of ($ergs/cm^3$ s).

When $T_d > T_g$, the heating rate, $\tilde{\Gamma}_g^d$, is positive, and heat is transferred from the dust to gas. For situations where $T_d < T_g$, $\tilde{\Gamma}_g^d$ is negative, the gas heats the dust. At very high density

(>10^4 cm^{-3}), $T_g \rightarrow T_d$, and the gas and dust temperatures approach equilibrium (Goldreich and Kwan, 1974; Scoville and Kwan, 1976).

In addition to heating by dust, gas within molecular clouds can be heated directly by cosmic rays at a rate (Goldsmith, 2001)

$$\Gamma_g^{cr} = 10^{-27}[n_{H_2}] \quad (3.29)$$

where

Γ_g^d = cosmic ray heating rate (ergs/cm^3 s)

n_{H_2} = the number density of molecular hydrogen molecules (cm^{-3})

Cosmic rays alone can heat the gas to ~10 K.

3.7 DUST POLARIZATION: ORIGIN AND MEASUREMENT

In the above discussion, we have assumed a spherical model for dust grains. However, the observed polarization of starlight, at the few percent level (Hall, 1949; Hiltner, 1949) at optical wavelengths, tells us this is not the case. For the starlight to suffer selective extinction in one polarization, and not the other, means that along a given line of sight some fraction of the dust grains are both elongated and aligned relative to each other. Assuming it is the prevailing interstellar magnetic field responsible for the alignment, one finds the presumed field geometry to be consistent with that derived from Zeeman splitting, synchrotron emission, and Faraday rotation (Hildebrand, 1983). Polarization of starlight is observed both in the diffuse ISM, and in relatively dense dust cores, where frequent collisions with gas molecules will work to disturb grain alignment.

If the dust grains are composed largely of ferromagnetic material, then one could perhaps think of them as "compass needles" aligned parallel to the prevailing field. However, it is alignment perpendicular to the field that is observed. Also, Spitzer and Tukey (1951) showed that given the weak nature of the field (~10^{-5} gauss), even the motions associated with the thermal energy of the surrounding gas (~$(3/2)kT$) are enough to prevent alignment of the "compass needles."

The most robust explanation for the observed grain alignment is the Davis–Greenstein mechanism (Davis and Greenstein, 1951). In this model, a dust grain with volume V is assumed to be elongated, immersed in a magnetic field, **B**, paramagnetic with an induced magnetization, **M**, and spinning at angular frequency, **ω**, due to collisions (see Figure 3.7). The grain will have a magnetic moment, V**M**, and experience a torque, **L**, in a direction to make **M** line up with **B**:

$$\mathbf{L} = \mathrm{V}\mathbf{M} \times \mathbf{B} \quad (3.30)$$

for

$$\mathbf{M} = \chi''\omega^{-1}(\omega \times \mathbf{B})$$

where χ'' is the imaginary part of the complex grain susceptibility.

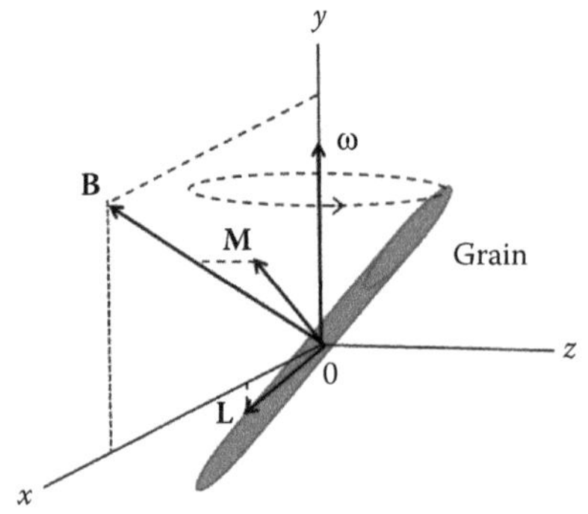

FIGURE 3.7 Davis–Greenstein effect. As the grain rotates about y, it experiences a torque, **L**, that works to align it perpendicular to **B**.

With each gyration of the grain about the axis, z, a small part of the rotational energy with spin vectors not aligned with **B** will be dissipated within the grain, due to paramagnetic absorption (an analog to dielectric loss). Given enough time, **M** will become aligned with **B**, with the result that the grain's long axis will be perpendicular to **B**, as is observed. As described by Davis and Greenstein, a familiar example of this type of "self-aligning" behavior is a rapidly spinning top on a table. The top may, at first, have a wobble, which prevents it from being aligned vertically with gravity, but very small frictional forces, between the top's pivot point and the table, work to dampen the precession and nutation motions with each spin, allowing the top to slowly right itself, until the top "sleeps."

In the original description of the mechanism, it would take ~10^5 years for a dust grain to come into alignment, a time scale much longer than the time taken for grain–molecule collisions. In order to dissipate perturbances from collisions, the grains need to spin much faster than speeds associated with pure stochastic processes. Proposed mechanisms for increasing the spin rate of grains involve anisotropic processes on the grains themselves, for example, through grain sites hosting exothermic H_2 formation (Hollenbach and Salpeter, 1971) producing suprathermal rotation or grain sites with "superparamagnetic" (e.g., ferrous) inclusions (Mathis, 1986).

If we assume the dielectric properties of subwavelength-sized dust grains are largely isotropic, then the E-vectors of their thermal black-body emission will take on the orientation of the bulk material, and be perpendicular to the magnetic field lines on which they find themselves. Polarized thermal emission from magnetically aligned dust grains was first observed at THz frequencies (i.e., the far-infrared) by Cudlip et al. (1982) and Hildebrand et al. (1984).

Linear polarization, P_ν, is measured by observing the flux density, F_ν, of an object, by first passing the incoming light through a polarization modulator (typically a half-wave plate) then through an analyzing element (typically a wire grid) located in front of the detector (see Figure 3.8). The sense of the incoming linear polarization changes as a function of the modulator's rotation angle, θ. The purpose of the modulator is to rotate the polarization angle of the incoming light relative to the grid, so that if some fraction of the light is polarized, the corresponding portion of the incoming flux will produce a maximum

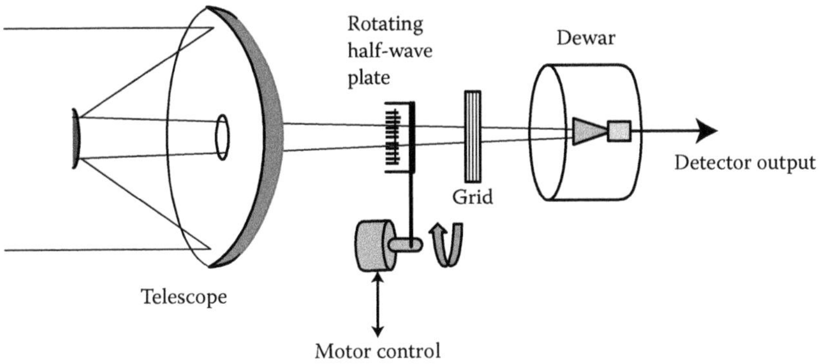

FIGURE 3.8 Single channel polarimeter utilizing rotating, half-wave plate. (After Glenn, J. et al., 1997, *Ap. J.*, 479, 325. With permission.)

and minimum response from the detector, as the modulator rotates. The peak-to-peak response can be used to estimate the percentage of the incoming signal that is polarized. Half-wave plates are made from either natural (e.g., quartz) or artificially formed birefringent material. In such materials, the index of refraction is different for different polarizations of light passing through it. At THz frequencies, a birefringent material can be produced by cutting closely spaced (typically a small fraction of a wavelength apart), parallel grooves in a dielectric material (e.g., Rexolite or Teflon). The groove depth and spacing are designed to produce a 180° phase shift between the electric field component of the incident wave parallel to the groove, relative to the electric field component of the incident wave orthogonal to the groove. The net effect is to rotate the polarization of the emergent signal by 90°. Kirschbaum and Chen (1957) provide a prescription for designing grooves for half-wave plates. The polarization axis aligned perpendicular to the grooves is referred to as the "fast optical axis." The polarization axis aligned parallel to the dielectric material forming the groove walls is called the "slow optical axis" (since light travels more slowly through a dielectric than free space). The operation of a half-wave plate is depicted in Figure 3.9.

With the polarimeter configuration of Figure 3.8, the flux density of the light incident on the detector is (Glenn, 1997)

$$F_\nu = 0.5F_0\{1 - P\cos[4(\psi + \phi) + 2\delta] + P_{sys}(\psi,\delta)\} \tag{3.31}$$

where

F_ν = flux incident on detector (Jy)
F_0 = flux from telescope entering polarimeter (Jy)
P_ν = fractional polarization (1 corresponds to 100% polarized)
ψ = fast axis position angle of plate (rad)
δ = constant phase offset of waveplate (rad)
ϕ = sum of other phase offsets of system, that is, parallactic angle, elevation angle (rad)
P_ν^{sys} = systematic polarization offset

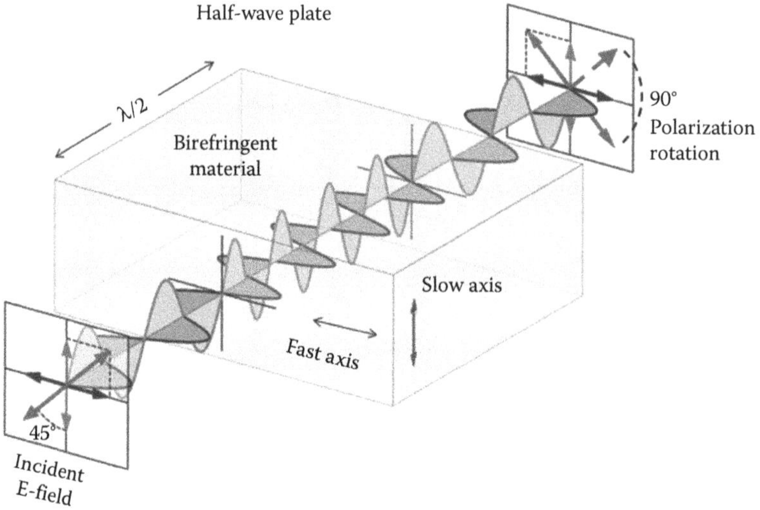

FIGURE 3.9 Operation of a half-wave plate. An electric field polarized at 45° is incident on a slab of birefringent material with a vertical slow axis and horizontal fast axis. As the wave propagates through the material, the vertical component of the field begins to lag behind the horizontal component, for a total phase delay of 180° (i.e., λ/2). The delay flips the sign of the vertical wave component. This sign reversal causes the polarization of the output electric field vector to be rotated 90° relative to what it was on incidence. (Adapted from Waveplate. Licensed under Creative Commons Attribution-Share Alike 3.0 via Wikimedia Commons—http://commons.wikimedia.org/wiki/File:Waveplate.png#mediaviewer/File:Waveplate.png. With permission.)

During observations, F_v is measured at four different angles, ψ, separated by 22.5°, both on and off source. From these flux measurements, the normalized Stokes parameters, Q and U, are derived (see Serkowski, 1974). Systematic offsets are determined by observing an unpolarized source (e.g., Jupiter). Once systematic offsets are removed, Q and U can be used to determine the linear polarization, P_v, at position angle, θ_v.

$$P_v = \sqrt{Q^2 + U^2} \tag{3.32}$$

$$\theta_v = 0.5\tan^{-1}\left(\frac{U}{Q}\right) \tag{3.33}$$

A 1.3 mm wavelength (230 GHz) linear polarization map of the star-forming region DR 21, made with the polarimeter design of Figure 3.8, is shown in Figure 3.10 (Glenn, 1997). The degree of linear polarization, P_v, is overplotted on the corresponding 1.1 mm dust continuum map from Richardson et al. (1989). The orientation of the magnetic field, **B**, is orthogonal to P_v.

For observing greater efficiency, and to reduce systematic measurement errors, two detectors (or detector arrays) can be employed. In this arrangement, the light just after the half-wave plate is split between the two detectors (or arrays) using a 1-D wire grid at 45° (Hildebrand et al., 1984; Platt et al., 1991; Schleuning et al., 1997; Dowell et al., 2010).

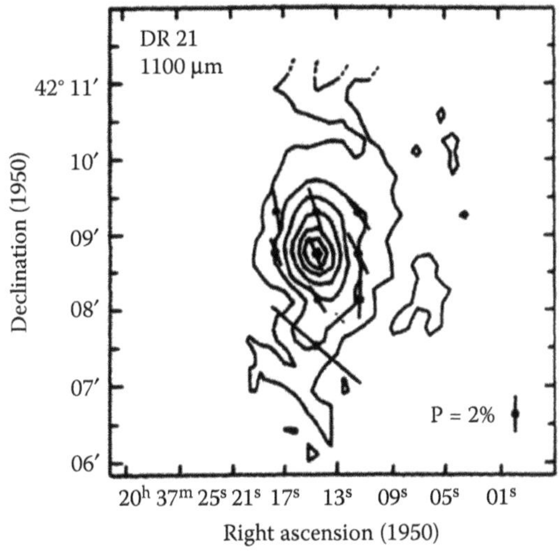

FIGURE 3.10 1.3 mm linear polarization map of DR21 overplotted on a contour plot of 1.1 mm thermal dust emission. (Adapted from Glenn, J. et al. 1997, *Ap. J.*, 479, 325.)

Polarization rotators have also been used in combination with heterodyne receivers for polarization studies of both line (Glenn et al., 1997) and dust emission. For single-dish observations, a polarimeter utilizing a wideband bolometer is more sensitive than one using a heterodyne receiver that has far less bandwidth (perhaps ~10× less). However, polarimetric observations with heterodyne receivers on a multielement interferometer can produce maps of magnetic field structure with both high sensitivity and angular resolution (see Rao et al., 2009; Hull et al., 2013).

The degree of polarization observed in THz emission, like in absorption in the optical, depends on the shape and dielectric properties of the grains, but where $\tau \ll 1$ (as is often the case) is independent of optical depth (Hildebrand, 1988). It is this property of THz polarized emission that makes it a powerful, unique probe of magnetic fields within regions of high visual extinction. In dense regions, where τ is nonnegligible, P_v will decrease with optical depth, according to

$$P_v = \frac{e^{-\tau}\sinh(P_0\tau)}{1 - e^{-\tau}\cosh(P_0\tau)} \tag{3.34}$$

where P_0 is the polarization when $\tau \to 0$ (Hildebrand et al., 1999).

CONCLUSION

Depending on the wavelength, dust grains will scatter or absorb starlight. Absorbed starlight is reemitted by dust grains at infrared and far-infrared wavelengths, thereby playing a key role in setting the energy balance within the ISM. Observations of dust emission and absorption can be used to derive dust/gas temperatures, column density, mass, and/or luminosity of interstellar clouds. The polarization of dust grains can be used to estimate the orientation and strength of magnetic fields within the ISM. In the next chapter, we will use what we have learned about line and dust emission/absorption to construct a simple radiative transfer model of a cloud core.

PROBLEMS

1. What is the average temperature of a dust core with a spectral energy distribution (SED) that peaks at 60 microns?

2. Using an interferometer, you measure the size of a dust core in the Taurus molecular cloud (distance ~ 140 pc) to be 8″ in diameter. The SED of the dust core suggests it has a mean temperature of 48 K and a spectral index of 1.5. At 230 GHz the measured flux density of the core is 2 Jy. What is the core's

   a. Dust optical depth?

   b. Dust mass?

   c. Gas mass?

3. At 230 GHz you measure a flux of 0.7 Jy toward the center of the galaxy NGC 253, using a 12 m radio telescope. NGC 253 is located ~3.3 Mpc from the Milky Way. Using observations from the literature, you construct an SED for NGC 253 from which you estimate its mean dust temperature and spectral index to be 29 K and 2, respectively. Estimate the total gas mass along your line of sight.

4. What is the heating rate between gas and dust particles in a protostellar core, if the gas and dust have temperatures of 27 and 40 K, and volume densities of 10^3 and 10^4 cm^{-3}, respectively? If the dust core is assumed to be a sphere 8″ in diameter, with a 30 L_ε protostar in the middle, what percentage of the protostar's luminosity is being used to heat the gas?

5. Using a polarimeter, you measure a fractional polarization in dust emission of 0.02 in the relatively low optical envelope of a protostellar core. Along the line of sight toward the center of the core, the fractional polarization drops to 0.015. Estimate the dust optical depth through the center of the core.

REFERENCES

Bernstein, M. P., Elsila, J. E., Dworkin, J. P., Sandford, S. A., Allamandola, L. J., and Zare, R. N., 2002, Side group addition to the polycyclic aromatic hydrocarbon coronene by ultraviolet photolysis in cosmic ice analogs, *Ap. J.*, 576, 1115.

Carlstrom, J. E. and Kronberg, P. P., 1991, H II regions in M82—High-resolution millimeter continuum observations, *Ap. J.*, 366, 422.

Condon, J., 1992, Radio emission from normal galaxies, *Annu. Rev. Astron. Astrophys.*, 30, 575–611.

Cudlip, W., Furniss, I., King, K. J., and Jennings, R. E., 1982, Far infrared polarimetry of W51A and M42, *MNRAS*, 200, 1169.
Davis, L. and Greenstein, J. L., 1951, The polarization of starlight by aligned dust grains., *Ap. J.*, 114, 206.
Dowell, C. D., Cook, B. T., Harper, D. A. et al. 2010, HAWCPol: A first-generation far-infrared polarimeter for SOFIA, *SPIE*, 7735E, 213.
Drain, B., 2003, Scattering by interstellar dust grains. I. Optical and ultraviolet, *Ap. J.*, 598, 1017.
Draine, B., 2011, *Physics of the Interstellar and Intergalactic Medium*, Princeton University Press, Princeton, New Jersey.
Draine, B. and Lee, H., 1984, Optical properties of interstellar graphite and silicate grains, *Ap. J.*, 285, 89.
Glenn, J., 1997, *Millimeter Wave Polarimetry of Star Formation Regions and Evolved Stars*, Ph.D. Dissertation, University of Arizona.
Glenn, J., Walker, C. K., and Jewell, P. R., 1997, HCO^+ spectropolarimetry and millimeter continuum polarimetry of the DR 21 star-forming region, *Ap. J.*, 479, 325.
Goldreich, P. and Kwan, J., 1974, Molecular clouds, *Ap. J.*, 189, 441.
Goldsmith, P., 2001, Molecular depletion and thermal balance in dark cloud cores, *Ap. J.*, 557, 736.
Gordon, K. D., Clayton, G. C., Misselt, K. A., Landolt, A. U., and Wolff, M. J., 2003, A quantitative comparison of the small magellanic cloud, large magellanic cloud, and milky way ultraviolet to near-infrared extinction curves, *Ap. J.*, 594, 279.
Hall, J., 1949, Observations of the polarized light from stars, *Science*, 109, 166.
Hildebrand, R., 1988, Magnetic fields and stardust, *QJRAS*, 29, 327.
Hildebrand, R. H., 1983, The determination of cloud masses and dust characteristics from submillimetre thermal emission, *QJRAS*, 24, 267.
Hildebrand, R. H., Dotson, J. L., Dowell, C. D., Schleuning, D. A., and Vaillancourt, J. E., 1999, The far-infrared polarization spectrum: First results and analysis, *Ap. J.*, 516, 834.
Hildebrand, R. H., Dragovan, M., and Novak, G., 1984, Detection of submillimeter polarization in the Orion nebula, *Ap. J.*, 284, 51.
Hiltner, W., 1949, Polarization of light from distant stars by interstellar medium, *Science*, 109, 165.
Hollenbach, D. and Salpeter, E. E., 1971, Surface recombination of hydrogen molecules, *Ap. J.*, 163, 155.
Hull, C., Plambeck, R., Bolatto, A. et al., 2013, Misalignment of magnetic fields and outflows in protostellar cores, *Ap. J.*, 768, 159.
Kirschbaum, H. S. and Chen, S., 1957, A method for producing broad-band circular polarization employing an anisotropic dielectric, *IRE Trans. Microwave Theory Tech.*, MTT-5, 199.
Klein, U., Wielebinski, R., and Morsi, H. W., 1988, Radio continuum observations of M82, *A&A*, 190, 41.
Mathis, J. S., 1986, The alignment of interstellar grains, *Ap. J.*, 308, 281.
Ossenkopf, V., 1993, Dust coagulation in dense molecular clouds: The formation of fluffy aggregates, *A&A*, 280(2), 617.
Ossenkopf, V. and Henning, T., 1994, Dust opacities for protostellar cores, *A&A*, 291, 943.
Platt, S. R., Hildebrand, R. H., Pernic, R. J., Davidson, J. A., and Novak, G., 1991, 100-micron array polarimetry from the Kuiper airborne observatory—Instrumentation, techniques, and first results, *PASP*, 103, 1193.
Predehl, P. and Schmitt, J. H. M. M., 1995, X-raying the interstellar medium: ROSAT observations of dust scattering halos., *A&A*, 293, 889.
Rao, R., Girart, J. M., Marrone, D. P., Lai, S.-P., and Schnee, S., 2009, IRAS 16293: A "Magnetic" tale of two cores, *Ap. J.*, 707, 921.
Richardson, K. J., Sandell, G., and Krisciunas, K., 1989, Small-scale structure in the DR 21/DR 21 (OH) region—A high resolution continuum study at millimetre and submillimetre wavelengths, *A&A*, 224, 199.
Schleuning, D. A., Dowell, C. D., Hildebrand, R. H., Platt, S. R., and Novak, G., 1997, HERTZ, A submillimeter polarimeter, *PASP*, 109, 307.
Scoville, N. Z. and Kwan, J. 1976, Infrared sources in molecular clouds, *Ap. J.*, 206, 718S.
Serkowski, K., 1974, Polarization techniques, *Methods of Experimental Physics*, N. Carleton, (ed.), Vol 12A: Astrophysics, 361.

Shu, F. H., 1977, Self-similar collapse of isothermal spheres and star formation, *Ap. J.*, 214, 488.
Shirley, Y., Huard, T., Pontoppidan, K., Wilner, D., Stutz, A., Bieging, J., and Evans, N., 2011, Observational constraints on submillimeter dust opacity, *Ap. J.*, 728, 143.
Spitzer, L. and Tukey, J. W., 1951, A theory of interstellar polarization, *Ap. J.*, 114, 187.
van der Marel, N., van Dishoeck, E., Bruderer, S. et al., 2012, A major asymmetric dust trap in a transition disk, *Science*, 340(6137), 1199–1202.
Walker, C. K., Adams, F. C., and Lada, C. J., 1990, 1.3 millimeter continuum observations of cold molecular cloud cores, *Ap. J.*, 349, 515.
Waveplate. Licensed under Creative Commons Attribution-Share Alike 3.0 via Wikimedia Commons—http://commons.wikimedia.org/wiki/File:Waveplate.png#mediaviewer/File:Waveplate.png.
Westbrook, W. E., Gezari, D. Y., Hauser, M. G., Werner, M. W., Elias, J. H., Neugebauer, G., and Lo, K. Y., 1976, One-millimeter continuum emission studies of four molecular clouds, *Ap. J.*, 209, 94.

4

SIMPLE RADIATIVE TRANSFER MODEL

PROLOGUE

By integrating what we have learned about line and dust radiative transfer, we are now in a position to construct a simple radiative transfer model for investigating the emergent radiation from an astrophysical object. As an example, in this chapter we will outline the steps to construct a model protostellar cloud core.

4.1 INTRODUCTION

A spherical protostellar cloud core is chosen for a first radiative transfer model because of its simple geometry and its wide range of physical conditions and velocity fields. The volumetric densities within a protostellar core are relatively high, making an initial simplifying assumption of local thermodynamic equilibrium or LTE (i.e., excitation dominated by collisions; $n = n_{crit}$) more valid than if we were to model a more diffuse structure (see Section 4.7 for a non-LTE approach). The decision about which computer language to use in realizing a radiative transfer code depends, to a large extent, on the desired numerical resolution of the model (with higher resolution advocating the use of a language that can support parallel computational strings), but, perhaps most importantly, on the taste and skill set of the coder.

4.2 GEOMETRY

Gravity prefers to sculpt objects into spheres. A protostellar core is no exception, so we will model our protostellar core as a series of nested, spherical shells. The greater the

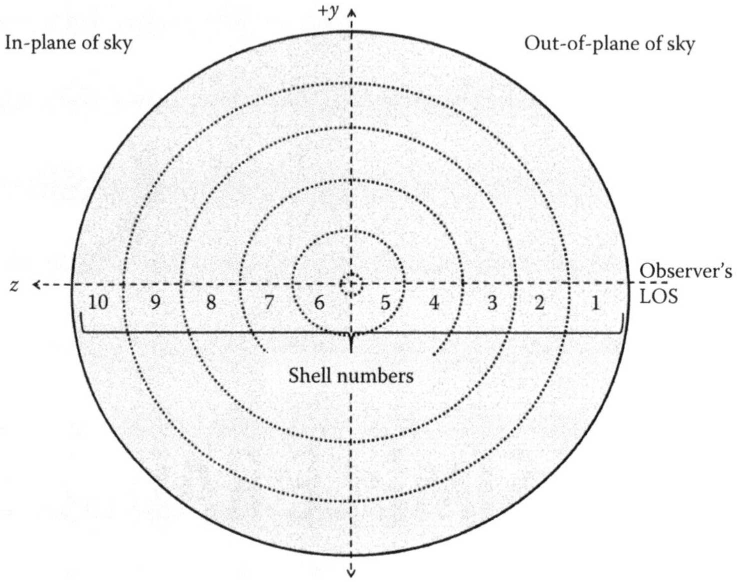

FIGURE 4.1 Geometry of spherical radiative transfer code. The object is divided into 10 hemispheres, five out of the plane of the sky and five in the plane of the sky. The observer views the sphere from the right. The geometric center of the object marks the origin of both Cartesian and spherical coordinate systems.

number of shells, the greater the model's spatial resolution will be, and the closer the model will be to reality. Let us place the center of the nested spheres at the origin (0, 0, 0) of a Cartesian coordinate system (x, y, z), with the plane of the sky corresponding to the x–y plane and lines of sight (LOSs) to the observer on or parallel to the z-axis. Such a structure, with $N_S = 10$ shells, is depicted in Figure 4.1. The observer is to the right of the core, with the first shell encountered along an LOS from the observer being designated as Shell 1. The larger the shell number, the further along the LOS you are. Using this methodology, each shell actually has two numbers, the smaller of the two numbers corresponding to the side of the shell that is out of the plane of the sky, and the larger number to the component of the shell that is within (or behind) the plane of the sky. For convenience, we will also make the center of the Cartesian coordinate system coincident with that of a spherical coordinate system (r, θ, ϕ), where r is radius, θ is the polar angle, and ϕ is the azimuthal angle.

4.3 SOURCE PHYSICAL CONDITIONS

A spherical geometry lends itself to defining physical conditions in terms of normalized, radial dependent expressions.

$$\text{Temperature: } T(r) = T_0\left(\frac{r}{r_1}\right)^{\alpha};= T_0 \quad \text{for } r \leq r_1 \tag{4.1}$$

$$\text{Density: } n(r) = n_0\left(\frac{r}{r_1}\right)^{\beta};= n_0 \quad \text{for } r \le r_1 \tag{4.2}$$

$$\text{Velocity: } v(r) = v_0\left(\frac{r}{r_1}\right)^{\gamma};= v_0 \quad \text{for } r \le r_1 \tag{4.3}$$

Typical values for the central velocity (v_0), number density (n_0), and temperature (T_0) expected for a collapsing, nonrotating, solar mass protostar within the infall region (r_1) are:

$$\begin{aligned} v_0 &= 0.8\,\text{kms}^{-1} \\ n_0 &= 10^6\,\text{cm}^{-3} \\ T_0 &\approx T_D \approx T_G = 40\,\text{K} \\ r_1 &= 4.7\times 10^{16}\,\text{cm} \\ \alpha &= -1 \\ \beta &= -2 \\ \gamma &= -\frac{1}{2} \end{aligned} \tag{4.4}$$

4.4 LINES OF SIGHT

Much of writing a radiative transfer code has to do with bookkeeping. Over each of our N_S shells, let us assume the model parameters defined by Equations 4.1 through 4.4 are held constant at an average shell value, as depicted in Figure 4.2. Now divide the LOS through the core into a number of discrete points, with the first point, z_1, being where the LOS first pierces the structure, and the last point, z_N, where the LOS exits the backside of the structure. The greater the number of points, N, along the LOS, the more accurate will be the model. In general, there should be ≥3 points sampling each shell. So, for $N_S = 10$ shells, the LOS should be divided into ≥300 points. The model resolution, Δz, along the LOS is then

$$\Delta z = z_{n+1} - z_n \tag{4.5}$$

4.5 MODEL EQUATION OF TRANSFER

Let us define a discrete version of the frequency dependent solution to the equation of transfer as

$$I_\nu^{N_{LOS}} = \sum_{z_1}^{z_N} S_\nu(z)(1 - e^{-\tau_\nu(z)})e^{-\tau_\nu^{tot}} \tag{4.6}$$

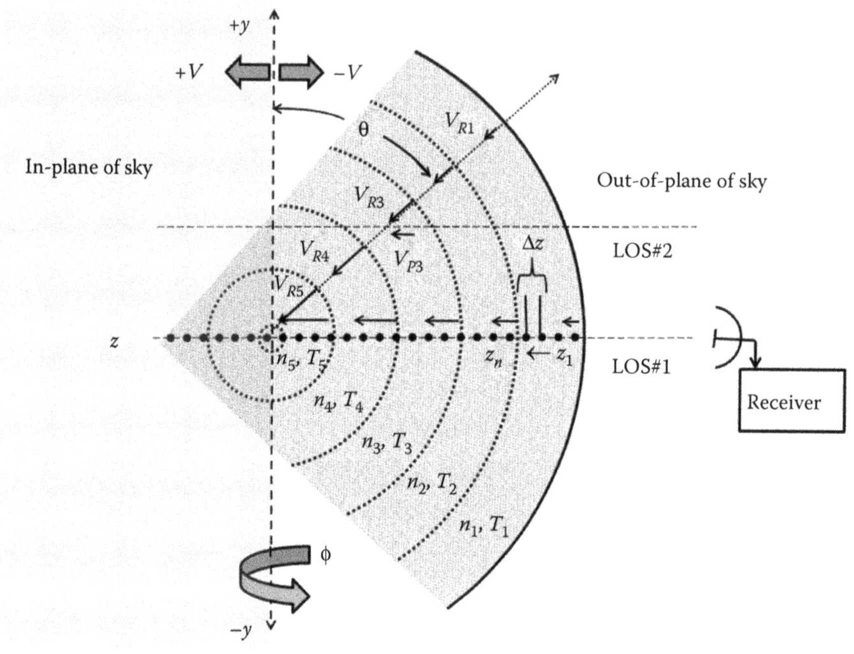

FIGURE 4.2 Cross-section of simple spherical radiative transfer code. Each shell is assigned a temperature, density, temperature, and velocity based upon its radius. The line of sight (LOS) through the sphere is divided into points $z_1 \rightarrow z_n$. Each point takes on the characteristics of the shell in which it finds itself. When computing the line profile function, $\phi_\nu(z)$, for each z_n the projection of the ambient velocity field (e.g., V_{P3}) along the LOS is used.

where

$I_\nu^{N_{LOS}}$ = integrated intensity along line of sight N_{LOS} at frequency ν
$S_\nu(z)$ = source function at point z and frequency ν
$\tau_\nu(z)$ = optical depth at point z and frequency ν
τ_ν^{tot} = total accumulated $\tau_\nu(z)$ back along LOS to observer
$z_1 = 1$ = first LOS point on front surface of model structure
z_n = last LOS point at exit of model structure
N = number of model points along LOS
N_{LOS} = designation of the LOS being considered (e.g., the (x,y) of the LOS on plane of sky)

At high densities, $n = n_{crit}$, where collisional excitation/deexcitation dominate over radiative transitions (see Equation 2.74), LTE is a good assumption and $S_\nu(z) \approx B_\nu(T_N)$. For lower density regions, a non-LTE source function based on statistical equilibrium or an escape probability formalism should be used (see Section 2.4).

Let us define an expression for optical depth as a function of z that includes the absorptive effects of both gas and dust using the approach of Yorke (1979).

$$\tau_\nu(z) = \alpha_\nu^{total}(z)\Delta z \tag{4.7}$$

where

$\alpha_\nu^{total}(z) = \alpha_\nu^g(z) + \alpha_\nu^d(z)$ = total absorption coefficient at sample point (cm^{-1})

$\alpha_\nu^g(z)$ = gas absorption coefficient at sample point (cm^{-1})

$\alpha_\nu^d(z)$ = dust absorption coefficient at a sample point (cm^{-1})

$$= \left(\frac{n_R m_R}{\rho}\right)\sigma_\nu^{a,R} + \left(\frac{n_V m_V}{\rho}\right)\sigma_\nu^{a,V} \tag{4.8}$$

ρ = gas mass density (g cm^{-3})

$\sigma_\nu^{a,R}$ = refractory dust cross-section (cm^{-2})

$\sigma_\nu^{a,V}$ = volatile dust cross-section (cm^{-2})

n_R = refractory dust particle density at sample point (cm^{-3})

n_V = volatile dust particle density at sample point (cm^{-3})

$m_R = 1.05 \times 10^{-15}$ g

$m_V = 3.36 \times 10^{-24}$ g

Refractory (R) dust particles are made from graphite and silicates. Volatile (V) dust particles are those covered with ices, such as water, ammonia, and methane. Therefore, the quantities (n_R/ρ) and (n_V/ρ) are temperature dependent.

$$\left(\frac{n_R}{\rho}\right) = \begin{cases} 1.43 \times 10^{12}\,\mathrm{g}^{-1}, & T_V < 150\,K \\ 1.907 \times 10^{12}\,\mathrm{g}^{-1}, & T_R \leq 1700\,K \text{ and } T_V \geq 150\,K \\ 0, & T_R \geq 1700\,K \end{cases} \tag{4.9}$$

$$\left(\frac{n_V}{\rho}\right) = \begin{cases} 4.77 \times 10^{11}\,\mathrm{g}^{-1}, & T_V < 150\,K \\ 0, & T_V \geq 150\,K \end{cases}$$

Values for the dust cross-sections, $\sigma_\nu^{a,R}$ and $\sigma_\nu^{a,V}$, rise steeply with frequency. From Figure 1 of Yorke (1979) we derive Table 4.1.

The gas opacity, $\alpha_\nu^g(z)$, comes from our earlier discussion (see Section 2.3.1, Equation 2.33).

$$\alpha_\nu^g(z) = \frac{c^2}{8\pi\nu_{ul}^2}\frac{g_u}{g_l}A_{ul}n_l\left(1 - e^{-(h\nu_{ul}/kT_{NS})}\right)\phi_\nu(z) \tag{4.10}$$

TABLE 4.1 Dust Cross-Sections

Frequency (THz)	$\sigma_\nu^{a,R}$	$\sigma_\nu^{a,V}$
1.0	1×10^{-15}	9×10^{-15}
2.0	5×10^{-15}	3×10^{-14}
3.0	2×10^{-14}	7×10^{-14}
4.0	3×10^{-14}	9×10^{-14}
5.0	4×10^{-14}	1.5×10^{-13}

where

ν_{ul} = line center frequency of chosen molecule (Hz)
$g_l = 2l + 1$ statistical weight of lower state, l
$g_u = 2u + 1$ statistical weight of upper state, u
A_{ul} = spontaneous emission coefficient from u to l of chosen molecule (s^{-1})

$$= \left(\frac{64\pi^4\nu_0^3}{3hc^3}\right)|\mu_{ul}|^2$$

$$|\mu_{ul}|^2 = \frac{\mu^2(l+1)}{(2l+1)}$$

μ = dipole moment of molecule (Debye)
n_l = number density of molecules in l (cm^{-3})
T_{N_S} = excitation temperature within shell N_S (K)

$$\frac{n_l}{n_{N_S}} = (2l+1)\exp\left(-\frac{l}{2}\frac{h\nu_{ul}}{kT_{ex}}\right)\left[\frac{hB}{kT_{ex}}\right]$$

n_{N_S} = total number density of molecules in shell N_S (cm^{-3})
B = rotational constant of molecule (Hz)

$$= \frac{\nu_{ul}}{2(l+1)}$$

$\phi_\nu(z)$ = line profile function at each point z_n along LOS (Equation 2.28).

If the goal is to generate a model line profile with spectral resolution $\Delta\nu$ between frequencies ν_1, ν_2, ν_3, to ν_n, then at each point, z_n, and frequency, ν_n, along the chosen LOS,

- ∞ Compute physical condition; T_{ex}, n^{mol}, $V_{r,\theta,\phi}$.
- ∞ Compute line profile function with velocities projected back along LOS to observer; $\phi_\nu(z)$.
- ∞ Compute total absorption coefficient; $\alpha_\nu^{total}(z)$.
- ∞ Compute source function; $S_\nu(z)$.

Starting with z_1, use Equations 4.6 and 4.7 to sum up the monochromatic integrated intensity, $I_\nu^{N_{LOS}}$, along the LOS for each channel frequency in the line profile. The computed channel intensities can be converted to radiation temperatures (T_R) using Equation 2.34.

Figure 4.3 is a suite of model $^{12}C^{32}S$ (main line) and $^{12}C^{32}S$ (isotope) line profiles toward the protostellar source IRAS 16293-2422 made using an LTE radiative transfer code based on the algorithm described above (Walker, 1988). Physical conditions similar to those of Equation 3.32 were assumed. (In this figure the broadband continuum emission from dust has been subtracted from the baseline.) The optically thick CS $J = 2 \rightarrow 1$ main line profile is heavily self-absorbed, with a small degree of blue asymmetry in the height of the two emission peaks. The presence of self-absorption is due to the core being hotter in the center than the edge, such that the cooler, overlying shells are able to absorb some fraction of the

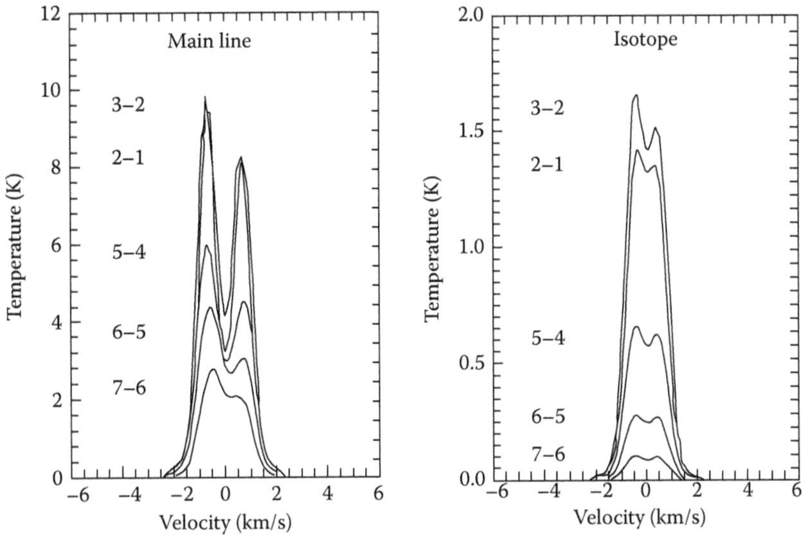

FIGURE 4.3 $C^{32}S$ (main) and $C^{34}S$ (isotope) spectra generated using an LTE spherical radiative transfer code of the type described in the text. Model conditions were chosen to reflect those expected in the vicinity of a low-mass protostar. The molecular core is centrally heated with radially dependent temperature, density, and velocity laws. The blueshifted self-absorption features are due to cool infalling (redshifted) gas absorbing photons from warmer, infalling gas underneath. Higher lying lines probe deeper into the protostellar core where infall velocities are greater. The model spectra are what would be observed toward the protostar with a 20″ telescope beam.

photons generated from the warmer gas underneath. If there were no systematic velocity field present, the self-absorption would be centered on the rest velocity of the cloud core, and there would be two symmetric emission peaks. The blue asymmetry is due to the presence of an ever-increasing infall velocity field causing redshifted gas in the outer layers of the core to preferentially absorb photons from warmer redshifted gas just beneath. The degree of blue asymmetry increases in higher lying lines, since these transitions (due to their higher critical density) probe deeper into the core where higher infall velocity fields and temperatures are present. Redshifted absorption continues along the LOS until the plane of the sky is reached. Beyond this point, the infalling layers appear blueshifted, that is, coming toward the observer. Since the blueshifted photons are generated in blueshifted gas, they cannot interact with the cooler, overlying redshifted gas (since they are offset in frequency/velocity space), and they travel unimpeded to the observer. The less abundant $^{12}C^{34}S$ line profiles also show evidence of a blue asymmetry, but to a far less degree. The line asymmetry is not necessarily confined to line core velocities. The asymmetry can occur at higher velocities, as long as the emission remains optically thick. In a collapsing cloud core, line profile features at higher velocities arise deep within the core, where infall velocities are greatest. Lower velocity features arise closer to the surface of the core, where infall velocities are less (see Figure 4.4).

A comparison of model and observed CS line profiles toward IRAS 16293 are shown in Figure 4.5 (Walker et al., 1986). Similar blueshifted asymmetric line profiles, suggesting the presence of infalling gas, have been observed toward a number of protostellar objects

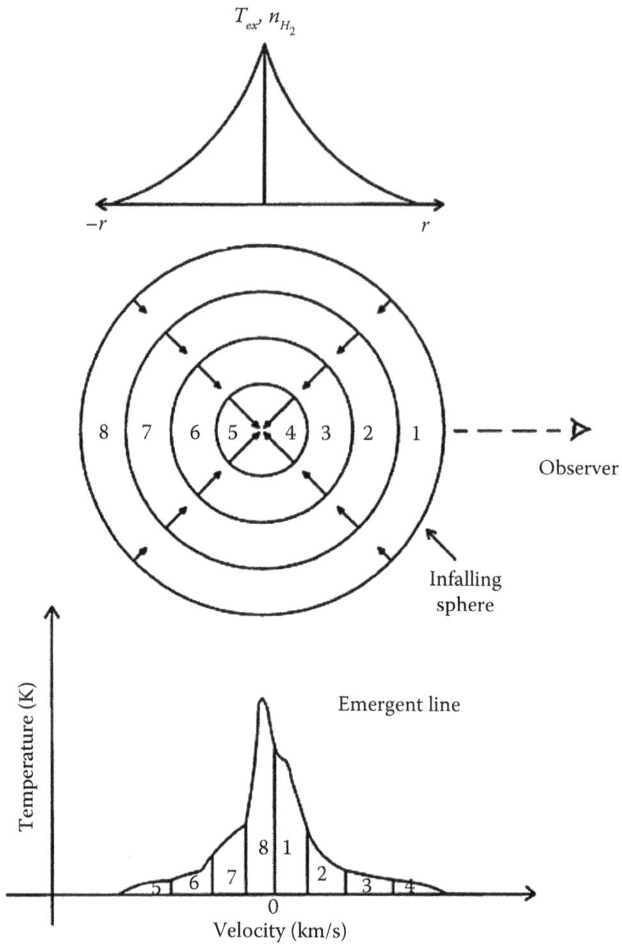

FIGURE 4.4 Anatomy of an infall spectral line profile. Here, the spherically symmetric, infalling cloud core is divided into layers numbered 1 through 8. As illustrated at the top of the figure, the temperature and density within the cloud core decrease with radius. Infall velocities also decrease with radius. Emissions from layers (1 through 4) are from the near side (out of the plane of the sky) of the protostar and appear redshifted to the observer. The emissions from the layers (5 through 8) from the far side of the protostar appear blueshifted. The emergent line from the collapsing core is shown at the bottom. The line is broken up into velocity bins. The layer responsible for the majority of the emission in a particular bin is indicated by its number. The emissions from the inner layers on the near (redshifted) side of the protostar will be partly absorbed by the cooler foreground layers before reaching the observer. Emission from the far (blueshifted) side of the protostar is not absorbed by foreground layers because (1) the foreground layers are hotter than the layer where the emission originated or (2) the foreground layer is at a substantially different velocity than the layer where the emission originated. Therefore, the emergent line appears self-absorbed and asymmetric, with the blueshifted (lower velocity) side of the line brighter than the redshifted side. The infall line asymmetry is not limited to line center velocities. The asymmetry can occur in the line wings as long as the emission remains optically thick. (From Walker, C. K. et al., 1994, *Astrophys. J.*, 431, 767. With permission.)

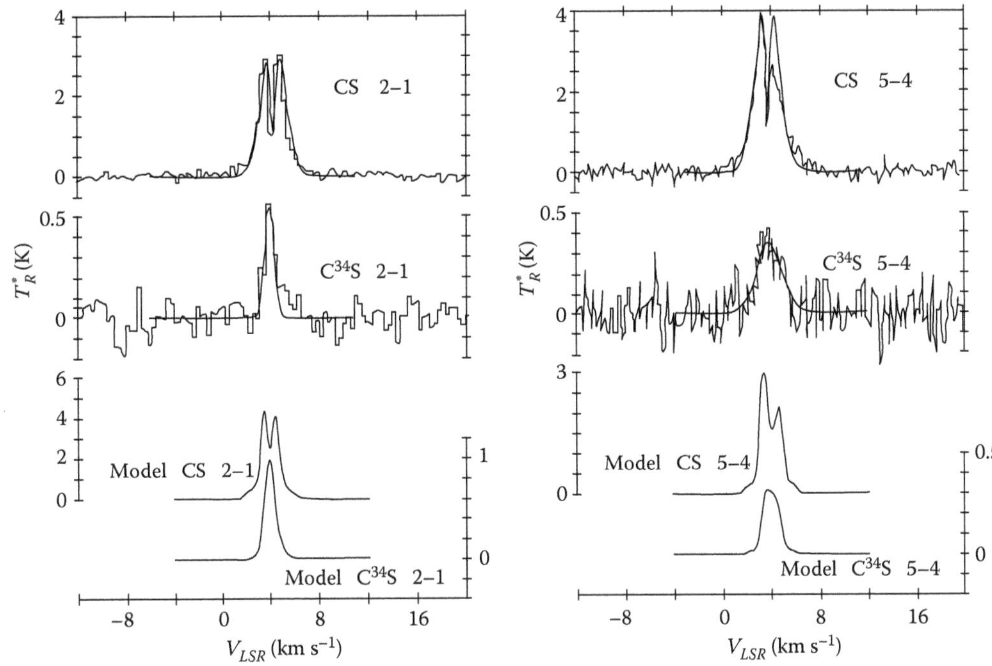

FIGURE 4.5 Observed $C^{32}S$ and $C^{34}S$ spectra (histogram) toward the IRAS16293-2422 protobinary system. Each spectrum is fitted with one or more Gaussians (smooth lines). Below the observed spectra are the corresponding model spectra generated using a non-LTE, Monte Carlo, radiative transfer code. The blueshifted asymmetry often associated with infall is clearly seen in the CS $J = 5 \rightarrow 4$ spectrum. (From Walker, C. K. et al., 1986, *Astrophys. J.*, 309L, 47. With permission.)

(e.g., Zhou et al., 1993; Lee et al., 1999; more recently, Reiter et al., 2011). Rotation can be readily added into the model through modification of the line profile function, $\phi_v(z)$, as described by Adelson and Leung (1988).

A model map can be generated by computing model spectra along multiple lines of sight and convolving the results. So as not to miss flux, the spacing between LOSs in the model map should be ≤1/2 the desired convolved, angular resolution. For example, if a final angular resolution of 30″ is desired, the model LOSs should be spaced no more than 15″ apart in x and y. The resulting grid of model spectra are then convolved in 2-dimensions, using a Gaussian beam function with a FWHM (full-width at half-maximum) of 30″.

EXAMPLE 4.1

The angular resolution of an observation can have a profound effect on the appearance of a spectral line. The $C^{32}S$ $J = 5 \rightarrow 4$ spectral lines in the right panel of Figure 4.5 were taken toward a protostellar source with an ≈30″ telescope beam. If the protostellar accretion disk has a diameter of ~100 AU and a mean temperature of ~1000 K, how bright would you expect the $C^{32}S$ $J = 5 \rightarrow 4$ emission from the disk to appear? Assume the line is optically thick, and the distance to the object is 160 pc. Is there any evidence for an underlying disk in the model spectra?

From Appendix A2, we have,

$$1\ \text{pc} = 3.09 \times 10^{18}\ \text{cm, and}$$

$$1\ \text{AU} = 1.5 \times 10^{13}\ \text{cm.}$$

The beam diameter in radians is $30'' \Rightarrow (30''/3600'')(3.141/180°) = 1.45 \times 10^{-4}$ rad. At the distance of the object this translates into a spatial resolution, s, of

$$\begin{aligned} s &= 160\,\text{pc}\left(3.09 \times 10^{18}\frac{\text{cm}}{\text{pc}}\right)1.45 \times 10^{-4}\ \text{rad} \\ &= 7.18 \times 10^{16}\ \text{cm} \end{aligned}$$

The diameter, D, of the disk in cm is

$$D = 100\,\text{AU}(1.5 \times 10^{13}\ \text{cm}) = 1.5 \times 10^{15}\ \text{cm}$$

When an intensity (such as a spectral line) is measured toward an object with a beam larger than the object itself, the intensity will appear diluted, that is, reduced, by an amount equal to the ratio of the source area, A_S, to the beam area, A_B, such that,

$$T^{obs} = \frac{A_S}{A_B}T^{obj},$$

or, in terms of beam solid angle, Ω (see Chapter 5), we could equivalently state,

$$T^{obs} = \frac{\Omega_S}{\Omega_B}T^{obj}$$

Substitution then yields

$$T^{obs}_{disk} = \left(\frac{1.5 \times 10^{15}\ \text{cm}}{7.8 \times 10^{16}\ \text{cm}}\right)^2 (1000\ \text{K}) = 0.37\ \text{K}$$

Examination of the $C^{32}S$ $J = 5 \rightarrow 4$ model spectra (lower right) does indeed reveal "shoulders" to the spectral line, which are consistent with the above value of T^{obs}_{disk}. The line broadening can be attributed to disk rotation. In interpreting the spectra toward a protostar spectral features associated with the disk can be masked by a protostar's molecular outflow. As the angular resolution of an observation is increased, the effects of beam dilution will decrease, yielding a line profile that more accurately reflects the temperature distribution within the unresolved source. In the case of the protostar, what are low-lying shoulders at 30″ angular resolution may grow to outshine the peaks in the unresolved line profile when higher angular/spatial resolutions are used.

4.6 LTE RADIATIVE TRANSFER WITH HYDRODYNAMIC SIMULATIONS

Spherical models can provide a good first look into the conditions in a variety of astrophysical objects. However, to account for nonspherical effects (e.g., rotation, magnetic fields, noncentralized density, and heating effects), or the presence of multiple bodies, 3-dimensional hydrodynamic simulations can be employed. There are various formalisms that can be used in creating such codes, including using finite difference solutions to the equations of hydrodynamics, gravity, and radiative transfer (e.g., Boss, 1993), and the smooth-particle hydrodynamics (SPH) approach (Gingold and Monaghan, 1977). The output of these simulations can be written such that the derived properties, for example, temperature, density, radiation field, magnetic field, and velocity field, at each point in space as a function of time, are identified in spherical coordinates.

Radiative transfer calculations to predict emergent spectra and/or continuum emission through a hydrodynamic simulation can be performed in a similar manner to what was outlined above for the simple spherical model. As an example, let us consider modeling the collapse of a cloud core yielding a protobinary star system, using the approach of Walker et al. (1994). First, a program reads in the temperature, density, and velocity field at each of the 3-dimensional hydrodynamic model grid points. The model protostellar core is then rotated to the desired viewing angle. Next, the average conditions at each point along a chosen line of sight (LOS) are determined. The number of points along a LOS is set by the desired spatial resolution. The greater the number of points, the greater the central processing unit (CPU) time required to digest the model nebula. For the model results discussed in the above referenced paper, a spatial resolution of 1.21×10^{15} cm was chosen. In order to take into account changes that may occur between sample points, a Gaussian weighting technique was used when computing the conditions at each LOS point. The sigma of the Gaussian function is set by the chosen spatial resolution (similar to what is done in the SPH approach). The conditions along each LOS are then stored. A second program reads in the Gaussian averaged conditions along each LOS and predicts the emergent spectrum for the desired molecular transition. The model spectrum is divided up into 100 frequency points. The equation of transfer is solved at each LOS point for each frequency. The calculated intensity at each velocity is then summed up along the LOS, following Equation 4.6.

Figure 4.6 is a spectral mosaic of $^{12}C^{32}S$ along a 3×5 grid of LOSs through the northern hemisphere of the model protostellar core 38,343 years (1.212 free-fall times) after the onset of gravitational collapse from the parent molecular cloud. The southern hemisphere is a mirror image of the northern hemisphere. In this figure, the bottom row of spectra is coincident with the core's equator, with the protobinary itself located in the center of

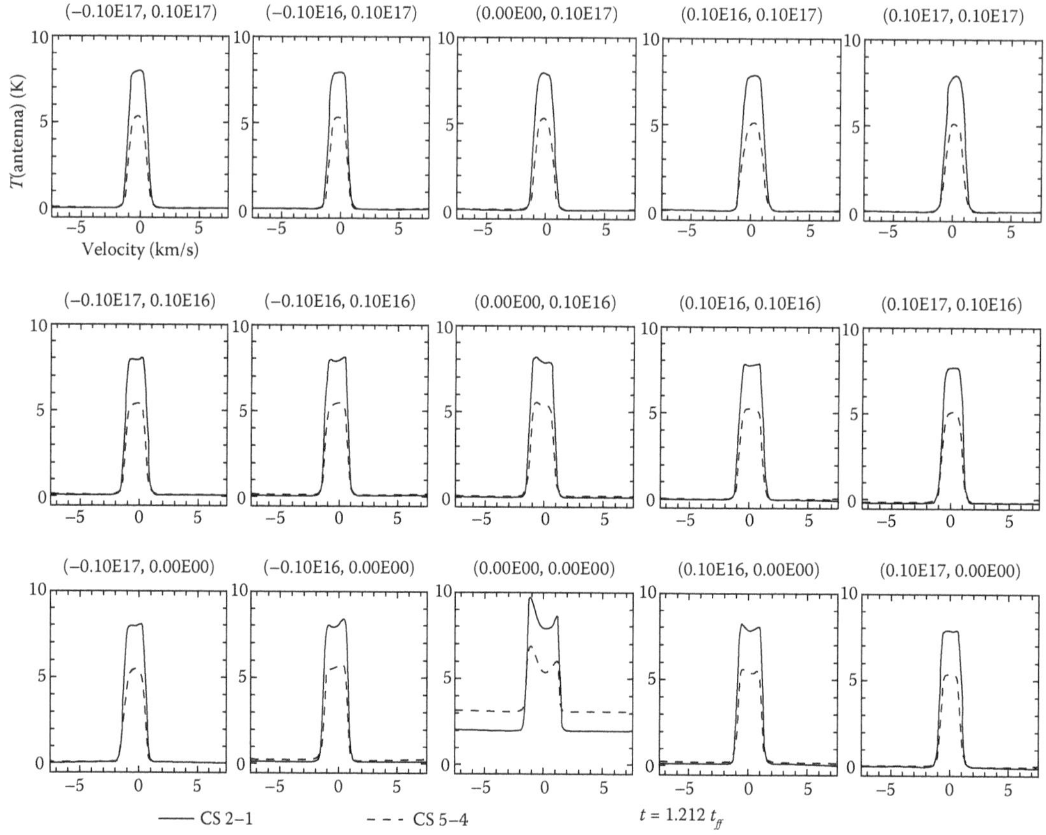

FIGURE 4.6 Model $C^{32}S$ and $C^{34}S$ spectra toward lines of sight through a collapsing, rotating protobinary system. Only northern hemisphere lines of sight are shown. The equator is along the bottom row, with the rotation axis down the middle. Along the LOS with the protobinary (bottom, center) both an elevated dust continuum and blueshifted infall asymmetry are seen. The line asymmetry is blueshifted to the right of the protostar and redshifted to the left, due to the presence of rotation. (From Walker, C. K. et al., 1994, *Astrophys. J.*, 431, 767. With permission.)

the row. The distance between adjacent LOSs is 1.0×10^{15} cm. The spectra are what would be observed with a 0.5″ telescope (e.g., ALMA, or Atacama large millimeter array) at the distance of Rho Ophiuchi (~160 pc). The distance between sample points along the LOS is 1.21×10^{15} cm (45 AU). The $^{12}C^{32}S$ $J = 2 \rightarrow 1$ emission lines are solid, and the $^{12}C^{32}S$ $J = 5 \rightarrow 4$ emission lines are dashed. Along the LOS to the protobinary, the baseline level is elevated due to continuum emission from a central dust core. A blue self-absorption asymmetry, due to the presence of infalling gas, is evident in both line profiles. Looking on either side of the central position, self-absorbed line asymmetries are also seen, with a slight blueshifted asymmetry on the right and a redshifted asymmetry to the left. This flip inline asymmetry across the position of the protobinary is due to rotation in the cloud core, with the rotation axis running down the middle of the figure. Where a blue-line asymmetry is observed (right side), the rotational velocities project into the page. Where the red-line asymmetry is observed (left side), is where the rotational velocities project out of the page. If an idealized "pencil beam" were used to observe the protostar, the rotational velocity fields would be perpendicular to the LOS, and not contribute to the shape of the line profile. But for beams (either theoretical or real) of finite size pointed toward the protostar, non-LOS rotational velocity components will be sampled. Here, the blue and redshifted rotational velocity fields will be evenly split across the beam, and tend to cancel out, leaving the infall velocity field to dominate the appearance of the line profile.

4.7 NON-LTE RADIATIVE TRANSFER WITH HYDRODYNAMIC SIMULATIONS

When considering low-density environments or when using atomic or molecular species with high Einstein A coefficients (i.e., with high electric diploe moments), the LTE approximation may no longer be valid. Under such non-LTE conditions, the source function must be calculated explicitly. For this calculation, an iterative procedure based on the Monte Carlo approach is often utilized (e.g., Bernes, 1979; van Zadelhoff et al., 2002). Using this technique, the steady state distribution of energy states among the atoms or molecules is found, and, once known, the source function, S_ν, can be calculated using Equation 2.15,

$$S_\nu = \frac{n_u A_{ul}}{n_l B_{lu} - n_u B_{ul}},$$

(recalling that in LTE, $S_\nu = B_\nu$, the Planck function). The explicit value of S_ν is then substituted into Equation 4.6 and the rest of the LOS calculation to generate a spectral line proceeds much as before. However, the complication is that the level populations now also depend on the mean intensity field at each grid point in the model,

$$J_\nu = \frac{1}{4\pi}\int I_\nu \, d\Omega, \tag{4.11}$$

which, in turn, depends on the intensity from other grid points. Therefore, it is necessary to initially guess the level populations (say, assuming LTE), solve for the mean intensity field, calculate updated level populations, and iterate until the level populations converge.

As discussed by Narayanan et al. (2006), the radiation field is modeled by "photon packets" that simulate many real photons. The number of photons represented by each packet is a function of the Einstein A_{ul} of the atom or molecule, and the number density, n_u, of the atom or molecule in the upper state of the transition being considered. The photon packets are isotropically emitted in a spontaneous manner with a line frequency set by the line profile function, ϕ_ν, (Equation 2.28). It is through ϕ_ν that the systematic and turbulent velocity fields express themselves in the formation of the spectral line. The photon packet takes a step, z, with the number of real photons represented by the photon packet diminished by a factor $e^{-\tau}$ due to absorptions. The photon packet continues to take steps in the same direction until it leaves the model grid, or the number of photons it represents goes to zero. During its journey, the number of photons a packet possesses at each model grid point is recorded. After all the model photons have cleared the model grid or been absorbed; the mean intensity is known throughout the model structure. Using the equations of statistical equilibrium (see Section 2.4), the level populations can be computed using standard matrix inversion techniques and updated. The process is iterated until level convergence is achieved. The time it takes to reach convergence increases with the number of model photons used and the optical depth encountered. This approach is very powerful and has been used to model a wide range of astrophysical structures, from protoplanetary disks, Figure 4.7, (e.g., Narayanan et al., 2006) to high z galaxies, Figure 4.8 (e.g., Narayanan et al., 2008).

PROBLEMS

Using a programming language of your choice, write a simple, spherically symmetric, LTE radiative transfer code for modeling the emission line produced along the line of sight through the center of a protostellar core. There are many ways to accomplish this task. Here, try using the approach outlined in Chapter 4. An effective philosophy to developing such a code is to break the task up into modules that can be individually tested and then, one by one, linked together. At each step, simple, representative calculations (like those presented in the example problems in Chapters 2 and 3) should be done by hand to serve as a "sanity check" against those generated by your model.

1. Construct a "preamble" module to your program in which all key parameters (e.g., physical constants, number of shells, line frequencies, selected atomic and/ or molecular parameters, as well as central temperatures, densities, velocities, and

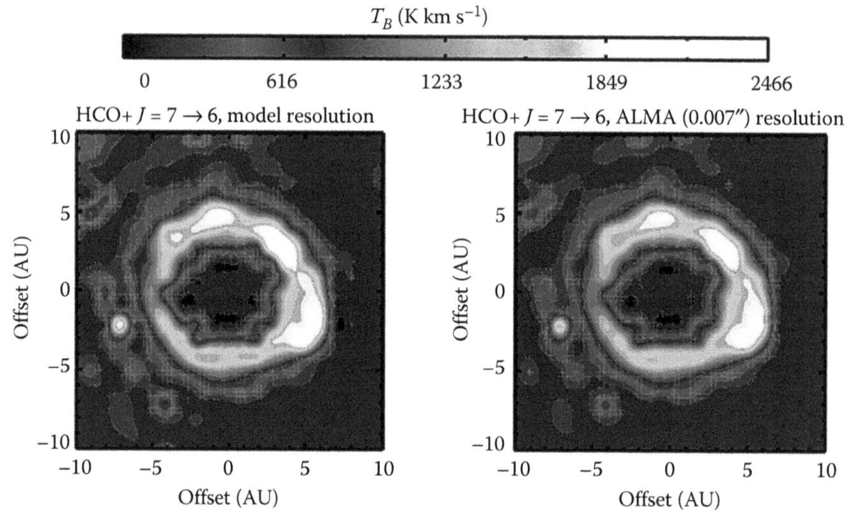

FIGURE 4.7 HCO+ $J = 7 \rightarrow 6$ image of a protoplanetary disk made by using a combination of hydrodynamic and non-LTE radiative transfer codes. To produce the image, the 3-D hydrodynamic simulation was first run and a file generated containing the physical conditions at each point in the model. A non-LTE radiative transfer code then read in the results of the model and generated spectra along multiple lines of sight. The integrated intensities from these model spectra were then used to make the images. Left: Image made at the model resolution. Right: Same image, but convolved to the highest expected ALMA angular resolution, 0.007″. (Images from Narayanan, D. et al., 2006, *Astrophys. J.*, 647, 1426. With permission.)

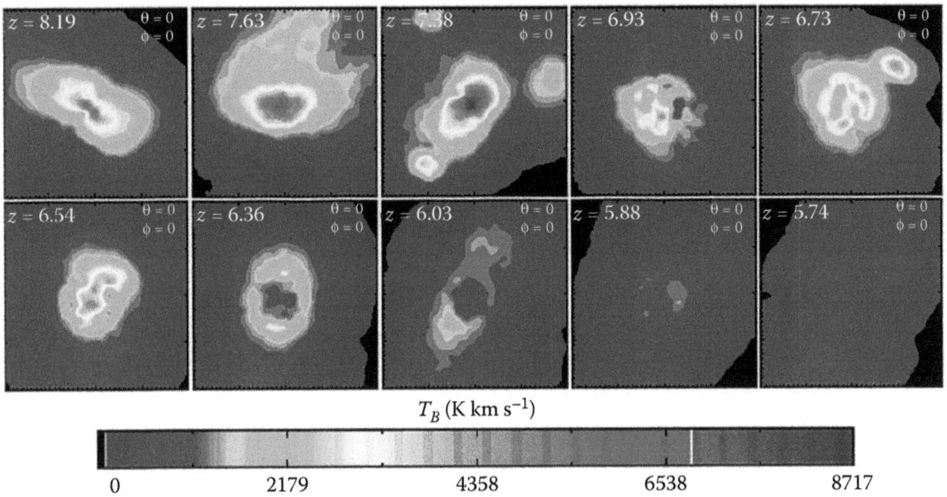

FIGURE 4.8 Model CO $J = 1 \rightarrow 0$ emission as a function of redshift toward the central 2 kpc of a quasar host galaxy. Image made by Narayanan et al. (2008) using the same non-LTE radiative transfer algorithm as in the case of the protoplanetary disk of Figure 4.5, but with a different hydrodynamic model solution.

associated power law exponents (see Equations 4.1 through 4.4) are defined. For this exercise, let us use model the $C^{32}S$ $J = 5 \rightarrow 4$ line.

2. Write a module that divides the internal structure of the model into a series of five nested spheres, with the model's outer radius normalized to one. A line of sight piercing the center of the sphere will encounter five shells on each side of the plane of the sky (see Figure 4.1). Using Equations 4.1 through 4.4, determine a value for the temperature, density, and radial velocity within each shell. A spherical coordinate system is convenient for defining shell values and rotating the object (if desired) relative to the observer. Cartesian coordinates are convenient for taking lines of sight through the object, once shell properties have been defined. Place the center of the protostar at the origin of the coordinate system with the z axis parallel to all LOSs.

3. Write a module that takes an LOS through the center of the module. Divide the LOS through the core into 300 discrete points, with the first point, z_1, being where the LOS first pierces the structure, and the last point, z_N, where the LOS exits the structure. This will give you ~3 sample points within each shell. Identify which shell (and thereby the associated physical parameters) for each LOS point.

4. Write an LTE radiative transfer model that calculates the source function (here the Planck function) and optical depth (Equations 4.7 and 4.10; for now let's ignore the dust emission) at each LOS point for each frequency in the line profile. To obtain sufficient frequency resolution, choose to have a 100 frequency points, separated by 50 kHz.

5. Next, write a module that utilizes the discrete version of the equation of transfer (here Equation 4.6) to sum up the emission along the LOS at each frequency.

6. Write a module that saves the resulting spectrum to a file.

7. Try running your code using different model values than those listed in Equation 4.4, and see how the appearance of the line profile changes.

GOING FURTHER

Your model generates an ideal, 1-dimensional, "pencil-beam" through the object along the z axis, with no spatial extent in x and y. To create a map of the object, you would need to modify your code to work "off-axis" from the protostar. This involves a bit more trigonometry and bookkeeping, but is straightforward to implement (e.g., Adelson and Leung, 1988). If you would like to compare your model to an observation with a telescope, first create a grid of model LOSs that fills the area that would be sampled by the telescopes's beam at the distance of the source. For full model sampling, the x,y spacing between LOSs in the grid should be ≤ the spacing between sample points along the LOS. Then, convolve each LOS model spectrum with a Gaussian function centered on the source (or other location of interest) that has the same FWHM as the telescope's beam. Example 4.1 examines the impact of angular resolution on the appearance of line profiles.

REFERENCES

Adelson, L. and Leung, C., 1988, On the effects of rotation on interstellar molecular line profiles, *MNRAS*, 235, 349.

Bernes, C., 1979, A Monte Carlo approach to non-LTE radiative transfer problems, *A&A*, 73, 67.

Boss, A., 1993, Collapse and fragmentation of molecular cloud cores. I—Moderately centrally condensed cores, *Astrophys. J.*, 410, 157.

Gingold, R. and Monaghan, J., 1977, Collapse and fragmentation of molecular cloud cores. I—Moderately centrally condensed cores, *MNRAS*, 181, 375.

Lee, C., Myers, P., and Tafalla, M., 1999, A survey of infall motions toward starless cores. I. CS (2-1) and N_2H^+ (1-0) observations, *Astrophys. J.*, 526, 788.

Narayanan, D., Kulesa, C. A., Boss, A., and Walker, C. K., 2006, *Astrophys. J.*, 647, 1426.

Narayanan, D., Li, Y., Cox, T. J. et al., 2008, The nature of CO emission from z ~ 6 Quasars, *Ap. JS.*, 174, 13.

Reiter, M., Shirley, Y., Wu, J., Brogan, C., Wootten, A., and Tatematsu, K., 2011, Evidence for inflow in high-mass star-forming clumps, *Astrophys. J.*, 740, 40.

van Zadelhoff, G., Dullemond, C., van der Tak, F. et al., 2002, Numerical methods for non-LTE line radiative transfer: Performance and convergence characteristics, *A&A*, 395, 373.

Walker, C., 1988, An observational study of the dynamics of molecular cloud cores, PhD dissertation, University of Arizona.

Walker, C. K., Lada, C. J., Young, E. T., Maloney, P. R., and Wilking, B. A., 1986, Spectroscopic evidence for infall around an extraordinary IRAS source in Ophiuchus, *Astrophys. J.*, 309L, 47.

Walker, C. K., Narayanan, G., and Boss, A. P., 1994, Spectroscopic signatures of infall in young protostellar systems, *Astrophys. J.*, 431, 767.

Yorke, H. W., 1979, The evolution of protostellar envelopes of masses 3 and 10 solar masses. I—Structure and hydrodynamic evolution, *A&A*, 80, 308.

Zhou, S., Evans, N., Koempe, C., and Walmsley, C., 1993, Evidence for protostellar collapse in B335, *Astrophys. J.*, 404, 232.

5

THz OPTICAL SYSTEMS

PROLOGUE

Light from space, whether it originates from a star, planet, galaxy, or the Big Bang itself, is the only physical evidence we have from which to deduce the nature of the distant Universe. Efficiently capturing, conveying, and analyzing this light is the purpose of all astronomical instrumentation. Since the time of Galileo, essentially all breakthroughs in astronomy have involved the use of a telescope. Telescopes are composed of a system of lenses and/or mirrors designed to capture photons over larger areas (and/or frequency ranges) than is possible with the unaided eye. In this chapter, we will investigate how light from distant objects couples to a THz telescope, and how it can be efficiently conveyed to a detection system.

5.1 INTRODUCTION: SOURCE–BEAM COUPLING

The primary lens or mirror of a telescope defines a window or aperture of diameter (D) through which the Universe is viewed. The geometry of the situation is depicted in Figure 5.1. The telescope aperture lies at the origin of a spherical coordinate system, with the telescope pointed toward the zenith. Light of intensity $B(\theta, \phi)$ from an object or source falls on the aperture, A, from a direction in space specified by the angles θ and ϕ. The infinitesimal power, dW, received from the object through a solid angle, $d\Omega$, on a surface area, dA, over a given bandwidth, $d\nu$, follows Lambert's cosine law (Kraus, 1966),

$$dW = B(\theta,\phi)\cos\theta \, d\Omega \, dA \, d\nu \tag{5.1}$$

where

dW = incident infinitesimal power, watts

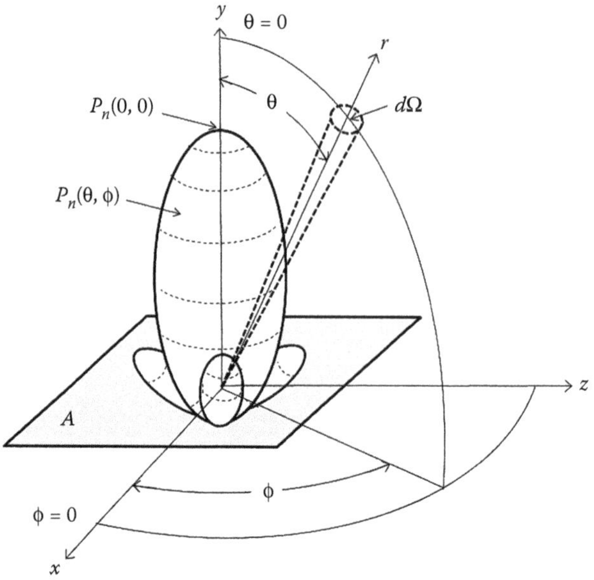

FIGURE 5.1 Source–beam coupling.

$B(\theta, \phi)$ = brightness of sky toward $d\Omega$, watts m^{-2} Hz^{-1} rad^{-2}
$d\Omega$ = infinitesimal solid angle of sky (=sin θ $d\theta$ $d\phi$), rad^{-2}
θ = angle between $d\Omega$ and the zenith
dA = infinitesimal area of surface, m^2
$d\nu$ = infinitesimal element bandwidth, Hz

Let us assume that the telescope is operating in its diffraction limit, meaning its response function to incoming photons is due to entirely its size and the wavelength of operation. The telescope response function can take on many forms. For single apertures of circular cross-section, the response function (for reasons discussed below) can be well approximated by a Gaussian with a central, main lobe, and one or more minor sidelobes (Figure 5.1). A telescope's response function (especially at THz and radio frequencies) is often expressed in terms of sensitivity to incoming power and referred to as a power pattern $P(\theta, \phi)$. A normalized power pattern can be obtained by dividing the pattern by its maximum value,

$$P_n(\theta,\phi) = \frac{P(\theta,\phi)}{P_m(0,0)} \tag{5.2}$$

where
$P_n(\theta, \phi)$ = normalized power pattern (dimensionless)
$P(\theta, \phi)$ = power pattern (W m^{-2})
$P_m(0, 0)$ = power pattern maximum (W m^{-2})

A power pattern can be thought of as a weighting function through which a telescope observes the Universe. Analytically, this is equivalent to saying

$$dW = B(\theta,\phi)P_n(\theta,\phi)d\Omega dA d\nu \tag{5.3}$$

where
- dW = incident infinitesimal power, watts
- $B(\theta, \phi)$ = brightness of sky toward $d\Omega$, watts m^{-2} Hz^{-1} rad^{-2}
- $d\Omega$ = infinitesimal solid angle of sky (=$\sin\theta d\theta d\phi$), rad^{-2}
- dA = infinitesimal area of surface, m^2
- $d\nu$ = infinitesimal element bandwidth, Hz

Here, $P_n(\theta, \phi)$ replaces the idealized cos θ term in Equation 5.1.

A telescope's effective collecting area, A_e, can be more or less than its physical collecting area, A_p. For example, the collecting area for a dipole antenna, from which low-frequency radio telescopes can be made, is $A_e = 0.133\lambda^2$, where λ is the wavelength of operation. For the single circular aperture antennas often used at high frequencies, the value of A_e is smaller than A_p. The ratio of a telescope's (i.e., antenna's) effective area to its physical area is referred to as its aperture efficiency, η_A,

$$\eta_A = \frac{A_e}{A_p} \tag{5.4}$$

where
- η_A = aperture efficiency, dimensionless
- A_e = effective (or electrical) area, m^2
- A_p = physical (or geometrical) area, m^2

An expression for the total power, W, captured by the telescope's aperture can now be obtained through integrating Equation 5.3,

$$\begin{aligned} W &= \iiint B(\theta,\phi)P_n(\theta,\phi)d\Omega dA_e d\nu \\ &= A_e\Delta\nu\iint B(\theta,\phi)P(\theta,\phi)d\Omega \end{aligned} \tag{5.5}$$

where
- W = captured power, watts
- $\Delta\nu$ = detector bandwidth, assumed to be uniform, Hz

If the source being observed is large and smooth (e.g., the cosmic background (CMB)) compared to the telescope's power pattern, Equation 5.5 can be further simplified:

$$\begin{aligned} W &= A_e\Delta\nu B\iint P_n(\theta,\phi)d\Omega \\ &= A_e\Delta\nu B\Omega_A \end{aligned} \tag{5.6}$$

where Ω_A is the telescope's antenna solid angle.

$$\Omega_A = \int P_n(\theta,\phi)d\Omega \tag{5.7}$$

The solid angle contained in the main lobe of a power pattern is Ω_M. The telescope main-beam efficiency is defined as

$$\eta_M = \frac{\Omega_M}{\Omega_A} \tag{5.8}$$

For a telescope with a surface roughness $<\lambda/30$, η_M is typically ≈ 0.7.

The flux density, S, of a source having brightness distribution $B(\theta, \phi)$ is expressed as

$$S = \iint_{source} B(\theta,\phi)\,d\Omega \tag{5.9}$$

where

S = flux density of source, watts m^{-2} Hz^{-1}

$B(\theta,\phi)$ = source brightness distribution, watts m^{-2} Hz^{-1} rad^{-2}

$d\Omega$ = element of solid angle ($=\sin\theta\, d\theta\, d\phi$), rad^2

and the integral extends over the solid angle of the source. As touched on earlier, when observed through a telescope, the brightness distribution of an object is multiplied by the normalized power pattern, $P_n(\theta, \phi)$, of the telescope before reaching the detector. The observed flux density, S_o, is then

$$S_o = \iint_{source} B(\theta,\phi) P_n(\theta,\phi)\,d\Omega \tag{5.10}$$

Since $P_n(\theta, \phi)$ is a weighting function, the observed flux density $S_o \leq S$.

Let us assume the telescope's optical system is well designed, and the contribution of Ω_m, the solid angle of the minor beams (or sidelobes), is negligible. Then, if the source's solid angle, Ω_S, is small compared to the telescope's main beam, Ω_M, we have

$$S_o \approx S, \quad \text{for } \Omega_S \ll \Omega_M. \tag{5.11}$$

If the source is extended relative to Ω_M, then

$$\begin{aligned} S_o &= B(\theta,\phi)\iint P_n(\theta,\phi)\,d\Omega \\ &= B(\theta,\phi)\Omega_M, \quad \text{for } \Omega_S \gg \Omega_M \end{aligned} \tag{5.12}$$

When the conditions of Equation 5.11 hold, the source is said to be *unresolved* by the beam. When Equation 5.12 holds, the source is said to be *extended* relative to the beam. Equation 5.11 is saying that when you are measuring the flux of an object (e.g., in dust continuum observations), as long as the beam of your telescope is larger than that of the object, and does not overlap with other unwanted sources, you will get a faithful measurement of the flux. However, as stated by Equation 5.12, if your beam is less than the size of the object, then you will only measure the fraction of the flux sampled by the beam.

When we are interested in observing the brightness distribution of a source, B_0, (e.g., in atomic or molecular line observations) instead of flux, S_0, then sources smaller than Ω_M will appear "beam diluted" according to

$$B_0 = \frac{S}{\Omega_M} = \frac{\Omega_s}{\Omega_M} B \tag{5.13}$$

If not properly taken into account, beam dilution can lead to significant underestimates of a line's true strength.

Usually, we find ourselves in the "in-between" case where the source is extended relative to our beam, while containing local maxima that are ≤ the telescope beam. This is often the situation when we observe molecular clouds (Figure 5.2). The telescope beam provides you with one pixel of information. In order to make a map of a source, a large number of positions need to be observed, either individually or by scanning. Angular information on a scale that is less than about half of a beam-width will be filtered out by the telescope's power pattern. The observed brightness distribution appears smoother than the true brightness distribution. Mathematically speaking, the source brightness distribution has been convolved with the telescope's power pattern. Namely,

$$S_0(\phi_0) = \int B(\phi) P_n(\phi - \phi_0) d\phi \tag{5.14}$$

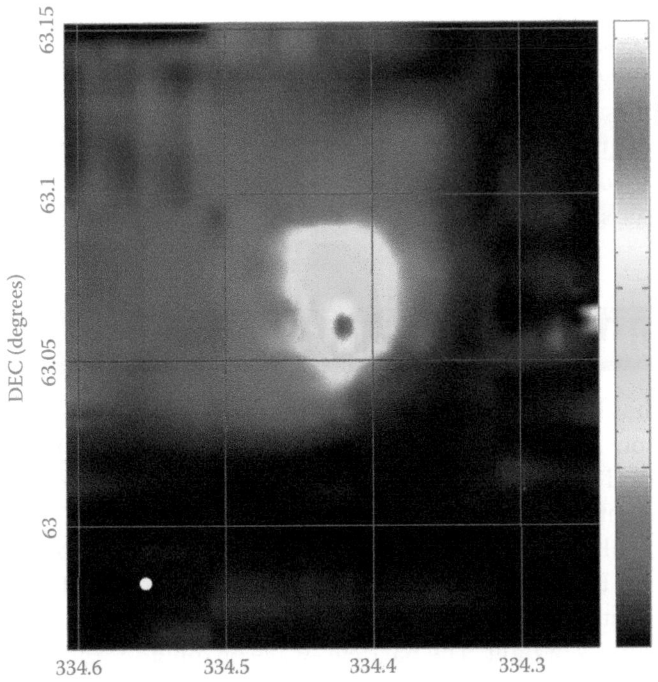

FIGURE 5.2 ^{12}CO $J = 3 \rightarrow 2$ image of the S140 star formation region taken at the Heinrich Hertz telescope (HHT) using the SuperCam heterodyne array. The white dot indicates the size of the telescope beam (~23″).

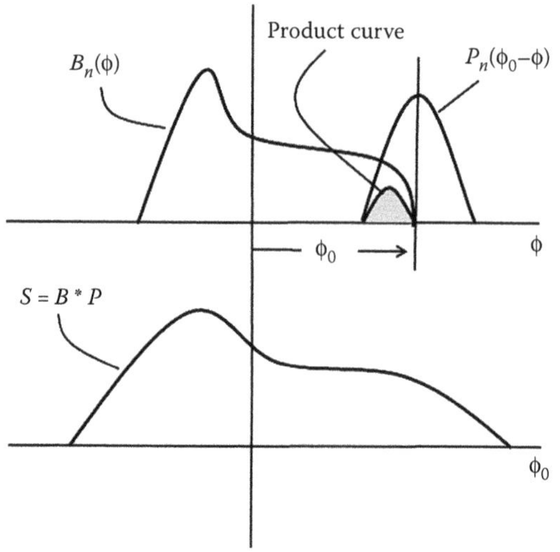

FIGURE 5.3 Convolution of an antenna power pattern with an extended source.

where

$S_o(\phi_o)$ = observed flux density, watts m^{-2} Hz^{-1}

$B(\phi)$ = source brightness distribution, watts m^{-2} Hz^{-1} rad^{-2}

$P_n(\phi - \phi_o)$ = telescope power pattern when it is displaced by ϕ_o

Convolution involves displacement, multiplication, and integration. A two-dimensional example is shown in Figure 5.3. If the telescope power pattern were infinitely sharp, then the observed brightness distribution would be identical to the true source distribution. Besides the benefit of greater collecting area and sensitivity, larger telescopes provide the higher angular resolution (see discussion below) required to accurately measure the structure of compact sources.

5.2 QUANTUM ELECTRODYNAMICS (QED) AND MAXWELL

The nature of light has been hotly debated since the time of Isaac Newton and Christian Huygens; Newton convinced it is composed of small particles, and Huygens convinced it is of a wave nature. The ability of the wave interpretation of light to explain everyday phenomena, such as interference, refraction, and diffraction, caused it to hold sway for over two centuries, culminating in the electromagnetic field equations of James Clerk Maxwell in the 1860s. However, in 1905, Einstein invoked the particle model to explain the photoelectric effect. Experimentation by Planck and others soon led to the idea of the particle–wave duality of light. The particle–wave or "photon" interpretation of light became a cornerstone of the rapidly evolving field of quantum electrodynamics (QED) and is how we view the nature of light today. One might ask, what role do Maxwell's classical wave equations have within this modern view of light? Fortunately for us, Maxwell's field equations do a superb job of describing and predicting the behavior of light when large numbers of photons are involved. In the parlance of quantum mechanics, one would say such situations correspond to phenomena with high photon occupation numbers. However, in situations where the number of photons is low (small occupation numbers), quantum mechanics is the way to go in describing the behavior of light.

At THz frequencies, photon energies are relatively low, so they can exist in prodigious numbers. At such high photon occupation numbers, Maxwell's equations are best used in the design of "optical" systems.

5.3 ORIGIN OF A SINGLE-APERTURE DIFFRACTION PATTERN

The region of space to which a telescope is sensitive is set by the number of photons of frequency, ν_p, that can fit side-by-side across its aperture, D. The size of the aperture determines the number of possible paths a photon can take to the detector. Each path has a probability amplitude, P_p, associated with it. It is the vector sum of the probability amplitudes of all possible paths that determines the likelihood of a photon from a given direction of space reaching a detector. To understand the origin of diffraction patterns in terms of probabilities, let us imagine we have a detector at the focus of a parabola of diameter D, as shown in Figure 5.4. For now let's assume the detector has no built-in prejudice to accept photons from one area of the parabola over another, it "uniformly illuminates" the telescope's aperture. In this case, the footprint of the telescope on the sky is solely determined by D. The shape of the parabola only serves to efficiently bring photons incident on the parabola's surface to a focus. For illustrative purposes, the aperture is divided into a number, n, of thin strips of width $\Delta d \ll \lambda$, where λ is the wavelength (or characteristic size) of the photon being detected. Each strip contributes

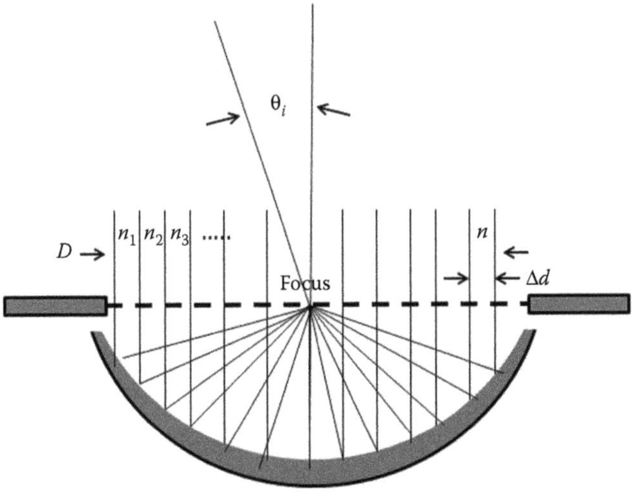

FIGURE 5.4 The diffraction pattern of a single aperture, such as the telescope shown here, is determined by the vector sum of all possible paths a photon can take in crossing the aperture.

to the probability that a photon will be received from a particular incidence angle θ_i. Let us denote this individual contribution to the probability by a position vector, where the position vector's length corresponds to the probability amplitude that a photon will pass through the strip, and the vector's position angle is proportional to the photon's arrival time difference between strips, $\Delta\phi$. For a photon arriving from large distances compared to D, the value of P_p and ϕ for each strip will be nearly equal. Therefore, for any given value of θ_i, the value of $\Delta\phi$ is constant. $\Delta\phi$ is zero on axis, and grows larger for greater values of θ_i.

The interaction of the incoming photon with the aperture is governed by the vector sum of these position vectors (Feynman, 1985). Graphically, one performs vector addition by arranging adjacent vectors head-to-tail, all the while rotating each position vector by its position angle, $\Delta\phi$. The probability of a photon entering the aperture from a given θ_i is found by drawing a resultant vector from the tail of the first position vector to the head of the last, and then squaring its amplitude. A graphical illustration of this process for four different values of θ_i is shown in Figure 5.5. For the case of $\theta_i = 0°$ (Figure 5.5a), photons arrive at all strips on the aperture simultaneously, so, $\Delta\phi = 0$. The position vectors fall on a straight line, resulting in a maximum probability amplitude P_m, suggesting photons prefer paths that take the shortest time. For values of $\theta_i > 0°$, adding the position vectors will produce an arc (see Figure 5.5b). The angle ϕ in Figure 5.5c is a measure of the total time difference between a photon arriving at one edge of the aperture compared to the other. From inspecting Figure 5.5c, one can write

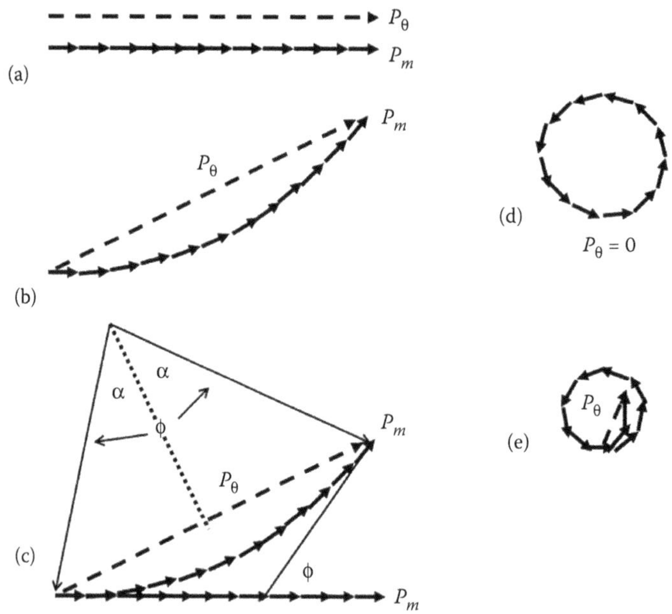

FIGURE 5.5 Diffraction pattern phasor sums. (a) Diffraction pattern peak ($\theta_i = \Delta\phi = 0$). (b) Off the peak $\Delta\phi \neq 0$ and the phasors form an arc. (c) Same as b, but showing geometry. (d) Diffraction pattern minimum, phasors wrap completely around and cancel out. (e) Secondary peak, phasors wrap one full time around, plus a bit more. (Adapted from Halliday, D. and Resnick, R., 1974, *Fundamentals of Physics*, John Wiley & Sons, Inc., New York.)

$$\sin\left(\frac{\phi}{2}\right) = \frac{P_\theta}{2R} \tag{5.15}$$

where

ϕ = phase difference between photons arriving cross the aperture

$\propto\tau$ = aperture crossing time

P_θ = the resultant vector amplitude

R = radii of curvature of vector sum

From the arc length formula, Figure 5.5c also shows

$$\phi \cong \frac{P_m}{R}; \quad \text{for small } \Delta\phi \tag{5.16}$$

Substituting Equation 5.16 into Equation 5.15 yields

$$P_\theta = \frac{P_m}{(\phi/2)}\sin\left(\frac{\phi}{2}\right)$$

Setting

$$\alpha = \frac{\phi}{2} \tag{5.17}$$

we find

$$P_\theta = P_m\frac{\sin\alpha}{\alpha} \tag{5.18}$$

Recalling ϕ is proportional to the time difference τ between photon arrival times at one end of the aperture versus the other, inspection of Figure 5.5 shows

$$\tau = \frac{D}{c}\sin\theta_i \tag{5.19}$$

and

$$\phi = \frac{\tau}{T} \tag{5.20}$$

where,

D = aperture diameter (m)

$T = \lambda/c$ = photon period (s)

λ = photon wavelength (m)

Equations 5.19 and 5.20 together show

$$\phi = \frac{D}{\lambda}\sin\theta_i \tag{5.21}$$

The probability, P_{θ_i}, of photon reception from any given θ_i is proportional to the square of Equation 5.18, namely,

$$P_{\theta_i} = P_m^2\left(\frac{\sin\alpha}{\alpha}\right)^2 \tag{5.22}$$

Equation 5.22 is an expression describing the familiar diffraction pattern of Figure 5.6. As can be seen in Figure 5.6, there are secondary peaks beyond the central one. These secondary peaks correspond to values of θ_i where the position vectors wrap part way around (see Figure 5.5d), and have a nonzero vector sum. Minima occur in the diffraction pattern where the position vectors arrange themselves in a circle, yielding a zero vector sum (Figure 5.5c).

The greater the value of D, the greater will be the number and time difference between off-axis photon paths. This will cause a faster wrapping of the position vectors, and a sharper peak in the probability distribution. The tighter the probability distribution, the tighter the central response peak of the telescope will be. In the nomenclature of classical antenna theory, the diffraction pattern of Figure 5.6 would be analogous to the "antenna beam pattern," with the central peak and secondary peaks referred to as the primary and secondary lobes. The minima correspond to nulls in the antenna pattern.

The above derivation was done considering the likelihood of a single photon passing through the aperture from an angle θ_i, but the result is the same regardless of the number of photons being considered. If one substitutes path-length (i.e., phase) differences for

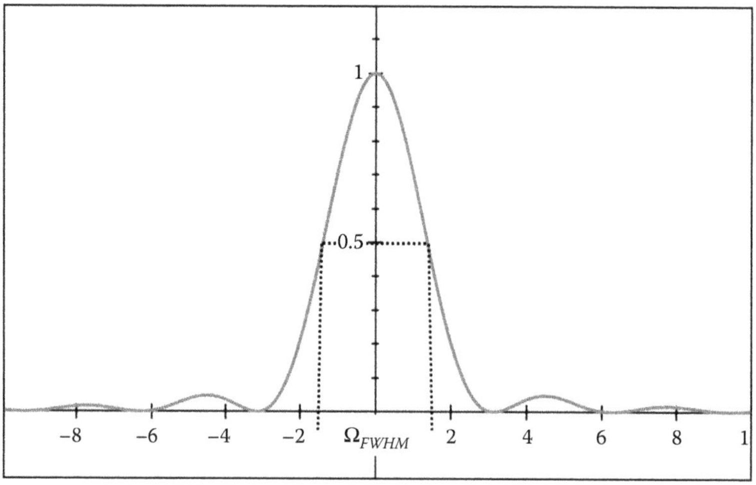

FIGURE 5.6 Diffraction pattern from a uniformly illuminated telescope with a circular aperture. The full-width-at-half-maximum (FWHM) of the pattern is $\Omega_{FWHM} \approx 1.2\lambda/D$.

time differences, and electric field amplitude for probability amplitude, the classical, wave-based form of the diffraction pattern is obtained.

5.4 GAUSSIAN BEAM OPTICS

5.4.1 GAUSSIAN BEAM BASICS

At visible and near infrared wavelengths, the sizes of lenses and mirrors are typically many thousands of wavelengths or more across. These systems can, for the most part, be analyzed or designed using ray tracing techniques. Light can be treated as being transmitted from place to place via plane waves. At THz frequencies, this is often not the case, with quasi-optical elements such as lenses and mirrors being only a few 10s or 100s of wavelengths across. Gaussian beam mode theory can be used to describe this situation.

A wave propagating through space can be described by the scalar version of the wave equation

$$\nabla^2\psi + k^2\psi = 0 \tag{5.23}$$

where

$k = 2\pi/\lambda$

ψ = scalar field function

λ = signal wavelength

For a beam moving paraxially in the z direction as depicted in Figure 5.7, define a function u in cylindrical coordinates such that (Allen, 1992)

$$\psi = u(r,\phi,z)\exp(-jkz)\exp(j2\pi ft) \tag{5.24}$$

The solutions to Equation 5.23 have the form

$$u(r,\phi,z) = \frac{C_{lp}^{LG}}{\omega(z)}\left(\frac{r\sqrt{2}}{\omega(z)}\right)^{|l|}\exp\left(-\frac{r^2}{\omega^2(z)}\right)L_P^{|l|}\left(\frac{2r^2}{\omega^2(z)}\right)\exp\left(ik\frac{r^2}{2R(z)}\right)\exp(il\phi) \\ \times \exp[-j(2p + |l| + 1)\zeta(z)] \tag{5.25}$$

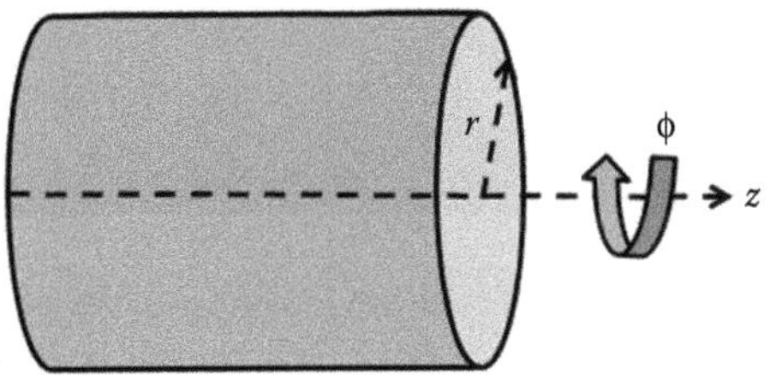

FIGURE 5.7 Paraxial beam in cylindrical coordinates.

where

L_p^l = generalized Laguerre polynomials

$p \geq 0$ = the radial index

l = azimuthal index

C_{lp}^{LG} = normalization constant

$$\zeta(z) = \arctan\left(\frac{\lambda z}{\pi\omega_0^2}\right) \tag{5.26}$$

$$\omega(z)^2 = \omega_0^2\left[1 + \left(\frac{\lambda z}{\pi\omega_0^2}\right)^2\right] \tag{5.27}$$

and

$$R(z) = z\left[1 + \left(\frac{\pi\omega_0^2}{\lambda z}\right)^2\right] \tag{5.28}$$

The solutions represent a set of propagating modes, as shown in Figure 5.8. For the vast majority of THz applications, the fundamental ($l = 0$, $p = 0$) mode is used. This mode has a single peak and (ideally) no sidelobes, making it the easiest to couple energy into and out of.

For the fundamental mode, the power distribution perpendicular to the axis of propagation is given by

$$P(r,z) = P_0 \exp\left(-\frac{r^2}{\xi(z)^2}\right) \tag{5.29}$$

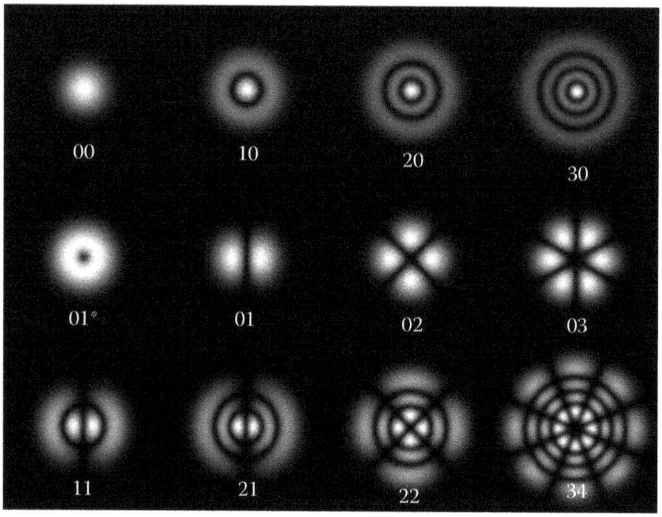

FIGURE 5.8 The intensity of Gaussian beam modes as viewed down the axis of propagation. The fundamental (00) mode is the one most commonly used at THz frequencies. (From Forget, S., 2007, Optical resonators and Gaussian beams, *Laser and Non-Linear Optics*, Université Paris. Available at: http://www.optique-ingenieur.org/en/courses/OPI_ang_M01_C03/co/Contenu_14.html. With permission.)

where

$$\xi(z) = \xi_0 \left[1 + \left(\frac{z}{k\xi_0^2} \right)^2 \right]^{\frac{1}{2}} = \frac{\omega(z)}{\sqrt{2}}$$

$$k = 2\pi/\lambda k = \frac{2\pi}{\lambda}$$

$$\xi_0 = \frac{\omega_0}{\sqrt{2}}$$

P_0 = on-axis (peak) power.

Substitution yields:

$$P(r) = P_0 \exp\left(-\frac{2r^2}{\omega(z)^2} \right) \tag{5.30}$$

A characteristic plot of Equation 5.30 is shown in Figure 5.9.

The quantity $\omega(z)$ is often referred to as the beam size or beam radius, and is the distance from the beam axis at which the field amplitude drops to $1/e$ of its on axis value. In practice, one measures power, not field amplitudes. In terms of power, the beam waist is the beam radius at which the power is down $(1/e)^2 \approx -8.69$ dB from its on-axis value. The quantity $R(z)$ is the field radius of curvature at point z.

Unlike with ray tracing, Gaussian beam optics does not assume the focus of a beam is point-like (see Figure 5.10). Instead, the focus is defined as being where the beam radius has its minimum value, called the *beam waist*, which is most often referred to as ω_0.

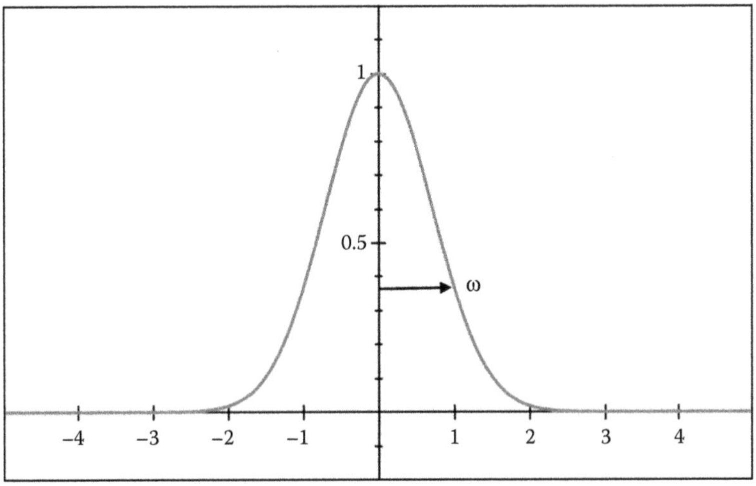

FIGURE 5.9 Gaussian beam profile.

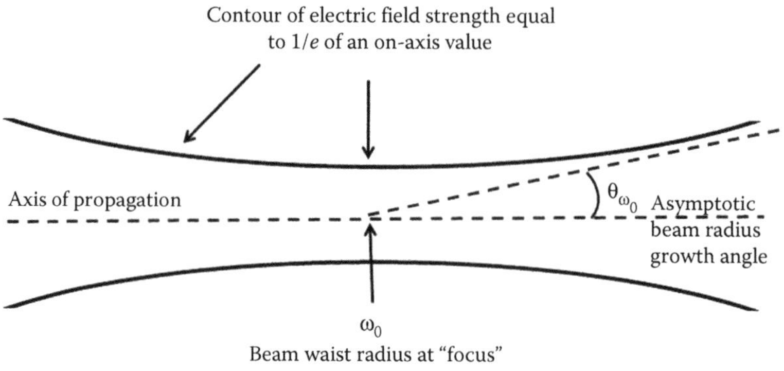

FIGURE 5.10 Focus of Gaussian beam. (Adapted from Goldsmith, P. F., 1998, IEEE Press/Chapman & Hall Publishers Series on Microwave Technology and RF, New York.)

The beam wave fronts are planar at ω_0. At large distances from ω_0, the beam has a radius of curvature equal to the distance from the beam waist. The asymptotic angle of growth of the beam waist is given by

$$\theta_{\omega_0} = \frac{\lambda}{\pi\omega_0}. \tag{5.31}$$

In the observational side of THz astronomy, we most often deal with the size of a beam at its full-width-half-maximum, Ω_{FWHM}, in terms of Gaussian beams, given by

$$\Omega_{FWHM} = 1.18\theta_{\omega_0} \tag{5.32a}$$

$$= 0.376\frac{\lambda}{\omega_0}. \tag{5.32b}$$

5.4.2 GAUSSIAN BEAM COUPLING

Much of the effort associated with designing THz optical systems is finding the best way to efficiently couple a Gaussian beam from one optical element (e.g., telescope, mirror, and lens) to another. Ideally, one would like to see the situation as shown in Figure 5.10, where a beam coming from the left side of the figure seamlessly flows into the beam from the right side. However, without proper foresight, the situation is likely to appear more like Figure 5.11.

Here, we have two Gaussian beams sharing the same axis of propagation, but with unequal beam waists that are not colocated. This situation can easily lead to suboptimal power coupling efficiencies, C_{12}, ≪1 (Goldsmith 1982).

$$C_{12} = \frac{4}{\left[([\omega_{01}/\omega_{02}] + [\omega_{02}/\omega_{01}])^2 + (\lambda/\pi\omega_{01}\omega_{02})^2 d^2\right]} \tag{5.33}$$

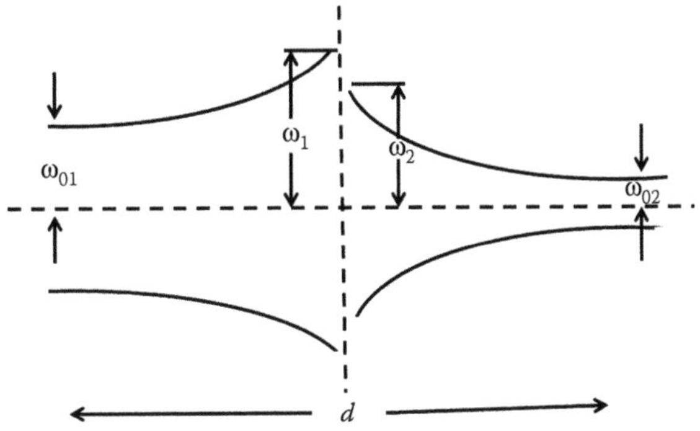

FIGURE 5.11 Coupling between two Gaussian beams. (Adapted from Goldsmith, P. F., 1998, IEEE Press/Chapman & Hall Publishers Series on Microwave Technology and RF, New York.)

For perfect beam coupling, $C_{12} = 1$, the two beams must have equal beam radii and be colocated. This is accomplished by using lenses and mirrors.

EXAMPLE 5.1

Estimate the loss associated with coupling two, on-axis Gaussian beams, with beam waists of 1.0 mm and 3 mm. Assume the wavelength is 0.35 mm (i.e., 0.86 THz), and the separation between beam waists is 1 mm.

Here, we have

$$\begin{aligned} \omega_{01} &= 1.0 \text{ mm} \\ \omega_{02} &= 2.0 \text{ mm} \\ d &= 1.0 \text{ mm} \\ \lambda &= 0.35 \text{ mm} \end{aligned}$$

The coupling efficiency, C_{12}, between the two beams can be found by substituting the above quantities into Equation 5.33.

$$\begin{aligned} C_{12} &= \frac{4}{\left[\left(\frac{1}{2}+\frac{2}{1}\right)^2 + \left(\frac{0.35}{\pi(1)(2)}\right)^2 (1)^2\right]} \\ &= 0.64 \end{aligned}$$

We see that 36% of the power is lost in the coupling process. At THz frequencies, even small beam misalignments can lead to significant losses in efficiency.

5.5 FOCUSING GAUSSIAN BEAMS

In astronomy, all objects of interest are far away, with the result that photons from them arrive on parallel trajectories. The purpose of a telescope is to collect these parallel photons, and focus them into a detector that is usually orders of magnitude smaller than the telescope's aperture. This is accomplished using a system of lenses and/or mirrors. Each of these focusing elements has its pros and cons. Systems that use lenses are more compact than those that use reflective optics, but are more lossy, due to reflection from and absorption by the dielectric material that forms the lens. It is often the case that the best solution is a hybrid approach, where you use mirrors where you can and lenses where you must.

5.5.1 LENSES

Lenses, like the parabolic antenna of Figure 5.4, are used to increase the likelihood that a photon will travel toward a designated point, for example, the detector. The input and output parameters of a lens described in terms of Gaussian beam optics are depicted in Figure 5.12.

As an example, let's assume light collected by a telescope is entering from the left and converging toward a minimum beam waist, ω_{01}, at distance d_1, along the axis of propagation. Unfortunately, d_1 does not turn out to be a convenient location for our detection system, which would rather be at d_2. It also turns out that for maximum power coupling, the detector would prefer that the light entering it should have a beam waist ω_{02}. These input and output parameters are related to each other through the following expression (Goldsmith, 1998):

$$\frac{1}{f} = \frac{1}{d_1\left[1+(\pi\omega_{01}^2/\lambda d_1)^2\right]} + \frac{1}{d_2\left[1+(\pi\omega_{02}^2/\lambda d_2)^2\right]} \tag{5.34}$$

where

ω_{01} = input beam waist
ω_{02} = output beam waist
d_1 = location of input beam waist
d_2 = location of output beam waist
f = focal length of lens

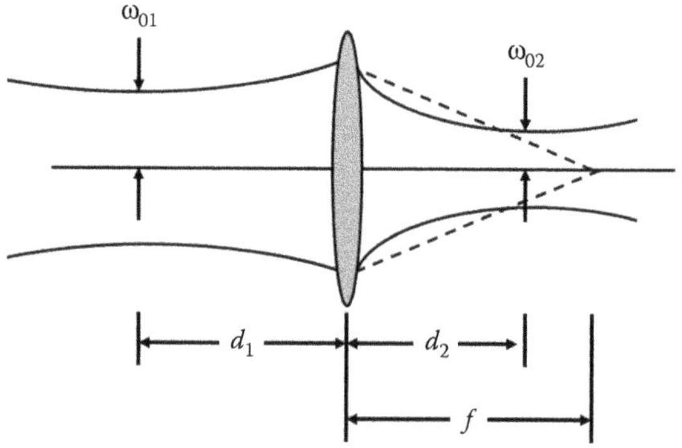

FIGURE 5.12 Focusing a Gaussian beam with a lens. (Adapted from Goldsmith, P. F., 1998, IEEE Press/Chapman & Hall Publishers Series on Microwave Technology and RF, New York.)

The above expression can be used to determine the focal length of the lens we need to match two Gaussian beams. If we have the situation,

$$d_1 = d_2 = f, \tag{5.35}$$

then we find

$$\frac{\omega_{02}}{\omega_{01}} = \frac{\lambda f}{\pi \omega_{01}^2}. \tag{5.36}$$

In the most basic optical system, the value of ω_{01} is set by the telescope (see Section 5.7), and the value of ω_{02} by the detector (see Section 5.6). Since these quantities are known, the value of f can be determined from Equation 5.36. A lens can only produce a beam waist, $\omega(z)$, to distances, z, such that (Lesurf, 1990)

$$z \leq \frac{\pi \omega(z)^2}{2\lambda}. \tag{5.37}$$

The next parameter to determine is the size of the lens. The size of the lens (or mirror) depends on the application, but here is a rule of thumb. At the anticipated location of the lens, use Equation 5.27 to calculate the beam radius, $\omega(z)$. If you have room in your system, make the *diameter* of the lens (or mirror) equal to $4\omega(z)$. By doing so, the power at the edge of the lens is ~3.5×10^{-4} (−35 dB) down from what it is at the center of the lens. If space in your optical system is tight, you can make the lens somewhat smaller (say, $3\omega(z)$) without seriously compromising performance. The amount of power, P_c, coupled through a lens/mirror of *radius*, r_{optic}, is given by

$$P_c = 1 - \exp\left(-2\left(\frac{r_{optic}}{\omega(z)}\right)^2\right) \tag{5.38}$$

EXAMPLE 5.2

What percentage of a Gaussian beam will make it through a lens of diameter $3\omega_z$?

A lens of diameter $3\omega_z$ has a radius $r_{optic} = (3/2)\omega(z)$. Substituting into Equation 5.38, we find

$$\begin{aligned} P_c &= 1 - \exp\left(-2\left(\frac{3}{2}\right)^2\right) \\ &= 1 - 0.011 \\ &= 0.989 \end{aligned}$$

With a lens *diameter* of $2r_{optic} = 3\omega(z)$, the above expression shows that ≈99% of the beam power will be coupled through the lens.

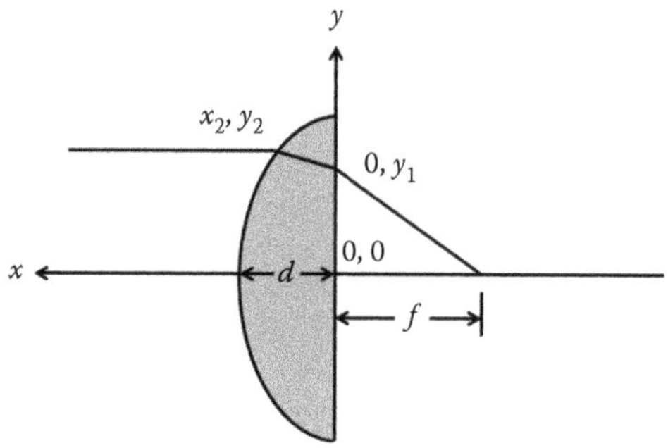

FIGURE 5.13 Plano-convex lens design.

Depending on the application, lenses can take on a variety of shapes, including biconvex, plano-convex, biconcave, plano-concave, and elliptical. Using modern optical design packages and numerical milling techniques, one-of-a-kind lenses can be created to provide optimum coupling between less than ideal beams (Hübers, 2004). However, the most common lens employed at THz frequencies is the plano-convex lens (see Figure 5.13). This type of lens is optimized to take photons on parallel trajectories, and bring them to a focus. A formula for specifying the curvature of a plano-convex lens is provided below (Jasik, 1961).

$$
\begin{aligned}
x_2 &= \left\{ \frac{(n-1)d + f - \sqrt{f^2 + y_1^2}}{n - (1 - (y_1^2/n^2(f^2 + y_1^2))^{1/2}} \right\} \cdot \left[1 - \frac{y_1^2}{n^2(f^2 + y_1^2)} \right]^{1/2} \\
y_2 &= y_1 \left[1 + \frac{x_2}{[n^2(f^2 + y_1^2) - y_1^2]^{1/2}} \right]
\end{aligned}
\tag{5.39}
$$

where

n = index of refraction of lens material
f = focal length
d = thickness
$x_2 y_2$ = coordinates on the curved surface of the lens
y_1 = vertical coordinate on flat surface of the lens

A biconvex lens can be formed by putting two plano-convex lenses back-to-back.

5.5.2 BEHAVIOR OF LIGHT AT DIELECTRIC INTERFACES

The index of refraction, n, of a material is a measure of how fast photons can travel through it, compared to free space.

$$
\begin{aligned}
n &= \frac{c}{v_D} = c\sqrt{\varepsilon\mu} \\
&= \sqrt{\frac{\varepsilon\mu}{\varepsilon_o\mu_o}}
\end{aligned}
\tag{5.40}
$$

where

ε = electrical permittivity of the material
ε_o = electrical permittivity of free space
μ = magnetic permeability of the material
μ_o = magnetic permeability of free space
v_D = speed of light in dielectric

$$n = \sqrt{\varepsilon_R \mu_R} \tag{5.41}$$

for $\varepsilon_R = \varepsilon/\varepsilon_o$ and $\mu_R = \mu/\mu_o$

In most dielectric materials of interest, it is safe to assume that $\mu = \mu_o$, so Equation 5.41 simplifies to

$$n \approx \sqrt{\varepsilon_R}, \quad n > 1 \tag{5.42}$$

Interestingly, it is possible to create materials where both ε_R and μ_R are negative, in which case, the index of refraction is said to be negative, meaning the photon paths have a negative angle of refraction. Such a material is referred to as a metamaterial, and can be formed from periodic structures with features smaller than the wavelength at which they will be used. Materials that have negative values of either ε_R or μ_R (but not both) are opaque (e.g., gold and silver).

When photons encounter a dielectric surface (whether it is a lens, beam splitter, dewar window, etc.), optical loss will be introduced in two ways:

1. Through reflections off the front and back surfaces, due to impedance mismatches at the material interfaces.
2. Through power dissipation in the dielectric material itself.

At normal incidence (see Figure 5.14) the reflectivity, r_1, from the surface is

$$r_1 = \frac{n_2 - n_1}{n_2 + n_1} \tag{5.43}$$

The fractional power lost, P_r, from the incident signal at the interface due to the reflection is

$$P_r = \left(\frac{n_2 - n_1}{n_2 + n_1}\right)^2 \tag{5.44}$$

There will be an equivalent amount of reflection experienced when the light exits the dielectric, r_2. The reflected photons from the front and back will interfere, with the total reflection loss, P_R, being a function of both the angle of incidence and material thickness (Lesurf, 1990).

$$P_R = \frac{(r_1 + r_2)^2 - 4r_1 r_2 \sin^2 \rho}{(1 + r_1 r_2)^2 - 4r_1 r_2 \sin^2 \rho} \tag{5.45}$$

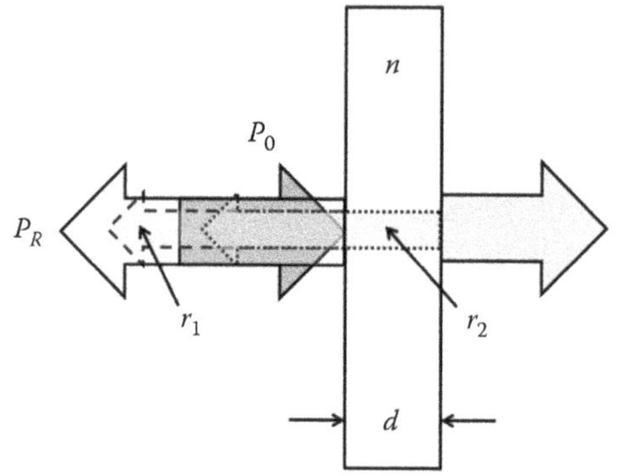

FIGURE 5.14 Light beam encountering a dielectric slab of thickness, d, and index of refraction n at normal incidence ($\theta = 0$). P_0 is the incident beam power. r_1 and r_2 are the reflected beam amplitudes from the material interfaces. P is the emergent beam power. P_R is the reflected power from the interference of the reflected beams.

where

$$\rho = \frac{2\pi d}{\lambda_0}\sqrt{n^2 - \sin^2\theta},$$

r_1 = reflectivity from interface 1
r_2 = reflectivity from interface 2
n = index of refraction of material
d = thickness of dielectric
θ = angle of incidence (measured from normal)
λ_0 = vacuum wavelength of interest

Equation 5.45 is particularly useful for designing vacuum windows and beam diplexers (often used for THz LO injection—see Appendix 7). For a dielectric slab at normal incidence in free space, $r_1 = r_2$ and Equation 5.45 reduces to

$$P_R^{\max} = \frac{4r^2}{(1+r^2)^2} \tag{5.46}$$

where
$P_R^{\max}$ = maximum fractional reflected power
$r = r_1$ or r_2
For a slab thickness d, the value of P_R will vary between $P_R^{\max}$ and 0 as a function of $\frac{d}{\lambda}$.

The absorptive power loss per unit length, α, through the material, is given by

$$\alpha = \frac{2\pi n \tan\delta}{\lambda_0} \tag{5.47}$$

where, tan δ is the loss tangent of the material,

$$\tan\delta = \frac{\omega\varepsilon'' + \sigma}{\omega\varepsilon'} \tag{5.48}$$

Here, σ is the electrical conductivity, and ε′ and ε″ are the real and imaginary parts of the complex permittivity. For nonconducting dielectrics, $\sigma = 0$ and Equation 5.48 reduces to

$$\tan\delta = \frac{\varepsilon''}{\varepsilon'} \tag{5.49}$$

The fractional power loss, L_d, due to dissipation through the material, is then

$$L_d = \left(1 - \frac{P}{P_o}\right) = (1 - e^{-\alpha d}). \tag{5.50}$$

EXAMPLE 5.3

What is the power loss from a 1.9 THz beam as it passes from free space through a 3 mm thick disk of high density polyethylene (HDPE) at normal incidence?

There will be both reflective and absorptive losses through the dielectric. The reflectivity, r_1, at the dielectric interface can be found from Equation 5.43. From Table 5.1 for HDPE we have, $n_2 = 1.47$ and for free space, $n_1 = 1$. Substitution then yields,

$$r_1 = \frac{1.47 - 1}{1.47 + 1} = 0.19.$$

Power will be reflected from both the front, r_1, and back, r_2, of the HDPE, where $r_1 = r_2$. For normal incidence, the total reflective power loss, P_R^{max}, can be found from substituting into Equation 5.46.

$$P_R^{max} = \frac{4(0.19)^2}{(1 + (0.19)^2)^2} = 0.13$$

Therefore, a maximum of ≈13% of the incident power is lost in reflections.

The absorptive power loss is computed using the absorption coefficient, α, of the HDPE (Equation 5.47) and the equation of transfer through the material, L_D, (Equation 5.50).

$$\alpha = \frac{2\pi n \tan\delta}{\lambda_0} = \frac{2\pi(1.47)(9.7 \times 10^{-4})}{0.158\ \text{mm}}$$
$$= 0.057\ \text{mm}^{-1}$$

$$L_D = 1 - e^{-\alpha d} = 1 - e^{-(0.057)3}$$
$$= 0.156$$

The absorptive loss through the 3 mm thick piece of HDPE is ≈16%. The total loss, L_{total}, through the dielectric is, then,

$$L_{total} = P_R + L_D \approx 0.28 \Rightarrow 28\%.$$

Dielectric disks such as this are often employed as vacuum windows in THz cryogenic systems. To reduce signal loss, the disks should be antireflection (AR) coated (if possible) and thin (as possible). The larger the window diameter, the thicker the disk needs to be to keep from blowing out due to the pressure differential across it.

TABLE 5.1 Commonly Used THz Dielectric Materials

Material	n	tan δ (×10^{-4})	Frequency (THz)	Reference
Polyethylene (LDPE)	1.5138 ± 0.0002	3–8	0.15–1.11	Birch et al. (1981)
Polyethylene (HDPE)	1.4711 ± 0.0003	9.7 ± 0.3	0.891	Qiu et al. (1992)
Silicon	3.416–3.419	2–12	0.6–4.2	Randall and Rawcliffe (1967)
Silicon (11,000 Ω-cm)	3.414	2.5 (lower Ω-cm has higher loss)	0.3	Afsar and Chi (1994)
Quartz	2.1133 ± 0.0004	2.49 ± 0.08	0.891	Qiu et al. (1992)
Quartz-o	2.1073–2.2072	—	0.6–6	Russell and Bell (1967)
Quartz-e	2.1541–2.2502	—	0.6–6	Russell and Bell (1967)
Fused silica	1.935 ± 0.001	35 ± 1	0.89	Tsuji et al. (1982)
Flurogold	1.624–1.636	70–310	0.15–1.02	Birch and Kong (1986)
Kapton	1.77 ± 0.009	46.5 ± 2.3	1.8	Smith and Lowenstien (1975)
Mylar	1.752 ± 0.002	237 ± 7	1.5	Smith and Lowenstien (1975)
Sapphire-o	3.0666 – 3.0649	4–9	0.09–0.4	Afsar (1987)
Sapphire-e	3.4056 – 3.4039	4–8	0.09–0.4	Afsar (1987)
Parylene C	1.62	132–168	0.45–2.8	Ji et al. (2000)
Parylene D	1.62	66–105	0.45–2.8	Ji et al. (2000)
Zitex (G125)	1.2 ± 0.07	1.45–26.8	0.4–1.6	Benford et al. (2003)
Germanium	3.9904 ± 0.0006	14.4 ± 0.4	0.891	Qiu et al. (1992)

Note: LDPE, low density polyethylene; HDPE, high density polyethylene.

The properties of commonly used THz dielectric materials are provided in Table 5.1. A more extensive listing of dielectric material properties can be found in Goldsmith (1998). Silicon is a favorite material amongst THz designers. For example, if we use a 3 mm thick lens of silicon at a wavelength of 158 μm, then using Equations 5.47 and 5.48, and a value for $n = 3.414$ from Table 5.1, we find 9.7% of the incident power will be lost through the lens. Using Equation 5.45, we find the associated reflective power losses will be 53%, more than five times that from absorption. One way to reduce reflective losses is to make the thickness of the lens such that the reflections from the front and back surfaces (r_1 and r_2) nearly cancel out. This is hard to do, since the lens' surfaces are not plane parallel. Even in the case of optical components that are plane parallel, such as a filter or window, the cancellation effect will come and go with frequency (see Equation 5.45). Since phase errors scale with thickness, in almost all instances, it is better to use a thin, antireflection (AR) coating (e.g., out of Parylene).

AR coatings work by providing an impedance match between the two media being considered. The impedance match is achieved by adding an intermediate layer in which the values of r_1 and r_2 associated with the layer cancel one another out. Examination of Equation 5.45 shows this happens for the case where the AR coating layer has an index of refraction, n_{AR}, such that

$$n_{AR} = \sqrt{n_1 n_2} \tag{5.51}$$

where

n_1= index of refraction of medium 1 ($n = 1$ for free space)
n_2= index of refraction of medium 2

The thickness, t, of the layer should be

$$t = \frac{(2\kappa + 1)\lambda_D}{4} \tag{5.52}$$

where

$\lambda_D = \lambda_0 / n_{AR}$ is the wavelength of the photons within the AR coating
κ = positive integer (lower values are better: less loss, higher bandwidth)

In other words, the thickness of the AR coating should be an odd number of $(\lambda_D/4)$'s thick (see Figure 5.15). Here is why. In terms of the phasor description of probability amplitudes discussed in Section 5.6, the position vector associated with a photon traveling the added $\lambda_D/4$ from surface 1 to surface 2 will pick up 90° of phase shift. The position vector will pick up an additional 180° of phase shift upon reflection at surface 2, and an additional 90° on the way back. Upon arrival back at surface 1, the position vector will have picked up a total of 360° of phase shift and be *in-phase* with the position vectors of photons entering the dielectric. However, it will be 180° *out-of- phase* with the position vector of photons that would reflect off of surface 1. The vector sum of the position vectors associated with r_1 and

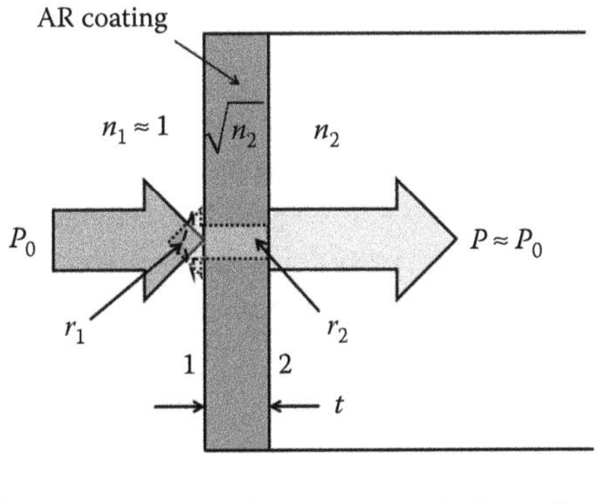

FIGURE 5.15 The addition of a proper AR coating can quench the reflected waves, r_1 and r_2, allowing a greater percentage of the incident wave to pass through.

r_2 is zero, meaning the probability of photons being reflected is zero, and the probability of photons entering the dielectric is one. In this ideal case, there is no power loss associated with reflection. To increase the useful frequency range (or bandwidth) of the coating, multiple layers can be deposited, each tuned to a slightly different frequency. For bandwidths of ~10%, a single $\lambda_D/4$ layer will be sufficient.

If no dielectric with the appropriate value of n_{AR} can be found, one can be simulated by sculpting the lens or dielectric slab with a series of grooves of depth $\lambda/4n_{AR}$, of pitch $\lambda/4n_{AR}$, and width $\lambda/4(1 + n_{AR})$. Such grooves will introduce a polarization dependency to the lens or slab. This polarization dependency can be somewhat mitigated by making the grooves concentric about the axis of propagation. To reduce polarization effects further, instead of a series of grooves, an array of holes can be drilled. Another alternative would be to etch deep triangular grooves several wavelengths deep into the dielectric (Jones and Cohn, 1955). However, due to the small size scales required, these machining methods become problematic to realize at THz frequencies.

Since the material transition looks the same to the photons leaving the dielectric as those entering, it is necessary to deposit, adhere, or machine an AR coating to both sides of the lens (or dielectric slab), to avoid reflective losses.

5.5.3 MIRRORS

As was pointed out by Isaac Newton, mirrors are an attractive alternative to refractive optics. A well-polished, conductive mirror has almost no reflective/absorptive loss, and can be machined into almost any shape. The three most commonly used mirror shapes are the flat, parabola, and ellipse. The flat mirror is used simply to redirect the photon stream. In designing a flat mirror, one needs to be sure it is smooth enough not to scatter the wavelengths, λ_0, of interest (i.e., with an rms surface roughness $\sigma_{rms} \leq \lambda_0/50$) and large enough to accommodate the beam (i.e., $\geq 3\omega_z$ in projection). The parabola and ellipse are mirrors with "power", meaning they can focus or defocus beams.

5.5.3.1 PARABOLA

For the purposes of THz astronomy, the parabolic surface of a telescope is often the first thing an incoming photon from deep space encounters. In Figure 5.4, a full parabola is used to collect parallel streams of incoming photons and bring them to a common focus. Unfortunately, we are almost never able to put our THz detectors at the prime focus of a parabola. Instead, there is a secondary mirror (typically a hyperbolic surface), followed by a system of mirrors or lenses that shepherd the photons to our detectors. When a clean, unobstructed beam is needed (e.g., in cosmic background studies), off-axis parabolas are often used for the primary mirror of a telescope. An example of an off-axis parabola is shown in Figure 5.16. A full parabola would be formed by simply revolving the mirror about the z-axis. The surface of Figure 5.16 has its vertex at the origin. The y-axis projects out of the page.

The formula describing the surface is

$$z = \frac{x^2 + y^2}{4f_p}, \tag{5.53}$$

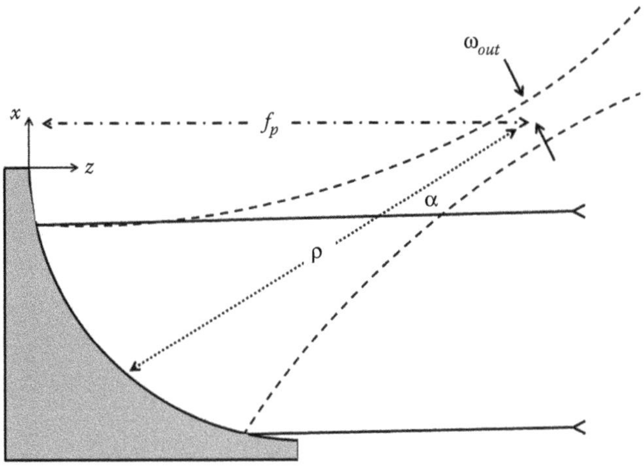

FIGURE 5.16 Off-axis parabola.

where

f_p = the distance from the parabola's vertex at $x = y = z = 0$ to the focus

ρ = the distance between a point on the surface and the focus

The expression for ρ as a function of f_p and α is

$$\rho = \frac{2f_p}{1 + \cos\alpha}. \tag{5.54}$$

In many instances, the designer may need to use only a segment of the parabolic surface to achieve the desired ends.

5.5.3.2 ELLIPSOID

The trouble with relying too heavily on the parabola alone to relay photons around, is that its parallel input beam (or output beam, depending on how it is used) will have an opening angle, Ω_{FWHM}, due to diffraction (see Figure 5.6) given by

$$\Omega_{FWHM} \approx 1.2\frac{\lambda}{D}, \tag{5.55}$$

where

λ = wavelength of operation

D= diameter of the optic presented to photons

If the optical system is relatively compact in length (<1000 wavelengths or so), diffraction may not be a big problem. For larger optical systems, a more conservative approach is to allow for beam expansion, and then refocus the beam down. Lenses can do this, or an elliptical mirror, like that shown in Figure 5.17, can be used.

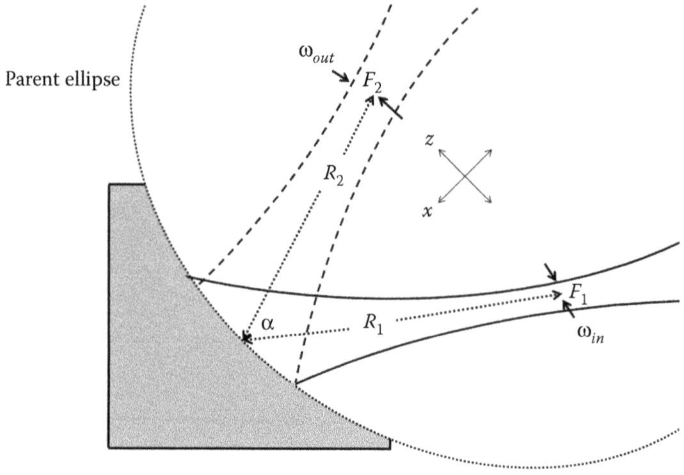

FIGURE 5.17 Off-axis ellipse.

The beauty of an ellipse is that it has two foci, one for the input beam (F_1) and other for the output beam (F_2). The equation for the surface of the parent ellipse is

$$\frac{x^2 + y^2}{b^2} + \frac{z^2}{a^2} = 1 \tag{5.56}$$

where

$x = y = z = 0$ is the center of the ellipse
a = length of semi-major axis
b = length of semi-minor axis

The eccentricity, e, of the ellipse is

$$e = \left(1 - \frac{b^2}{a^2}\right)^{1/2} \tag{5.57}$$

The distances between the two foci and a point on the surface are related by

$$2a = R_1 + R_2. \tag{5.58}$$

The eccentricity can also be defined in terms of R_1, R_2, and the angle α between input and output beams (Goldsmith, 1998),

$$e = \frac{(R_1^2 + R_2^2 - 2R_1R_2\cos\alpha)^{1/2}}{R_1 + R_2}. \tag{5.59}$$

The ellipse is equivalent to having two lenses (like that of Figure 5.13), or two parabolas (like that of Figure 5.15), arranged back-to-back (with zero separation), connected by a parallel beam (see Figure 5.18). In this model, R_1 and R_2 are the radii of curvature of the

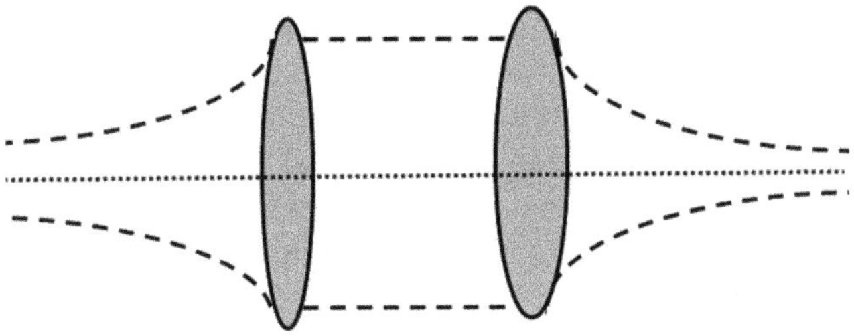

FIGURE 5.18 Two-lens model of ellipse.

input and output optical elements (lenses or mirrors). In optics, when you stack two or more lenses, one says the combination has an effective focal length, f_e. For an ellipse, the effective focal length is defined to be (Goldsmith, 1998)

$$f_e = \frac{R_1 R_2}{R_1 + R_2}. \tag{5.60}$$

The ratio between the output, ω_{out}, and input, ω_{in}, beam waists, gives us the magnification factor, M, of the ellipse.

$$M = \frac{\omega_{out}}{\omega_{in}}. \tag{5.61}$$

Physical constraints on an optical system will usually force us to have a particular input beam waist, magnification factor, M, and input distance, d_{in}. We can use this knowledge to solve for the required focal length of the system using the expressions (Goldsmith 1998)

$$f = z_c\left(\frac{d_{in}}{\zeta}\right)\left[1 \pm \left(1 - \zeta\left[1 + \left(\frac{d_{in}}{z_c}\right)^{-2}\right]\right)^{0.5}\right] \quad \text{for } M \neq 1 \tag{5.62}$$

and

$$f = z_c \frac{1 + (d_{in}/z_c)^2}{(2d_{in}/z_c)} \quad \text{for } M = 1 \tag{5.63}$$

where ζ and the confocal distance, z_c, are given by

$$\zeta = 1 - M^{-2} \tag{5.64}$$

and

$$z_c = \frac{\pi \omega_0{}^2}{\lambda}. \tag{5.65}$$

Assuming for the case of our ellipse, $d_{in} = R_1$ and $f = f_e$, we can solve for R_2, using Equation 5.60. The input to output beam angle, α, works into our prescription of the ellipse through Equation 5.59, which, together with Equations 5.58 and 5.57, can be used to solve for the ellipse parameters a and b (see Equation 5.56). The last thing to figure out is the required size of the ellipse. This is done by projecting the input beam onto the ellipse by substituting ω_{in} for ω_0, R_1 for z, and solving for $\omega(z)$, using Equation 5.27. The mirror diameter should be made $\geq 3\omega_z$ in projection.

5.5.3.3 MIRROR REFLECTIVE LOSSES AND BEAM DISTORTION

Reflection from the surface of a mirror can introduce signal loss in two ways: absorption within the material itself, and by surface scattering. When it comes to metallic mirrors, scattering is the dominant loss mechanism.

Most telescopes designed for operation at optical and infrared wavelengths, use optics made of glass. In order to make it an efficient reflector of THz photons, the surface must be coated with a thin layer of metal, typically through the vacuum deposition of aluminum. How thick to make the coating is a function of the skin depth of the metal being used. The skin depth, δ, is the distance from the surface of a conductor to where the electric field associated with the incoming light drops to e^{-1} of its original value. At THz frequencies, the skin depth is well approximated in free space by

$$\delta = \sqrt{\frac{2\rho}{\omega\mu_0}} = 503\left(\frac{\rho}{f(\mathrm{THz})}\right)^{1/2} \tag{5.66}$$

where

ρ = resistivity (Ωm)

$\omega = 2\pi f$

$\mu_0 = 4\pi \times 10^{-7}\ \mathrm{V\cdot s/A\cdot m}$

For example, aluminum has $\rho = 2.74 \times 10^{-8}\ \Omega$m. The resistivities of more expensive silver, gold, and copper are within a factor of two of this value. For aluminum, at 2 THz, we find $\delta = 0.06\ \mu$m. It only takes a few skin depths of these materials ($\approx 0.2\ \mu$m for aluminum) to make a surface essentially a perfect conductor at THz frequencies. Goldsmith (1998, Figure 5.23) shows that the reflective power loss due to conduction over these surfaces is, in most instances, negligible.

When photons encounter a surface with features of the order of their wavelength, they will most likely scatter upon reflection. The smaller the size scale of the surface features compared to the photon's wavelength, the smaller the effects of scattering will be. The efficiency, η_R, by which a metallic surface with roughness $\varepsilon_{\mathrm{rms}}$ reflects power, is described by Ruze (1966) to be

$$\eta_R \approx \frac{P_R}{P_0} \approx \exp\left[-\left(\frac{4\pi\varepsilon_{\mathrm{rms}}}{\lambda}\right)^2\right] \tag{5.67}$$

where

ε_{rms} = root-mean-square surface roughness (microns)
λ = wavelength (microns)
P_R = reflected power (watts)
P_o = incident power (watts)

Assuming we wish the fractional power reflected off a surface to be ≥90% , then $\varepsilon_{rms} \leq \lambda/40$.

Reflection from an off-axis mirror will introduce distortion in the amplitude distribution of a Gaussian beam, and cross-polarization. The amplitude distortion will result in some of the beam's power being transferred into unwanted, nonfundamental Gaussian beam modes. The fractional power loss from the (desired) fundamental Gaussian beam mode upon reflection, L_S, is a function of the angle of incidence, θ_i, and beam size on the mirror surface (Murphy, 1987; Goldsmith, 1998),

$$L_S = \frac{\omega_M^2 \tan^2\theta_i}{8f^2} \propto \left(\frac{\lambda}{D}\tan\theta_i\right)^2 \tag{5.68}$$

where

θ_i = angle of incidence
f = mirror focal length
ω_M = incident beam waist

As might be expected, Equation 5.68 argues for small values of λ/D and θ_i. For most optical systems, a value of $\theta_i \leq 45°$ provides low loss (of order 1%) performance. If the Gaussian beam is polarized, reflection from an off-axis mirror will move some of the beam's power into the orthogonal polarization. For polarization-sensitive detector systems (e.g., those using feedhorns and/or wire grids), this will result in a loss of power, L_P, given by

$$L_P = \frac{\omega_M^2 \tan^2\theta_i}{4f^2}. \tag{5.69}$$

From comparison of Equations 5.68 and 5.69, we see that the power loss from the incident beam due to cross-polarization is twice that due to loss into unwanted modes, an important point when working with polarizations sensitive detection systems (e.g., those using wire grids, detectors in waveguide, and quasi-optical detector systems using slot antennas).

In many optical systems, multiple mirrors are required. In these instances, it is possible to mitigate beam distortion by arranging pairs of mirrors so that the second mirror compensates for the distortions introduced by the first (Lesurf, 1990). Figure 5.19a illustrates the optimum arrangement of a mirror pair if the input and output beams are going in opposite directions, and Figure 5.19b if they are going in the same direction.

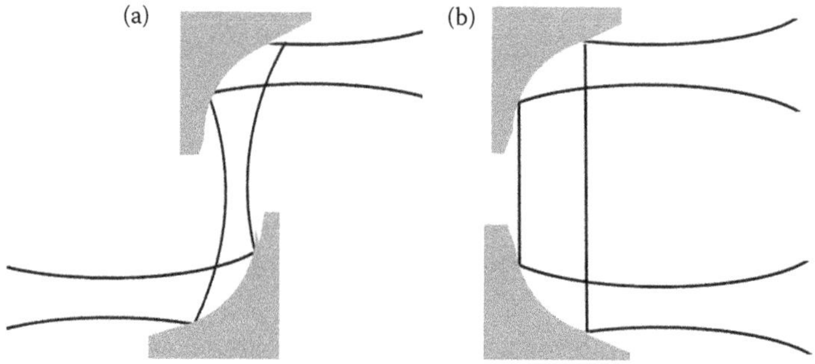

FIGURE 5.19 Mirror pairs arranged to cancel distortion. (a) Best configuration when input and output beams are in the same direction. (b) Best configuration when input and output beams are in the same direction. (Adapted from Lesurf, L., 1990, *Millimetre-wave Optics, Devices & Systems*, Adam Hilger, Bristol New York.)

5.5.4 GAUSSIAN BEAM TELESCOPE

The focus of an optical system designed using Equation 5.34 will have a wavelength dependency. The positions of the input and output beams will change with frequency which is, in general, undesirable. The way to circumvent this problem is to put two lenses (or mirrors) that satisfy Equation 5.35 in series (see Figure 5.20), with the input beam waist of the second lens , ω_{in2}, equal to the output beam waist of the first lens, ω_{out1}. The distance between the lenses, d, is

$$d = f_1 + f_2. \tag{5.70}$$

With this arrangement, the wavelength dependence of the beam waists' locations drops out, giving

$$d_{out2} = \frac{f_2}{f_1}\left(f_1 + f_2 - \frac{f_2}{f_1}d_{in1}\right). \tag{5.71}$$

The magnification factor, M, of the telescope is defined to be

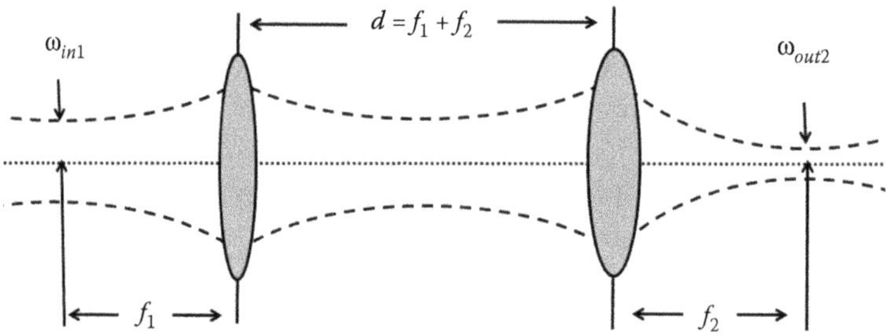

FIGURE 5.20 Gaussian beam telescope.

$$M = \frac{\omega_{out2}}{\omega_{in1}} = \frac{f_2}{f_1}, \tag{5.72}$$

which is also independent of the wavelength. With an $M = 1$, we find

$$f_1 = f_2 \tag{5.73}$$

and we have succeeded in shifting the focal plane along the optical path by an amount d. A series of lenses so ordered forms a Gaussian beam waveguide.

5.5.5 WIRE GRID POLARIZERS

Each photon has an intrinsic polarization, be it linear or elliptical, imparted to it by the emission mechanism that produced it. The polarization of a photon can provide clues to its origin. In astrophysics, knowledge of the photon's original polarization is often lost due to interactions with materials (e.g., dust grains) and/or fields (e.g., magnetic) that the photon experiences on its way to our solar system. In these instances, by analyzing the polarization of received photons, we can "back out" the properties of the medium through which the photon has traveled. With a few notable exceptions (e.g., synchrotron and maser emission), microscopic collisions between emitting atoms and/or molecules in astrophysical sources randomize their physical orientation, with the result that the emitted photons appear, as a whole, unpolarized. Due to the stochastic nature of black-body radiation (say from a gas or solid), it too appears unpolarized.

A photon can be modeled as the superposition of orthogonal time varying electric and magnetic fields. The amount of photon energy stored in each is identical; however, the electric field strength is far greater. For this reason, in physics and engineering, one typically describes the behavior of photons and their detection in terms of electric fields. At any given point in time, the position vector describing the electric field, E, of an incoming photon, can be described in terms of the field components E_x (horizontal) and E_y (vertical) projected on to the x and y axes of a Cartesian coordinate system (see Figure 5.21). Some detection systems, for example a bolometer fed by a Winston cone, are sensitive to both. Others, such as a detector mounted in a rectangular waveguide, are not. To maximize sensitivity, or to measure the degree of polarization impressed on the incoming stream of photons, a detection system should, if possible, be made sensitive to both vertical and horizontal field components. A common way to achieve this is to use a one-dimensional wire grid oriented at 45° with respect to the incoming beam of light (see Figure 5.22). Here, the wire grid serves the role of a polarization diplexer. The grid splits the incoming photon stream into two, with vertically polarized photons going to one detector system and horizontally polarized photons to another. Here, the grid is said to act as a polarization analyzer.

The grid of Figure 5.21 has a power reflectivity (Lesurf, 1990) for photons with polarization parallel to the wires, $R_{\|}$, of

$$R_{\|} = \left[1 + \left(\frac{2S}{\lambda}\right)^2 \ln\left(\frac{S}{\pi d}\right)^2\right]^{-1} \tag{5.74}$$

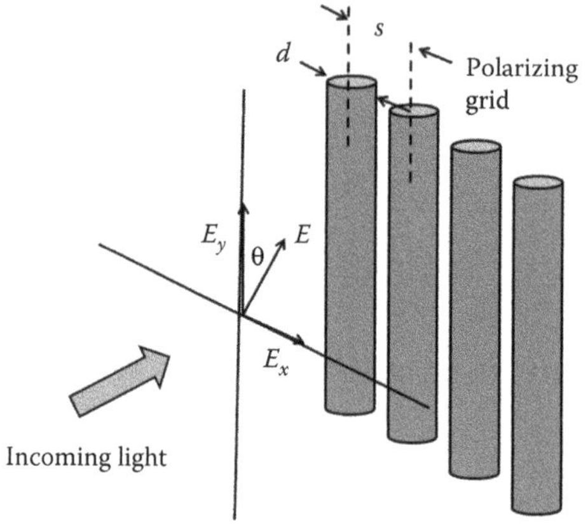

FIGURE 5.21 Wire-grid polarizer geometry.

and perpendicular to the wires, $R_\perp$, of

$$R_\perp = \frac{(\pi^2 d^2)^2}{(2\lambda S)^2\left[1 + (\pi^2 d^2/2\lambda S)^2\right]} \tag{5.75}$$

where

S = center to center spacing between wires
d = wire diameter
λ = wavelength of interest

At a given wavelength, the ratio of wire thickness to spacing, S/d, can be adjusted to allow a desired amount of power to flow through or be reflected from the grid. In most applications, you will want to make all the photons associated with one polarization go one

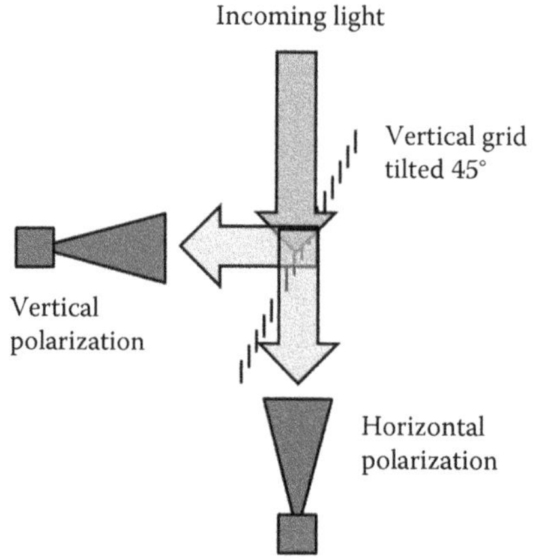

FIGURE 5.22 Wire grid used as a polarization diplexer. Grid holder is tilted 45° to incident beam. The wires in the grid are rotated so that they are aligned with the polarization of one feedhorn and orthogonal to the other.

direction, and all the photons associated with the other polarization, the other. The rule of thumb for achieving this is to make $S/d \approx 2$, with $d \leq \lambda/10$. For an input angle, $\theta = 45°$, such a grid will act as a 50/50 power splitter.

Excellent grids can be fabricated to work to ~5 THz by winding 5 μm diameter tungsten wire at 10 μm spacing (often referred to as the "pitch") on a rectangular, metal frame. Such grids have a power reflectivity, $R_{\parallel}$, of ~96% at 3 THz, for example. Once wound, an epoxy is used to secure the wires to the frame. If the grid is to be used at low temperatures, it is best to make the frame out of a material such as stainless steel, that has a small coefficient of thermal contraction. Otherwise, the wires may lose tension (and therefore lose parallelism) at low temperatures. Polarizing grids can also be fabricated by using photolithographic techniques to deposit wires on a thin, transparent dielectric sheet. When designing or specifying such a polarizer, one must take into account the reflective properties of the underlying dielectric material.

5.6 INTERCEPTING GAUSSIAN BEAMS

To this point, we have learned how to focus and manipulate Gaussian beams, but, ultimately, our goal is to have the photons deposit their energy onto a detector. How best to do this can depend both on the type of detector we are using, and the wavelength of operation. Incoherent detectors, such as bolometers, are designed for broadband operation. For these devices, an "open-structure" detector mount is often the preferred choice. For more narrowband operation, typical of coherent detection systems, waveguide detector mounts are commonly used. Above ~2 THz, where machining waveguide mounts becomes problematic, coherent detection systems have commonly used open structure optics.

5.6.1 OPEN-STRUCTURE DETECTOR OPTICS

An example of an open-structure (quasi-optical) detection system is shown in Figure 5.23. Here, a lens with a low focal-length-to-diameter ratio (e.g., $f/d \sim 2$) focuses the incoming light onto an absorbing layer on which is mounted one or more broadband detectors (e.g., bolometer and transition edge sensor). In the past, a broadband light concentrator, such as a Winston cone, was often used instead of a lens to convey the incoming light to the detector assembly (see Chapter 7). The absorber/detector assemblies were handmade, and located in a cavity up to a few wavelengths in diameter just behind the concentrator. To increase sensitivity and response time, the detector assembly is thermally isolated from its surroundings by thin silicon nitride or Kevlar supports. In terms of Gaussian beam optics, Winston cones are multimode, designed to provide uniform illumination across the telescope primary. However, as we shall soon learn, at THz frequencies, single mode operation can yield higher telescope efficiency. For this reason, the Winston cone is sometimes replaced with a single Gaussian mode feedhorn (see below). More recently, detector assemblies are made using photolithographic processes on thin silicon or quartz substrates. Here, the detector is located at the feed point of a planar antenna structure, such as a spiral or twin-slot (see Figure 5.23). The antenna structures are typically 1–3 wavelengths in diameter, producing a hemispherical power pattern extending into the substrate material. The substrate carrying the detector is epoxied to the back of a hemispherical or elliptical lens

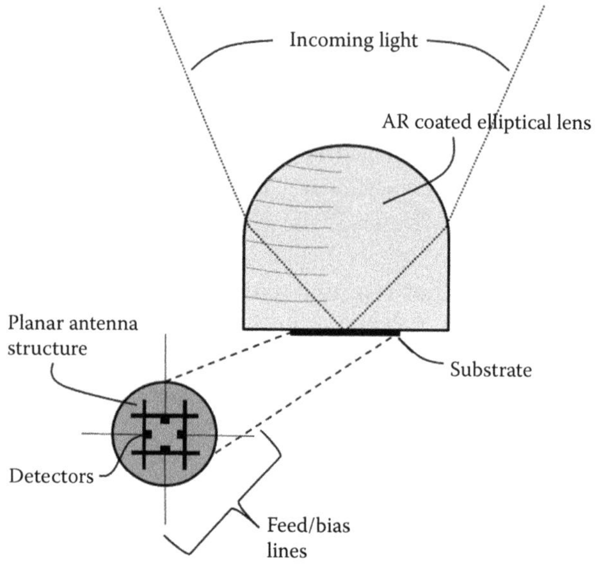

FIGURE 5.23 Open-structure (quasi-optical) detector optics.

made of the same material as the substrate, typically silicon, with a Parylene antireflection coating. The higher permittivity of the substrate and lens material yields less resistance to the passing of photons than free space, and effectively guides the incoming photons to the detector. To ensure that the detector's Gaussian beam axis is coincident with the central axis of the lens, the center of the planar antenna structure should be located within $\approx\lambda/10$ of the center of the lens. Use of photolithographic techniques allows many detector assemblies to be fabricated simultaneously, side-by-side, on the same substrate wafer, making the realization of low-cost, large focal plane arrays of detectors possible (see Figure 5.24).

5.6.2 TRANSMISSION-LINE-MOUNTED DETECTORS

In many THz systems, transmission line is used to convey photons to the detector, either from an antenna (e.g., Figure 5.4) or from a feedhorn (see Section 5.6.3). Transmission lines achieve this by confining the flow of photons between two or more conductors. A lossless transmission line can be thought of as a series of inductors and shunt capacitors, as shown in Figure 5.25. Photons are parcels of electromagnetic energy. The photon's energy is shared between its electric and magnetic fields, analogous to what occurs in a transmission line. A schematic of a lumped circuit equivalent photon model is shown in Figure 5.26, with an effective photon capacitance, C_p, and inductance, L_p. Within a transmission line, photons continue to share their energies between electric and magnetic fields, with the values of C_p, L_p, and λ_p (the photon's wavelength) modified by the physical properties (i.e., dielectric constant, magnetic permeability, and geometry) of the transmission line itself. Since the transmission line is composed of distributed capacitors and inductors, it lends itself to being configured into a variety of circuits.

The most basic waveguide is composed of two conductors running in parallel. The effective impedance, Z, of the waveguide (i.e., the ratio between the voltage and current

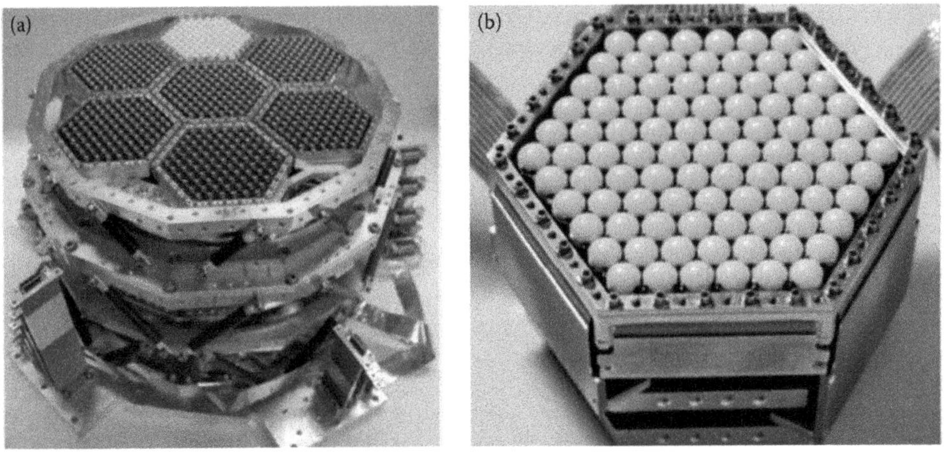

FIGURE 5.24 (a) "POLARBEAR" focal plane array composed of 637 antenna coupled bolometers. The array is constructed from 7 × 91 pixel subarrays designed for a spectral band centered on 148 GHz. Six subarrays use silicon lenses. One subarray uses alumina lenses. (b) Close-up of 91 pixel alumina lenslet subarray. (Adapted from Arnold, K. et al. 2012, *Proc. SPIE* 8452, Millimeter, Submillimeter, and Far-Infrared Detectors and Instrumentation for Astronomy VI, 84521D (September 24, 2012), doi:10.1117/12.927057. With permission.)

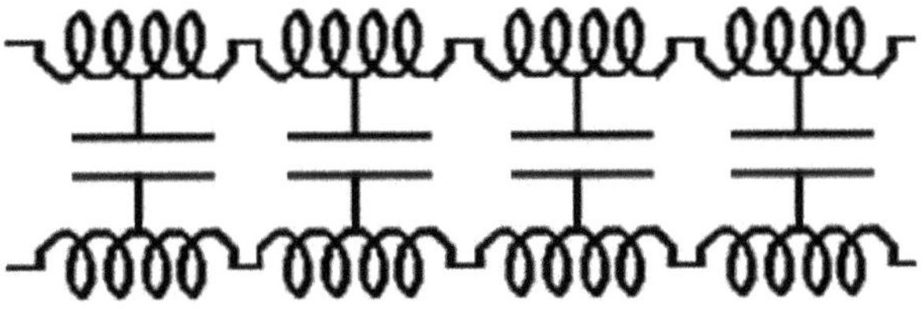

FIGURE 5.25 Lossless transmission-line model.

conveyed by the electric and magnetic fields) of the photons traveling between the conductors is set by the spacing between them; a wider spacing corresponds to higher impedance, and a smaller spacing to lower impedance.

In general, Z is a complex number, expressed in the same units as resistance, the Ohm (Ω).

$$Z = R \pm jX \tag{5.76}$$

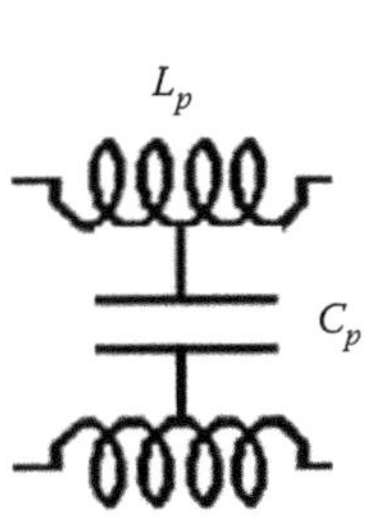

FIGURE 5.26 Lumped circuit model for a photon.

where R = resistance and X = reactance. (In electrical engineering j is used instead of i to denote imaginary numbers, so as not to be confused with current.) R is a measure of the ability of a component or circuit to dissipate energy, while X is a measure of its ability to store energy. Inductors have a positive reactance, X_L, and capacitors a negative reactance, X_C, given by

$$\begin{aligned} X_L &= \omega L \\ X_C &= \frac{1}{\omega C} \end{aligned} \tag{5.77}$$

where

$\omega = 2\pi f$

f = frequency of operation

For optimal sensitivity, one desires the impedance of a transmission-line mount to be a conjugate match to that of the detector itself. In other words, you match the real part of the impedances and cancel the imaginary component of the impedance. The imaginary component of the impedance is cancelled by providing an environment where the opposite type of reactance is presented by the detector. When reactances cancel in this way, we have a resonant circuit. Since the reactances of Equation 5.77 are frequency-dependent, resonance occurs at a frequency, f_r.

$$f_r = \frac{1}{2\pi\sqrt{LC}} \tag{5.78}$$

In a resonant circuit that is truly lossless, resonance occurs at only one single frequency. However, in physical circuits, there is always a resistance, R, either intentional or not, that dissipates some amount of energy. It also will have the effect of broadening out the frequency response of the circuit. For a parallel resonant circuit, we define a quality factor, Q, such that

$$Q = \frac{R}{X}, \tag{5.79}$$

where

R = parallel load resistance in Ω

X = reactance in Ω of either the inductor or capacitor

The higher the Q, the more dissipative energy loss there will be in a circuit compared to the stored reactive energy. There is a direct relationship between the Q of a circuit and its usable bandwidth, BW.

$$BW = \frac{f_r}{Q} \tag{5.80}$$

BW is the fractional frequency range over which the resonant circuit operates with greater than 50% power efficiency.

For series resonant circuits, Equations 5.78 and 5.80 are still valid, but Equation 5.79 is traditionally inverted. At resonance, parallel resonant circuits present the greatest impedance to the passage of photons through them, while series resonant circuits present the least.

EXAMPLE 5.4

What is the effective capacitance and inductance of a lumped circuit model photon in free space at 1 THz? What is its effective impedance, Z_p?

The energy stored in our model "photonic" capacitor and inductor is simply the energy of the photon itself, as described by Planck's Law.

$$E_p = hf_p = h\frac{v_p}{\lambda_p},$$

where h is Planck's constant and λ_p and f_p are the wavelength and frequency of the photon. The photon's speed, v_p, and wavelength, λ_p, are properties of the material through which it travels. A single photon has an effective voltage, V_p, given by the ratio of the photon's energy to a multiple of electric charge quanta, Ne. Here, N is the number of charge quanta associated with the generation, propagation, or interception of the photon.

$$V_p = \frac{hf_p}{Ne}$$

Maxwell's equations predict $N = 2$ (one positive and one negative charge) for an oscillating electric dipole in free space. In which case, for our model photon, $Ne = 2e \approx 3.2 \times 10^{-19}$ C.

From Gauss's Law, it can be shown that the energy, E_C, stored in a capacitor, C_p, with an effective voltage, V_p, is

$$E_C = \frac{1}{2}C_p V_p^2.$$

Similarly, from application of Faraday's Law, the energy, E_L, stored in an inductor, L_p, through which an effective current, I_p, flows, is found to be

$$E_L = \frac{1}{2}L_p I_p^2.$$

At any given time, half of a photon's energy is in E_C, and the other half in E_L. Therefore,

$$E_c = \frac{1}{2}E_p = \frac{1}{2}C_p V_p^2.$$

Substituting in the expression for V_p, the capacitance of a photon is found to be

$$C_p = (Ne)^2 \frac{1}{hf_p}.$$

An expression for L_p can be derived from making use of the fact that the capacitive reactance, X_C, and inductive reactance, X_L, are, in this case, equal,

$$X_L = X_C = \omega L_p = \frac{1}{\omega C_p},$$

where $\omega = 2\pi f_p$. Solving for L_p, we find

$$L_p = \frac{1}{\omega^2 C_p}.$$

Substitution of C_p into the above equation yields an expression for the effective inductance of a photon

$$L_p = \left(\frac{1}{2\pi}\right)^2 \frac{h}{f_p (Ne)^2}.$$

For a 1 THz photon, substitution yields

$$C_p = (2(1.6 \times 10^{-19}\mathrm{C})^2 \frac{1}{6.626 \times 10^{-34}\mathrm{m^2 kg\ s^{-1}}(1 \times 10^{12}\ \mathrm{Hz})} = 1.54 \times 10^{-16}\mathrm{F} = 0.154\mathrm{fF}$$

$$L_p = \left(\frac{1}{2\pi}\right)^2 \frac{6.626 \times 10^{-34}\mathrm{m^2 kg\ s^{-1}}}{1 \times 10^{12}\ \mathrm{Hz}\ (2(1.6 \times 10^{-19}))^2} = 1.639 \times 10^{-10}\mathrm{H} \approx 16\mathrm{nH}.$$

The effective impedance of our lossless, model photon is then, simply,

$$Z_p \approx \sqrt{\frac{L_p}{C_p}} \approx 1030\Omega.$$

The derived value of Z_p is greater than $\eta_o = \sqrt{\mu_o/\varepsilon_o} \approx 377\Omega$, the wave impedance of a plane electromagnetic wave in free space. The difference is either a byproduct of the assumptions associated with our simple photon model, and/or the application of Maxwell's wave model to a single photon.

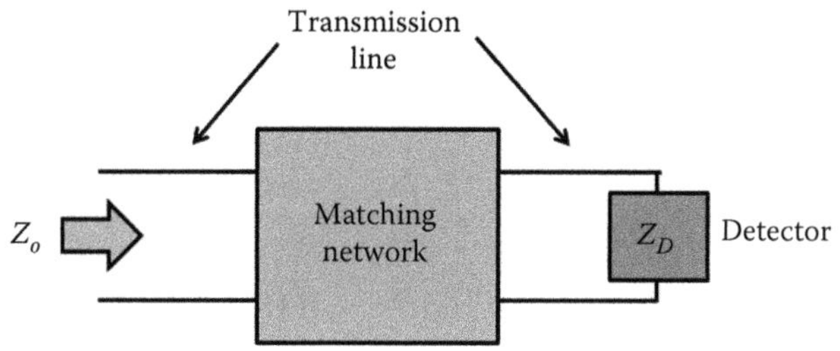

FIGURE 5.27 Matching network can enable an efficient conjugate match between transmission line and the detector.

Often, we find the detector's impedance (sometimes referred to as a load) does not match the characteristic impedance of the transmission line with which we are working. In this situation, instead of all the signal power going into the load, some of it bounces back and can create standing waves on the line. When this occurs, a matching network can be added (see Figure 5.27) to transform the impedance of the line to what the detector wants to see for maximum power transfer. Below ~20 GHz, matching networks are usually composed of lumped circuit elements (i.e., discrete parts; resistors, capacitors, and inductors). Above ~20 GHz, pieces of the transmission line can be cut to a length that, for the frequency of operation, makes the line appear either capacitive or inductive, and/or can adjust the real (resistive) component of the line impedance presented to the load.

Microstrip and stripline are examples of waveguides that are bounded on two sides (see Figure 5.28). These transmission-line types are in regular use at microwave and millimeterwave frequencies. Both can be made using photolithographic techniques. The impedance of these lines is set by the width, W, to height, h, ratio of the line, the thickness, t, of the central conductor, and the substrate material on which it is made. Losses in the substrate material can limit the use of both types of line at THz frequencies. Due to its ease of fabrication and the ability to attach discrete components directly to it, the microstrip line is more commonly used. However, the fields within the stripline are more tightly confined, and, therefore, will result in less crosstalk on a tightly packed circuit board. An approximate expression for the impedance, Z_m, of the microstrip is given by Wheeler (1977),

$$Z_m = \frac{Z_0}{2\pi\sqrt{2(1+\varepsilon_r)}}\ln\left(1+\frac{4h}{w_{eff}}\left(\frac{14+(8/\varepsilon_r)}{11}\frac{4h}{w_{eff}}+\sqrt{\left(\frac{14+(8/\varepsilon_r)}{11}\frac{4h}{w_{eff}}\right)^2+\pi^2\frac{1+(1/\varepsilon_r)}{2}}\right)\right) \tag{5.81}$$

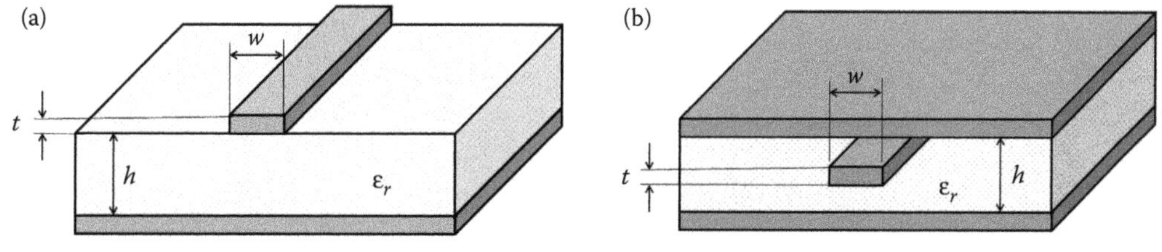

FIGURE 5.28 (a) Microstrip transmission line. (b) Stripline transmission line.

where Z_0 is the impedance of free space (377 Ω) and ε_r is the relative impedance of the substrate material being used. The effective width, w_{eff}, of the central strip can be derived from

$$w_{eff} = w + t\frac{1+(1/\varepsilon_r)}{2\pi}\ln\left(4e\left(\sqrt{\left(\frac{t}{h}\right)^2 + \left(\frac{1}{\pi}\frac{1}{(w/t)+(11/10)}\right)^2}\right)^{-1}\right).$$

A hollow waveguide, if it can be fabricated, is an intrinsically low-loss transmission line. The waveguide is essentially a bounded version of two wire transmission lines. Being bounded, it is less susceptible to unwanted pick-up from neighboring circuits than the microstrip or stripline. To see how the waveguide works, let us begin with the parallel wire transmission line on the left side of Figure 5.29. The incoming stream of photons is confined to travel between two wires of diameter, d. The impedance of the transmission line is set by the spacing, b, between them. A smaller spacing results in a higher concentration of energy and lower line impedance, Z_0.

$$Z_0 = 276\log_{10}\frac{2b}{d} \tag{5.82}$$

For constant impedance, the spacing between the lines must be maintained. A wire bracket can be used for this purpose if it is made λ/4 long. The choice of this bracket length follows the same reasoning described above in Section 5.5.2 for antireflection coatings. In terms of the phasor description of probability amplitudes discussed in Section 5.6, the position vector associated with a photon traveling the added λ/4 down the bracket, will pick up an additional 90° of phase shift. The phasor will pick up an additional 180° of phase shift upon reflection at the bottom of the bracket, and another 90° on the return trip. Upon arrival back at the bracket entrance, the position vector will have picked up a total of 360° of phase shift and be *in phase* with the position vectors of photons passing by. However, it will be 180° *out of phase* with the position vector of any photons that

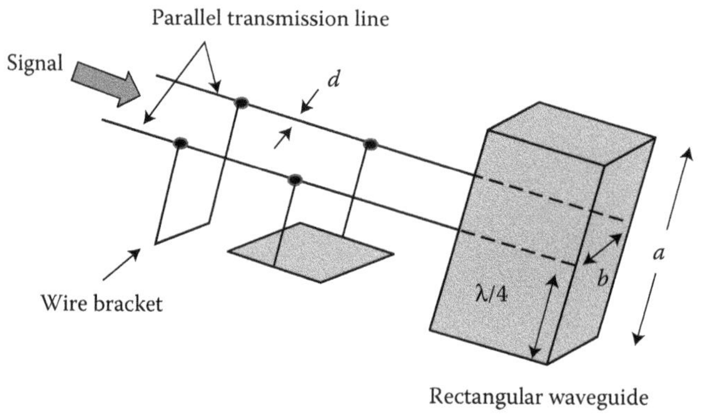

FIGURE 5.29 Waveguide evolution from parallel transmission line.

would reflect off the bracket entrance. The vector sum of the two position vectors is zero, meaning the probability that photons will actually travel down the bracket is zero, and the probability of photons continuing on down the transmission line unperturbed by the presence of the bracket is, in this ideal case, one. Therefore by using a $\lambda/4$ metal bracket, the parallel transmission line can be rigidly held in place at any desired separation, b. If this is true, there is no reason why the bottom of the bracket cannot be a plate instead of a wire. Now, if we imagine a continuous bracket instead of a discrete one, and extend it both above and below the parallel line, we now have a rectangular waveguide. The waveguide input impedance is given by

$$Z_g = 754\frac{b}{a}\frac{\lambda_g}{\lambda_o} \tag{5.83}$$

where

$\lambda_g = \lambda_o/\sqrt{1-(f_c/f)^2}$ = guide wavelength

λ_o = free space wavelength

b = guide height

a = guide width

$f_c^{10} = (c/2a)$ = guide cut-off frequency for TE_{10}

c = speed of light

The cutoff frequency, f_c, is the frequency below which the photons have gotten too big to efficiently propagate through the waveguide.

Like Gaussian beams, photons can propagate down a waveguide in a number of different modes, but detector systems are almost always designed using a waveguide in the dominant mode, *that is*, the *lowest mode* that can propagate and also the one with the simplest field configuration. For the rectangular waveguide of Figure 5.30, this would be the transverse electric mode, abbreviated TE_{10}. The electric and magnetic field configurations associated with photons propagating in this mode are shown in Figure 5.30. The electric field is seen to "bunch up" in the waveguide center, making it a convenient location across which to mount a detector. A schematic drawing of a waveguide-mounted detector probe is shown in Figure 5.31. As we go up in frequency, the photons get correspondingly smaller,

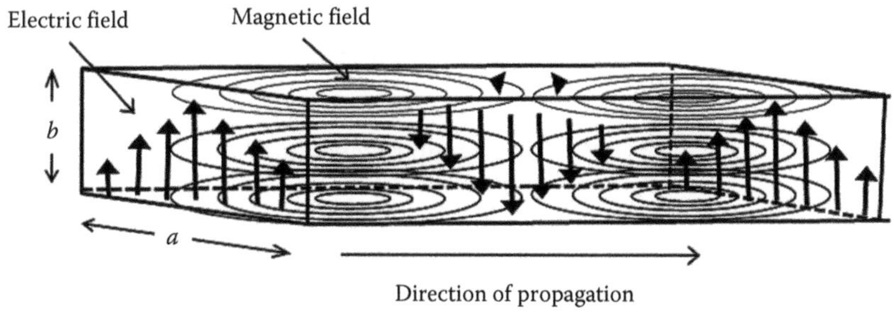

FIGURE 5.30 TE_{10} electric and magnetic field configuration in rectangular waveguide.

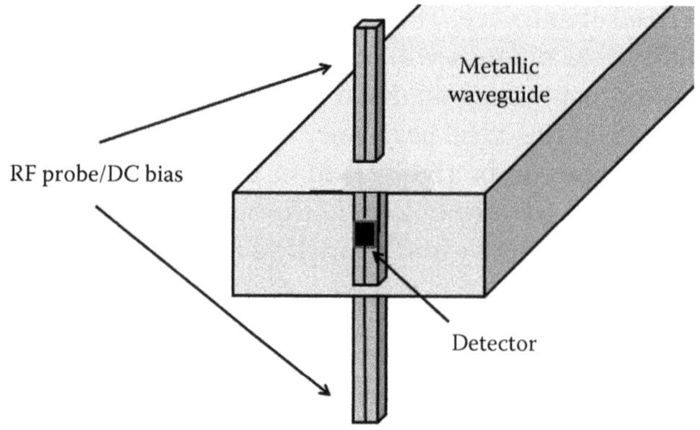

FIGURE 5.31 Waveguide mounted detector.

until a point is reached where two sets of field components can fit "side-by-side" down the waveguide. This mode of operation is referred to as the TE_{20} mode. The cutoff frequency of this mode is $f_c^{20} = 2f_c^{10}$. Therefore, a rectangular waveguide can operate in the TE_{10} mode over a fractional bandwidth of ~50%.

Circular waveguide (*a.k.a.*, cylindrical waveguide) is also in use at THz frequencies, principally as the input to conical feedhorns. The dominant mode for circular waveguide is the TE_{11} mode, with a cutoff frequency of

$$f_c^{11} = \frac{c}{1.7d} \tag{5.84}$$

where d = the waveguide diameter.

The field configuration of the TE_{11} mode in a circular waveguide looks very similar to that of the TE_{10} mode in a rectangular waveguide, with the electric field peaking in the middle and trailing off at the sidewalls. The next mode in a circular waveguide is the TM_{01} mode, starting ~23% higher in frequency.

In both circular and rectangular waveguides, power losses go up near the edges of the operating band, on the low end, due to the increasing difficulty that photons have squeezing through the waveguide near cutoff and, on the high end, due to photons slipping into higher, unwanted modes. In general, the best place to locate the operating frequency, f_{op}, is midband, such that,

$$f_{op} = 1.25 f_c^{10} \quad \text{for a rectangular waveguide} \tag{5.85}$$

$$f_{op} = 1.17 f_c^{11} \quad \text{for a circular waveguide} \tag{5.86}$$

Besides a greater fractional bandwidth, a rectangular waveguide has the added advantage that its impedance can be adjusted by stepping down its height (Equation 5.83). Since a circular waveguide has only one characteristic dimension, it does not have this flexibility.

The power transmitted through the waveguide, P_T, is given by

$$P_T = \frac{|E_0|^2 ab}{4Z_0}, \tag{5.87}$$

where E_0 is the magnitude of the associated electric field.

So, for a given impedance, Z_0, the greater the waveguide's cross-sectional area, ab, the higher the transmitted power (and the lower the loss) will be.

5.6.3 WAVEGUIDE FEEDHORNS

As discussed in Section 5.1, the main beam of a single-aperture telescope is well modeled by a fundamental mode Gaussian beam. Subsequent relay optics, if required, are designed to convey the stream of photons contained in this beam to the detector. Feedhorns and antennas (as depicted in Figure 5.32) can be thought of as transition mediums between free space optics and the transmission line of Figure 5.25.

For horns, one can define "Gaussicity" and "copolarization" metrics for comparing performance (Johansson, 1995). Gaussicity is the maximum achievable power coupling of the beam produced by the feedhorn to a linearly polarized Gaussian beam. Copolarization is the percentage of the photons within the horn's beam that share the same polarization. Another useful parameter is "ω/a," the ratio of the horn's projected Gaussian beam radius at its aperture, ω, to its physical aperture radius, a. The corrugated, Potter, diagonal, Neilson (a.k.a. profile) horns are (for now) the most commonly used designs at THz frequencies. In general, the output beam waist of these horns is, $\omega_0 \approx 0.3d$, where d is the diameter (or longest dimension) of the horn aperture. Table 5.2 lists additional parameters for these horn types.

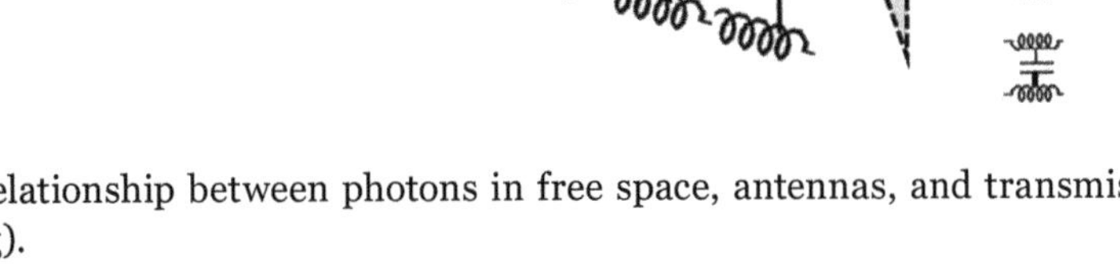

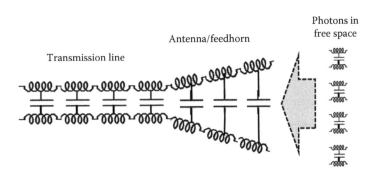

FIGURE 5.32 Relationship between photons in free space, antennas, and transmission line (here shown receiving).

TABLE 5.2 Horn Design Parameters

Feed type	Gaussicity (%)	Copolarization (%)	Matching (ω/a)	Bandwidth (%)
Corrugated horn	98.1	~100	0.644	35
Pickett–Potter horn	≥96.3	~100	0.556	20
Diagonal horn	84.3	90.5	0.60	55
Neilson horn	99.3	~100	0.62	20

5.6.3.1 CORRUGATED HORN

Corrugated horns have been the "gold standard" for horn performance for decades. A cross-sectional view of an early corrugated horn (Chu and Barrow, 1939) fed by a circular waveguide, is depicted in Figure 5.33. A "good" horn is one that produces a Gaussian looking main beam with the lowest possible sidelobes. When photons enter the aperture on the right of Figure 5.33, the first thing they see are a series of closely packed, $\lambda/4$ deep grooves that have been etched or milled into the inside of a conical surface. These grooves act as an antireflection coating (see Section 5.5.2), preventing the flow of currents on the metal surface in the vicinity of the horn's aperture. This has the effect of optically "softening" the appearance of the aperture to incoming (or outgoing) photons, thereby significantly reducing the occurrence of unwanted sidelobes produced by hard edges (see Figure 5.6). As expected, the larger the horn aperture, the tighter will be the horn's beam. The cone cross-section is designed to taper down to the diameter of the dominant mode, circular waveguide (see Equation 5.84). On the way there, the depths of the grooves are increased to $\lambda/2$. At this depth, the path-length down the groove, and back up again, is one λ. To the passing photons, this path-length effectively brings the conical surface up to them, and currents will flow just as if it was a smooth-walled surface, thereby providing a seamless transition to the circular waveguide that follows.

The taper angle of the horn affects both the angular size of the main beam and the location of the horn's phase center, that is, the effective location of its beam waist. Following Thomas (1978), let us define a dimensionless parameter, Δ, which is the difference (in units of wavelength) between the location of the horn's spherical wave front and plane aperture (see Figure 5.33).

$$\Delta = \frac{a}{\lambda_o}\tan\frac{\theta_o}{2} = \frac{R}{\lambda_o}\sin\theta_o \tan\frac{\theta_o}{2} \tag{5.88}$$

where

λ_o = free space wavelength

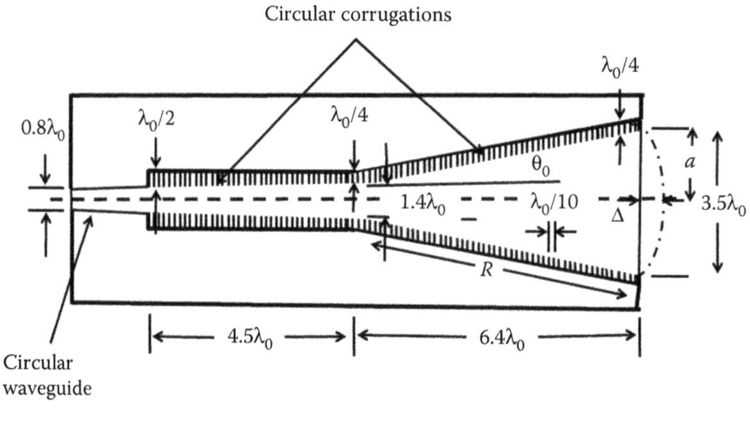

FIGURE 5.33 Corrugated horn design. Dimensions are in units of the free space wavelength. The input and output apertures are round. (Adapted from Kraus, J. and Marhefka, R., 2002, *Antennas for all Applications*, 3rd edn., McGraw Hill, New York.)

R = horn slant length
a = *radius* of horn aperture
θ_0 = taper angle

When $\Delta < 0.4$, the horn's beam-width is $\propto a/\lambda_0$, with the phase center located near the horn aperture. Such horns are narrow band and have narrow beam profiles. The Gaussian beam waist of these horns is given by (Goldsmith, 1998)

$$\omega_0 = 0.644\,a \tag{5.89}$$

For horns with values of $\Delta > 0.75$, the resulting beam-widths are $\propto\theta_0$, with the phase center located near the horn's apex (Thomas, 1978). Both the beam-widths and operating bandwidths of these horns are larger than for narrow-band horns. These wide-band horns (sometimes referred to as scalar horns) are appropriate for use in applications where the required θ_0 is relatively large (e.g., as the feedhorn for a prime focus system).

Corrugated horns can be fabricated up to ~2 THz using micromachining or platelet techniques (see Walker et al., 1997). However, the necessity of using specialized equipment in the fabrication and assembly process limits their popularity at frequencies ≥1 THz.

5.6.3.2 PICKETT–POTTER HORN

If one is willing to trade operating bandwidth for ease of fabrication, the dual-mode, Pickett–Potter horn is an attractive alternative to the corrugated feedhorn. The dual-mode horn was first analyzed by Potter (1963). The basic design begins with a smooth-walled, conical feedhorn being fed by a circular waveguide. Potter introduces a step at the waveguide–conical horn transition that generates a TM_{11} waveguide mode in addition to the usual TE_{11} mode. By adjusting the relative phase and amplitude of these two modes, as they propagate down the horn, he was able to generate a symmetric, circular beam at the horn aperture, and significantly lower the sidelobe level within the beam. The Potter horn is a highly tuned structure and, therefore, has a significantly lower operational bandwidth (~5%) than a corrugated horn (~35%). Later, Pickett et al. (1984) were able to significantly simplify the design and fabrication of the Potter horn and, in the process, increase its operating bandwidth to ~20%, suitable for most THz applications. The design of the Pickett–Potter horn is shown in Figure 5.34. It can be readily fabricated by drilling

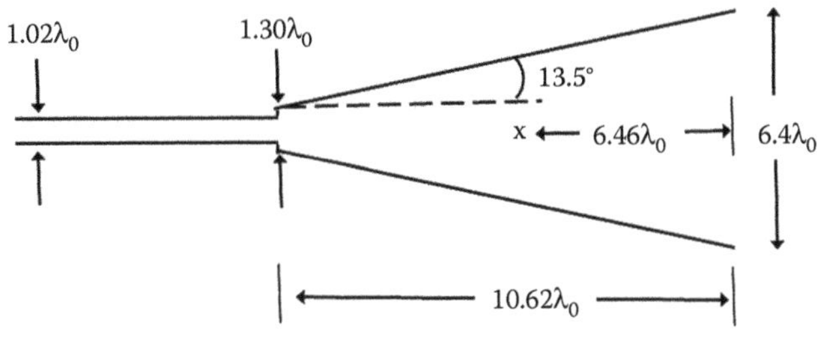

FIGURE 5.34 Pickett-Potter horn design. Dimensions are in units of the free space wavelength. "X" marks the location of the horn's phase center. The input and output apertures are round. (Adapted from Pickett, H. M., Hardy, J. C., and Farhoomand, J., 1984, *IEEE Trans. Microwave Theory Tech.* MTT-32(8), 936–937.)

a suitable metal (typically copper or aluminum) with a custom endmill, made to be the "negative" of Figure 5.34.

5.6.3.3 NEILSON HORN

The Neilson horn is another example of a dual-mode feedhorn, where both the TE_{11} and TM_{11} modes propagate. However, instead of a step generating the TM_{11} mode, here a sinusoidal variation in the shape of the horn's wall is used to generate the mode. The sinusoidal taper should have a period equal to the beat period between the TE_{11} and TM_{11} modes. A simple "arc-taper" (see Figure 5.35) between the radius of the input waveguide and the desired aperture (output) radius can be used as a starting point in the design, with optimization of the curve shape being done with an E&M simulation program. The amount of TM_{11} power generated depends on the shape of the curve. A TM_{11} to TE_{11} power ratio of ~0.4 will taper the propagating field energy to near zero at the horn's wall (as corrugations do for a corrugated feedhorn), and yield a Gaussian beam coupling factor of 0.96. A power ratio of ~0.2 will increase the coupling factor to 0.98 at the expense of a higher sidelobe level (Neilson, 2002). The operating bandwidth can be greater than 20%. The main advantage of the Neilson design over the Pickett–Potter design is its ability to accommodate larger horn output diameters. A larger horn output diameter (i.e., aperture) means a larger horn input beam waist, which is usually a good thing when trying to match the beam from a telescope. This advantage comes at the expense of needing to run a computer simulation to optimize the horn design and the increased difficulty of fabrication. The shape of the Neilson horn is conducive to fabrication by electroforming or by precision drilling with an endmill made to be a negative of the desired horn. Other examples of smooth profile horns can be found in Granet et al. (2004). A description of multiple flare-angle, dual-mode horns with excellent performance is given by Leech et al. (2011).

5.6.3.4 DIAGONAL HORN

The diagonal horn was invented by Li (1959) and used to illuminate a test spherical reflector at Bell Telephone Laboratories. It was later analyzed and further developed by Love (1962). The design of a typical diagonal horn is shown in Figure 5.36a (Johansson and Whyborn, 1992). Like the other horns discussed here, the horn's beam is circularly symmetric. However, unlike the other horns, the diagonal horn has a square aperture. Indeed,

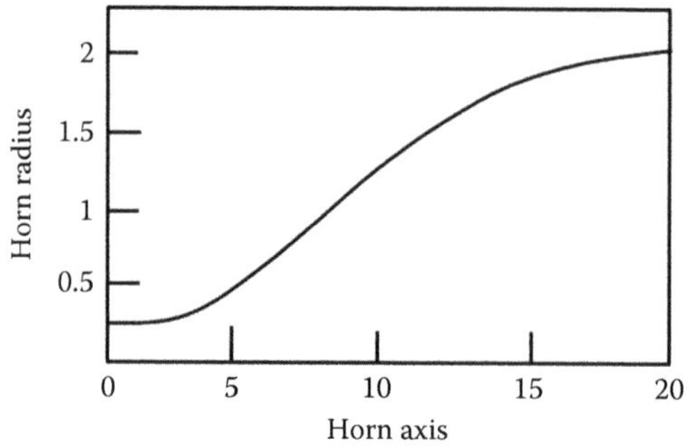

FIGURE 5.35 Neilson horn arc-taper profile for a 650 GHz horn. A full horn design can be generated by revolving the curve about the horizontal axis. (After Neilson, J., 2002, *IEEE Trans. Antenn. and Propag.*, 50, 1077–1081.)

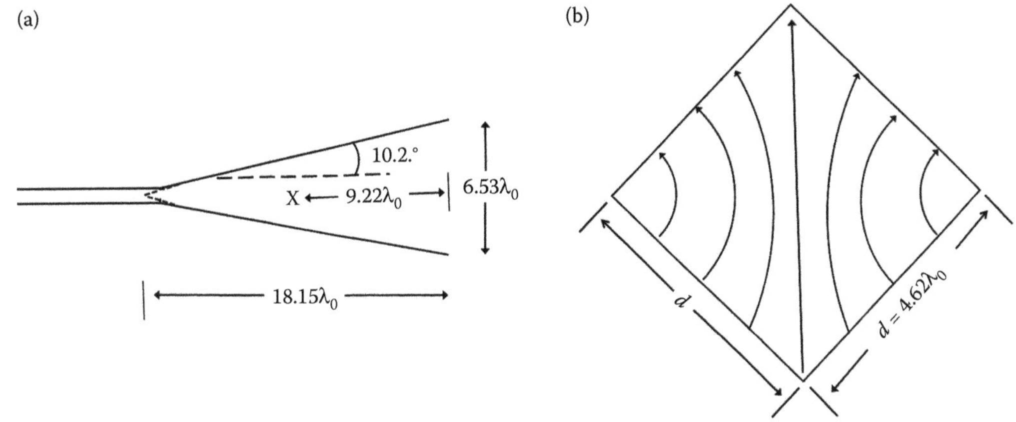

FIGURE 5.36 (a) Diagonal horn design. Dimensions are in units of the free space wavelength. "X" marks the location of the horn's phase center. The input and output apertures are square. The electric field configuration at the horn aperture is shown in (b). (Derived from Johansson, J. and Whyborn, N., 1992, *IEEE Trans. Microwave Theory Tech.*, 40(5), 795.)

all cross-sections through the horn are square, which makes it comparatively easy to fabricate using direct machining if a "split-block" architecture is employed, that is, the waveguide block is assembled from two halves, mirrored about the axis of propagation (see for instance, Groppi et al., 2011). At the horn aperture, the electric field associated with incoming photons has the configuration shown in Figure 5.36b and represents the superposition of two orthogonal TE_{10} modes, with power equally distributed between them. The maximum electric field vector is oriented across the diagonal of the horn's square aperture; hence, the name. The photons maintain this field distribution as they propagate down the horn. Most of the time, waveguide detectors are mounted across a dominant mode, rectangular waveguide, as shown in Figure 5.31. At THz frequencies, a smooth waveguide transition is typically used to match the square aperture at the throat of the horn to the rectangular aperture of the waveguide (Love, 1962; Johansson and Whyborn, 1992). In order to reduce the generation of standing waves, the transition should be 2 or more λ_g (see Equation 5.83) long. At lower frequencies (where machining is easier), the length of the transition can be reduced by employing a multisection quarter-wave, step transformer (see, for instance, Young, 1963).

Diagonal horns have large (55%) bandwidth and a high ω/a value. However, due to their field structure, they do have a significant (~10%) cross-polarized field component to their beams. If the horn is being used at the output of a transmitter (e.g., a local oscillator source), this means 10% of the power will be lost to a polarization sensitive detector and should be accounted for when performing link budget calculations (unless a diagonal horn is being used with the detector as well). However, in most THz astronomy applications, the light being collected from astrophysical sources is unpolarized, meaning there are approximately equal numbers of photons in both horizontal and vertical polarizations. What the diagonal horn may miss in one polarization component, it will pick up in the other, with an overall efficiency close to 100%. Due to their ease of fabrication and high packing density, diagonal horns are well suited for array applications (see Figure 5.37 and Groppi et al., 2011; Groppi and Kawamura, 2011; Kloosterman, 2014).

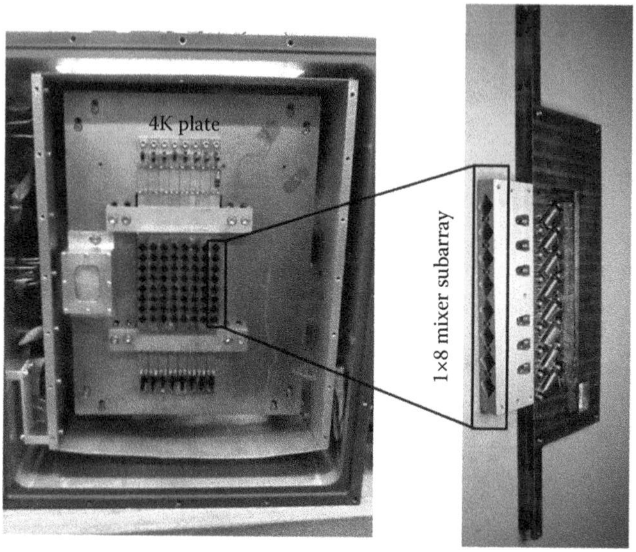

FIGURE 5.37 64 pixel SuperCam 345 GHz SIS Mixer Array. Right: SuperCam mixer array installed in hybrid cryostat. The array consists of eight, 1 × 8 mixer subarrays. Left: A single 1 × 8 SuperCam mixer subarray. Each pixel has its own diagonal feedhorn, SIS waveguide mixer, IF amplifier, and electromagnet, followed by an IF processor and digital spectrometer. The pixel to pixel spacing is 10 mm, corresponding to $3\omega_0$, where ω_0 is the beam waist of the feedhorns. This spacing reduces the cross-talk between adjacent horns to the 0.1% level, and yields a beam spacing on the sky of ~$2f\#\ \lambda$ (every other θ_{FWHM} beam). Here, the $f\#$ is the effective f/D ratio of the telescope being used (see Equation 5.94 and Figure 6.43). If each feedhorn is repointed (either mechanically or optically) toward the center of the secondary, a typical on-axis Cassegrain telescope can support a ~30 × 30 array of $3\omega_0$ spaced pixels with <1% gain loss at the edges. (Shillue 1997, MMA Memo 175: Gain Degradation in a Symmetrical Cassegrain Antenna Due to Laterally Offset Feeds, http://legacy.nrao.edu/alma/memos/html-memos/alma175/memo175.html.)

5.7 ILLUMINATING THz TELESCOPES

Astrophysical sources are so far away that the photons arriving from them strike all parts of the Earth's surface and the collecting surfaces of our telescopes uniformly. How many of these photons arrive at the detector(s) depends not just on the size of the telescope's aperture, but also on how it is illuminated. Most THz telescopes use either a Cassegrain (on-axis) or Gregorian (off-axis) design. The optical and structural simplicity of the Cassegrain design makes it somewhat more common, even if it does have somewhat lower efficiency and greater beam distortion due to the central blockage of the secondary.

The Gaussian beam illumination of a Cassegrain telescope is depicted in Figure 5.38. The telescope consists of a primary and secondary mirror. The parabolic shape of the primary is designed to bring the photons incident on its surface to a primary focus, f_p, but before that point is reached, a hyperbolic-shaped secondary mirror intercepts the photons and brings them to an effective focus, f_e. In this system, it is the characteristics of the Gaussian beam projected on to the secondary by the detection system that determines the

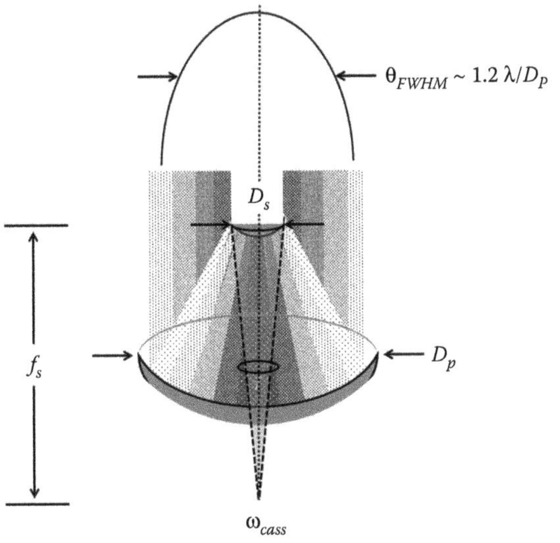

FIGURE 5.38 Cassegrain telescope illumination with a Gaussian beam.

photon weighting or "edge taper" across the primary. Edge taper, T_e, is often expressed in decibels (dB) as the ratio between the power at the edge of the aperture to the power at the center.

$$T_e(\text{dB}) = 10\log_{10}\left(\frac{P_{edge}}{P_{center}}\right) \tag{5.90}$$

A smaller edge taper means the aperture is more uniformly sampled. A larger edge taper means that photons are preferentially collected nearer the center of the aperture, making the edge of the primary appear "softer" to incoming photons. A softer edge means unwanted diffraction effects will be reduced, yielding a lower sidelobe level than would be observed with a uniformly illuminated aperture (see Figure 5.6). If the telescope is ground-based, there is the additional benefit of reducing the likelihood of picking up photons due to the illuminating beam from the secondary partially falling onto the warm ground behind the primary. This type of stray radiation is referred to as "warm-spillover." However, there is a trade. Using a large edge taper means the diameter of the telescope is effectively being reduced, which will result in a reduction in aperture efficiency, and a somewhat broader main beam. For the case of an unblocked aperture, the increase in the full-width at half-maximum, or FWHM, (i.e., 3 dB point) of the telescope beam is given by (Goldsmith, 1998).

$$\begin{aligned}\Delta\Omega_{FWHM} &= \left[1.02 + 0.0135\,T_e(\text{dB})\right]\frac{\lambda}{D_p} \\ &\approx 1.2\frac{\lambda}{D_p}\end{aligned} \tag{5.91}$$

where

Ω_{FWHM} = Full-width at half-maximum beam size (rad)
T_e (dB) = edge taper in dB
D_p = telescope primary diameter
λ = wavelength of observation

In practice, a value of T_e (dB) between 10 and 14 dB is often found to represent a good compromise between the sidelobe level and aperture efficiency.

The telescope beam waist, ω_{cass}, can be derived from the following very helpful expression (Goldsmith, 1998)

$$\omega_{cass} = 0.22\left[T_e(\text{dB})\right]^{1/2} \frac{f_e}{D_p}\lambda \tag{5.92}$$

The ratio f_e/D_p is often referred to as the telescope's $f\#$, and typically runs between 10 and 15. For large telescopes, the location of ω_{cass} can be assumed to be coincident with the geometric focus of the telescope. For more modest sized telescopes (~1 m or less at THz frequencies), there can be a Gaussian beam correction, Δz, toward the primary given by (Goldsmith, 1998)

$$\Delta z = f_e - z \cong f_e\left(\frac{\lambda f_e}{\pi\omega_a^2}\right)^2 \tag{5.93}$$

where ω_a is the Gaussian beam waist on the primary.

When designing an instrument, the goal is to match the size and location of the input beam waist of the detector system to the output beam waist, ω_{cass}, of the telescope (see Example 5.5).

EXAMPLE 5.5

You have a waveguide receiver operating at 492 GHz that uses a feedhorn to launch a Gaussian beam. The diameter of the feedhorn is 3 mm. You want to efficiently illuminate a 10 m telescope with the receiver. The $f\#$ of the telescope is 13.8. Assuming you can place the receiver near the Cassegrain focus, design the required optical system.

Our goal is to efficiently transfer the power collected by the telescope to the receiver. This is accomplished by designing an optical system that transforms the telescope's Gaussian beam waist to that of the receiver's feedhorn.

The beam waist of the feedhorn can be found from Equation 5.89.

$$\omega_h \approx 0.66a \approx 0.66(1.5\text{ mm}) \approx 1\text{ mm}.$$

Our wavelength of observation, λ_o, is

$$\lambda_o = \frac{3\times 10^8\text{ (m/s)}}{492\times 10^9\text{ Hz}} = 0.610\text{ mm}.$$

A good edge taper to use on THz telescopes is $T_E = 14$ dB. Substitution into Equation 5.92 then yields

$$\begin{aligned}\omega_{cass} &= 0.22[14\ \mathrm{dB}]^{\frac{1}{2}}(13.8)(0.610\ \mathrm{mm}) \\ &= 6.93\ \mathrm{mm}.\end{aligned}$$

We will need a lens or mirror to match ω_{cass} to ω_h. A system using a lens is typically more compact and easier to align than an optical system using an off-axis mirror, but has the disadvantage of having both reflective and absorption losses. Since no physical restrictions are specified, let us choose to use an elliptical mirror.

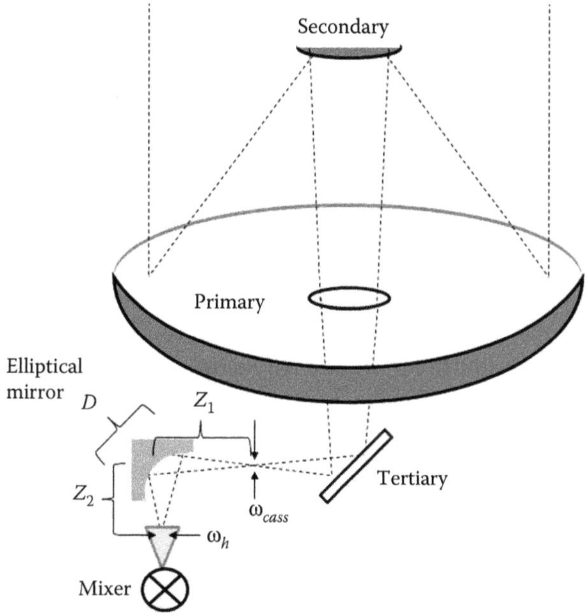

Optical schematic (not to scale)

Next, draw a picture of the proposed optical system.

Choose a location, z_1, for the mirror where the beam has passed through the telescope focus and expanded to several times ω_{cass}, such that,

$$\omega_{z_1} \approx 3\omega_{cass} \approx 20\ \mathrm{mm}.$$

ω_{z_1} is the Gaussian beam waist on the elliptical mirror.

Beam growth is given by Equation 5.27. Solving Equation 5.27 in terms of path-length z we find

$$z = \frac{\pi\omega_0^2}{\lambda_0}\left(\left(\frac{\omega_z}{\omega_0}\right)^2 - 1\right)^{1/2}$$

In the above expression, ω_0 is the beam waist radius and ω_z is the beam radius at a distance z from the beam waist, assumed to be at the telescope focus. For this example, ω_0 is ω_{cass} and ω_z is ω_{z_1}. Substituting into the expression, we find

$$z_1 = \frac{\pi(6.93\,\text{mm})^2}{\lambda_0}\left(\left(\frac{20\,\text{mm}}{6.93\,\text{mm}}\right)^2 - 1\right)^{1/2}$$
$$= 669.7\,\text{mm}$$

Next, we need to find the distance from the feedhorn to the mirror, z_2, such that the waist of the expanding beam from the feedhorn matches that of the beam from the telescope, ω_{z_1}.

$$z_2 = \frac{\pi(1.0\,\text{mm})^2}{\lambda_0}\left(\left(\frac{20\,\text{mm}}{1.0\,\text{mm}}\right)^2 - 1\right)^{1/2}$$
$$= 102.9\,\text{mm}\,.$$

z_2 is measured from the front surface of the elliptical mirror to the phase center of the feedhorn. For horns with narrow flare angles, this will be just behind the horn aperture (see Equation 5.88, here ≈0.9 mm). The elliptical mirror should have a projected diameter, $D \geq 3\omega_z$, and should account for continued beam growth from the top to bottom. If space permits, choose

$$D \approx 4\omega_{z_1} = 4(20\,\text{mm}) = 80\,\text{mm}.$$

CONCLUSION

In this chapter, we have investigated how photons interact with a telescope, and what tools, techniques, and components can be used to efficiently convey them to a detection system. The Gaussian beam approach to designing quasi-optical systems is found to be particularly powerful, since, at THz frequencies, we often operate in the diffraction limit. Coherent and incoherent detectors are used to transform the light collected by an optical system into voltages that can be recorded. In the next two chapters, we will investigate the physics of operation and sensitivity limits of these two families of detectors.

PROBLEMS

1. The sky brightness distribution $B(\theta,\phi)$ at a frequency of 230 GHz is uniform over the portion of the sky being observed and over a bandwidth of 2 GHz. The value of 1×10^{-28} (watts/m^{2}Hz rad^2). The telescope diameter is 10 m with aperture efficiency of 0.7.

a. What is the flux density, S_v, of the source?

b. What is the spectral power and total power collected by the telescope?

2. An idealized, one-dimensional, rectangular-shaped antenna power pattern, P, is shown in Figure P5.1 together with a one-dimensional source brightness distribution, B. The antenna power pattern is 10″ wide and the source has a 5″ wide central peak containing a flux density, S, seen in the middle of a broader, 30″, plateau of emission with a flux density of $0.5S$.

a. Draw a graph of the observed flux density as the antenna is scanned across the source in steps of 5″.

Hint: This is a convolution process. After each 5″ shift, multiply P with B and find the area under the product curve. The area corresponds to the observed flux density at that position. Try doing this problem the old-fashioned way, by using graph paper and counting squares under each product curve.

b. What is the maximum observed flux density, S_o, during the scan?

c. Now imagine the source distribution is plotted as a temperature, instead of a flux, with a central peak source brightness temperature of T, and a plateau source brightness temperature of $0.5\ T$. What would be the peak observed temperature, T_o, during the scan?

3. What is the FWHM beam size of a 10 m telescope at 1 THz in radians and arcseconds?

4. In the lab you measure the FWHM of a 1 THz Gaussian beam to be 14°. Assuming it is a Gaussian beam, what is its beam waist, ω_o, at its focus?

5. You need to design a lens that transforms a 0.81 THz Gaussian beam with a waist of 4 mm to one with a beam waist of 1 mm.

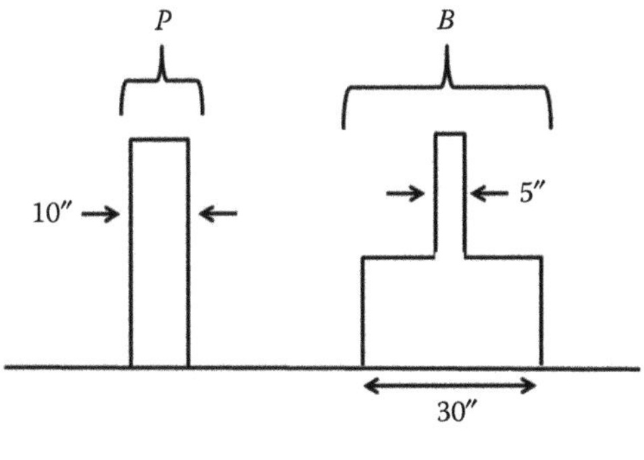

FIGURE P5.1 Antenna pattern and source brightness distribution.

a. Assuming you want to make the input and output beam waist locations equidistant from the lens, what focal length, *f*, should you design the lens to have?

b. Adopting this focal length, what diameter, *D*, should the lens have?

c. Using Equation 5.39, design a lens made out of LDPE with the appropriate diameter, *D*, and focal length, *f*. Make a plot showing the lens profile. *Hint:* You will need to iterate the value of the lens thickness, *d*, until you reach the target diameter.

6. Following the procedure outlined in Example 5.3, what would you estimate the signal loss through the center of a 5 mm thick HDPE lens to be at 0.81 THz?

7. You are given the opportunity to build a 1 THz receiver system for the telescope described in Example 5.5. You choose to use an optical system similar to that shown in the example's figure. Here, the mixer's feedhorn has a diameter of 1.41 mm. Assume the phase center of the horn is at its aperture. Calculate values for z_1 and *D*, assuming $z_2 = 75$ mm. Show all work.

REFERENCES

Afsar, M. N., 1987, Precision millimeter-wave dielectric measurements of birefringent sapphire and ceramic alumina, *IEEE Trans. Instrum. Meas.*, IM-36, 554.

Afsar, M. N. and Chi, H., 1994, Millimeter wave complex refractive index, complex dielectric constant, and loss tangent of extra high purity and compensated silicon, *Int. J. of Infrared Milli. Waves*, 15, 1181–1188.

Allen, L., 1992, Orbital angular momentum of light and the transformation of Laguerre-Gaussian laser modes, *Phys. Rev. A* 45(11), 8185.

Arnold, K., Aide, P., Anthony, A., and 52 others, 2012, The bolometric focal plane array of the POLARBEAR CMB experiment *Proc. SPIE* 8452, Millimeter, Submillimeter, and Far-Infrared Detectors and Instrumentation for Astronomy VI, 84521D (September 24, 2012), doi:10.1117/12.927057.

Birch, J., Dromet, J., and Lesurf, J., 1981, The optical constants pf some low-loss polymers between 4 and 40 cm^{-1}, *Infrared Phys.*, 21, 225–228.

Birch, J. and Kong, F., 1986, Birefringence and optical measurements in fluorogold at near-millimeter wavelengths, *Infrared Phys.*, 26(2), 131–133.

Chu, L. J. and Barrow, W. L., 1939, Electromagnetic horn design, *Trans. AIEE*, 58, 333–337.

Benford, D. J., Gaidis, M. C., and Kooi, J. W., 2003, Optical properties of Zitex in the infrared to submillimeter, *Appl. Opti.*, 42(25), 5118.

Feynman, R., 1985, *QED: The Strange Theory of Light and Matter*, Alix G. Mautner Memorial Lectures, Princeton University Press, Princeton, NJ.

Forget, S., 2007, Optical resonators and Gaussian beams, *Laser and Non-Linear Optics*, Université Paris. Available at: http://www.optiqueingenieur.org/en/courses/OPI_ang_M01_C03/co/Contenu_14.html.

Goldsmith, P. F., 1982, Quasi-optical techniques at millimeter and submillimeter wavelengths, in *Chapter 5, Infrared and Millimeter Waves*, 6, K. J. Button (ed.), Academic Press, New York. P. 277.

Goldsmith, P. F., 1998, *Quasioptical Systems*, IEEE Press/Chapman & Hall Publishers Series on Microwave Technology and RF, New York.

Granet, C., James, G., Bolton, R., and Moory, 2004, A smooth-walled spline-profile horn as an alternative to the corrugated horn for wide band millimeter-wave applications, *IEEE Trans. Antenn. Propag.*, 52(3), 848.
Groppi, C., Walker, C. K., Kulesa, C. et al., 2011, SuperCam: a 64 pixel heterodyne imaging spectrometer, in *Millimeter and Submillimeter Detectors and Instrumentation for Astronomy IV*, W. D. Duncan, W. S. Holland, S. Withington, and J. Zmuidzinas (eds.), Proceedings of SPIE 7020, SPIE, Bellingham, WA 2008, 702011.
Groppi, E. and Kawamura, J., 2011, Coherent detector arrays for terahertz astrophysics applications, *IEEE Trans. Terahertz Sci. Technol.*, 1(1), 85.
Halliday, D. and Resnick, R., 1974, *Fundamentals of Physics*, John Wiley & Sons, Inc., New York.
Hübers, H., Semenov, A., Richter, H. et al., 2004, Heterodyne receiver for 3–5 THz with hot electron bolometric mixer, in *Proc. Conf. Millim. Submillim. Detectors and Instrumentation for Astron. II*. SPIE, Glasgow, U.K., 5498, 579–586.
Jasik, H., 1961, *Antenna Engineering Handbook*, McGraw-Hill Book Company, New York.
Ji, M., Musante, C., Yngvesson, S., Gatesman, A., and Waldman, J., 2000, Study of Parylene as antireflection coating for silicon optics at THz frequencies, in *ISSTT Proceedings 2000*, Ann Arbor, MI.
Johansson, J., 1995, A comparison of some feedhorn types, *ASP Conf. Ser.*, 75, 82.
Johansson, J. and Whyborn, N., 1992, The diagonal horn as a sub-millimeter wave antenna, *IEEE Trans. Microwave Theory Techn.*, 40(5), 795.
Jones, E. M. T. and Cohn, S. B., 1955, Surface matching of dielectric lenses, *J. Appl. Phys.*, 26(4), 452.
Kloosterman, J., 2014, Heterodyne Arrays for TeraHertz Astronomy, PhD Dissertation, University of Arizona.
Kraus, J., 1966, *Radio Astronomy*, McGraw Hill, New York.
Kraus, J. and Marhefka, R., 2002, *Antennas for all Applications*, 3rd edition, McGraw Hill, New York.
Leech, J., Tan, B.-K., Yassin, G. et al., 2011, Multiple flare-angle horn feeds for sub-mm astronomy and cosmic microwave background experiments, *Astron. Astrophys.*, 532, A61.
Lesurf, L., 1990, *Millimetre-Wave Optics, Devices & Systems*, Adam Hilger, Bristol New York.
Li, T., 1959, A study of spherical reflectors as wide-angle scanning antennas, *IRE Trans. Antenn. Propag.*, 7, 223.
Love, A. W., 1962, The diagonal horn antenna, *Microwave J.*, V, 117–122.
Murphy, J., 1987, Distortion of a simple Gaussian beam on reflection fron off-axis elliptical mirrors, *Int. J. Infrared Milli. Waves*, 8, 1165–1187.
Neilson, J., 2002, An improved multimode horn Gaussian mode generation at millimeter and submillimeter wavelengths, *IEEE Trans. Antenn. Propag.*, 50, 1077–1081.
Pickett, H. M., Hardy, J. C., and Farhoomand, J., 1984, Characterisation of a dual-mode horn for submillimetre wavelengths, *IEEE Trans. Microwave Theory Tech.* MTT-32(8), 936–937.
Potter, P. D., 1963, A new horn antenna with suppressed sidelobes and equal beam widths, *Microwave J.*, 6, 71–78.
Qiu, B., Liu C., Huang, J., and Qiu, R., 1992, An automatic measurement for dielectric properties of solid material at 890 GHz, *Int. J. Infrared Milli. Waves*, 13, 923–931.
Randall, C. M. and Rawcliffe, 1967, Refractive indices of germanium, silicon, and fused quartz in the far-infrared, *Appl. Opt.*, 6, 1889–1894.
Russell, E. E. and Bell E.E., 1967, Measurement of the optical constants of crystal quartz in the far-infrared with the asymmetric Fourier-transform method, *J. Opt. Soc. Am.*, 57, 341.
Ruze, J., 1966, Antenna tolerance theory—a review, *Proc. IEEE*, 54, 633.
Shillue, B., 1997, MMA Memo 175: Gain Degradation in a Symmetrical Cassegrain Antenna Due to Laterally Offset Feeds, http://legacy.nrao.edu/alma/memos/html-memos/alma175/memo175.html
Smith, D. and Loewenstein, E., 1975, Optical constants of far infrared materials 3. plastics, *Appl. Opt.*, 14, 1335.
Thomas, B., 1978, Design of corrugated conical feedhorns, *IEEE Trans. Antenn. Propag.*, AP-26, 367–372.
Tsuji, M., Shigesawa, H., and Takiyama, K., 1982, Submillimeter-wave dielectric measurements using an open resonator, *Int. J. Infrared Milli. Waves*, 3, 801–815.

Walker, C., Narayanan, G., Knoepfle, H. et al., 1997, Laser micromachining of silicon: A new technique for fabricating high quality terahertz waveguide components, in *8th International Symposium of Space THz Technology*, Harvard University, Cambridge, Massachusetts.

Wheeler, H. A., 1977, Transmission-line properties of a strip on a dielectric sheet on a plane, *IEEE Trans. Microwave Theory Techn.*, 25(8), pp. 631–64.

Young, L., 1963, Practical design of a wide band quarterwave transformer in waveguide, *Microwave J.*, 6(10), 7679.

6

THz COHERENT DETECTION SYSTEMS

PROLOGUE

In order to probe the composition, chemistry, and dynamics of the ISM, detector systems capable of resolving THz emission and/or absorption lines of atoms and molecules are needed. For individual clouds, this often requires sub-km/s velocity resolution or, equivalently, frequency resolution on the order of ~1 MHz. To achieve these spectral resolutions, coherent detection systems are employed. Coherent detection systems have been in use for nearly a century and are utilized by the majority of the world's population in a variety of items on a daily basis. Such items include radios, TVs, cellphones, wireless networks, etc. In this chapter, the basic physics and technology behind coherent detection systems will be discussed, along with their implementation for THz astronomy.

6.1 INTRODUCTION

The emission from celestial sources is largely incoherent, originating in random atomic and molecular processes in the interstellar medium, stars, planetary bodies, or the cosmic background. If converted to audio frequencies, the emission would sound like static, similar to the noise generated within our detection systems. The total power received from a 1 Jansky source over a 10 m telescope aperture in 10 min is just 6×10^{-22} watts, far less than the noise power generated in a detection system. At THz frequencies, the situation is even worse, since, for ground-based systems, we must also contend with the noise power injected into our cosmic photon stream by atmospheric water vapor.

An antenna presents a noise power to the detection system as if originating from a resistor, R_A, at temperature T_A. R_A is referred to as the antenna radiation resistance and T_A is the temperature the antenna "sees" through its power pattern (Kraus, 1966). It is not the actual physical temperature of the antenna or telescope structure. Let us define S_ν^o at the output of our antenna (i.e., telescope) such that

$$S_\nu^o = \frac{2kT_A}{A_e} \tag{6.1}$$

where

S_ν^o = flux density at antenna terminals, that is, telescope output (watts m^{-2} Hz^{-1})
T_A = antenna temperature (K)
A_e = effective aperture of antenna (m^2)
k = Boltzmann's constant (1.38×10^{-23} joule K^{-1})

S_ν^o and T_A include optical and Ohmic (i.e., resistive) losses in the antenna itself and, for ground-based systems, the noise flux contribution from the atmosphere.

The total system noise temperature, T_{sys}, referenced to the antenna terminals is

$$T_{sys} = T_A + T_{RX}, \tag{6.2}$$

where

T_{RX} = equivalent black-body receiver noise temperature (K).

The purpose of a receiver system is to increase the observed source flux density, S_ν^o, to a level where it can be detected and, most often, digitized. In designing a receiver system, we should do our best to reduce unwanted noise power from our detectors, optics, and the atmosphere (e.g., by locating the telescope on a high, dry site, on a high-altitude balloon, or out of the atmosphere completely). However, no matter what we do, the instantaneous noise flux in our detection system, S_ν^{sys}, is likely to swamp the signal, S_ν^o, from the astronomical source we are trying to observe. The value of S_ν^{sys} depends on the noise contribution from each element in the detection system. For a signal-to-noise ratio (SNR) of one, S_ν^o is equal to the rms noise level of the system, ΔS_ν^{rms}, such that

$$S_\nu^o \approx \Delta S_\nu^{rms}, \quad \text{for SNR} = 1$$

ΔS_ν^{rms} is related to the system noise flux, S_ν^{sys}, frequency resolution, B_{pd}, and integration time, $\Delta\tau_{int}$, through the radiometer equation,

$$\Delta S_\nu^{rms} = \frac{2k}{A_e} \frac{K_s S_\nu^{sys}}{\sqrt{\Delta\tau_{int} B_{pd}}} \tag{6.3}$$

where

K_s = sensitivity constant (~1 to 2 depending on receiver type and observing strategy, dimensionless)

Equation 6.3 can be rewritten in terms of temperature as (Kraus, 1966)

$$\Delta T_{\text{rms}} = \sigma_{\text{rms}} = \frac{K_s T_{sys}}{\sqrt{\Delta\tau_{\text{int}} B_{pd}}} \tag{6.4}$$

where

ΔT_{rms} = rms noise temperature of system (K)
σ_{rms} = 1 standard deviation in the Gaussian noise floor (K)
$\approx (1/3)T_{p\text{-}p}$
$T_{p\text{-}p}$ = peak-to-peak variation observed in output power (K)
T_{sys} = system noise temperature (K)
$\Delta\tau_{\text{int}}$ = postdetection integration time on source (s)
B_{pd} = postdetection bandwidth, that is, frequency resolution (Hz)
K_s = sensitivity constant (~1 to 2 depending on receiver type and observing strategy, dimensionless)

The quantity T_{sys} is the equivalent black-body noise temperature of the receiver system, that is, the equivalent noise injected into the signal you are trying to observe by the receiver. The greater the value of T_{sys}, the longer you will need to integrate to reach the target value of ΔT_{rms}. Equation 6.4 says you can integrate this noise down by the square root of the integration time, as long as it remains uncorrelated, white noise.

To detect an astronomical source of temperature $T_\nu^s (\propto S_\nu^s)$, we must integrate over a time, $\Delta\tau_{\text{int}}$, and predetection bandwidth, B_{pd}, until T_ν^s is above the rms noise floor, of our system, such that

$$\frac{T_\nu^s}{\Delta T_\nu^{\text{rms}}} > 1 \tag{6.5}$$

In practical systems, the maximum allowed integration time "ON" source, $\Delta\tau_{\text{int}}^{\text{max}}$, is set by the gain stability, $\Delta G/G$, of the overall system and is usually restricted to between ~10 and 30 s before the noise stops integrating down according to Equation 6.4. Gain instabilities can arise from individual components, from the interaction between components, or, for ground-based systems, from the atmosphere itself. We will discuss these issues in Section 6.5. In order to normalize out the instrumental response (and atmospheric absorption effects, if they are significant) an "OFF" position free of source emission is observed for an equivalent amount of time. This mode of observing is referred to as "position switching." An alternative approach to obtaining an "OFF" is to tune the receiver to an emission-free frequency and integrate there. This approach is referred to as "frequency switching." The observation of an ON and an OFF constitutes a "scan." The OFF measurement is then subtracted from the ON measurement, and the difference spectrum calibrated by multiplying by the receiver's output power-to-temperature conversion factor or "Cal." The Cal is obtained by having the receiver observe a load at a known temperature. Multiple scans can be averaged together to meet the total integration time requirement of the observation. When the same amount of time is spent ON and OFF source, the value of K_s in Equation 6.4 is 2. More information on THz observing techniques is provided in Chapter 8.

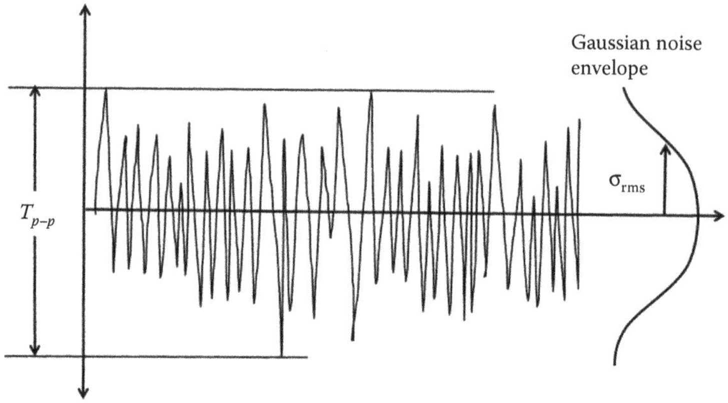

FIGURE 6.1 Relationship between receiver output noise, T_{p-p}, and σ_{rms}. Since the mean of the noise is zero, ΔT_{rms} has the same value as the standard deviation σ_{rms}, where $\sigma_{rms} \approx (1/3)T_{p-p}$.

During observations or laboratory measurements, it is often useful to estimate T_{sys} or T_{RX} from the peak-to-peak noise in the baseline of an observed spectrum or in the strip chart recording of the receiver's total power output. The relationship between σ_{rms} and T_{p-p} is shown graphically in Figure 6.1. It is the same whether you are looking at total power variations on the output of the receiver as a function of time (e.g., a strip chart reading), or the noise in an observed spectrum (e.g., Figure 2.6).

EXAMPLE 6.1

Below is a ^{12}CO $J = 2 \rightarrow 1$ spectrum taken toward an external galaxy (UGC 2936). The T_{sys} during the observations was 341°K and the frequency resolution (i.e., postdetection bandwidth) was 7.7 MHz. How long did it take (ON + OFF time) to reach the observed noise level?

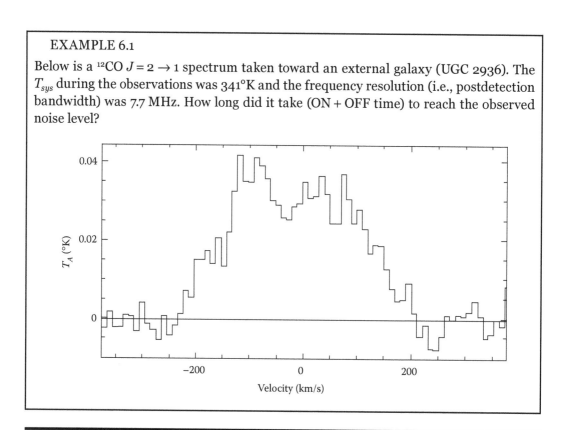

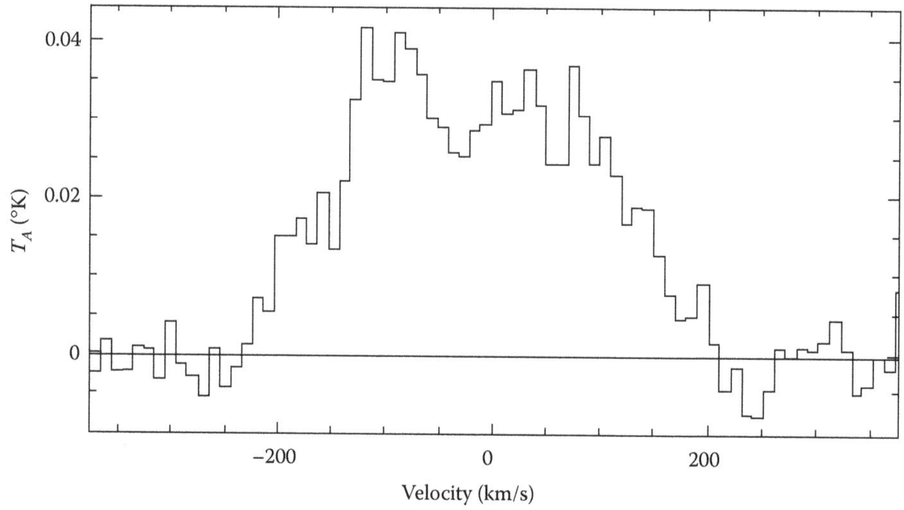

Solving for $\Delta\tau_{int}$ in Equation 6.4, we find,

$$\Delta\tau_{int} = \frac{(KT_{sys}/\Delta T_{rms})^2}{B_{pd}}$$

From examination of the baseline, we can estimate $T_{p\text{-}p} \approx 0.01$ K, therefore,

$$\Delta T_{rms} \approx \sigma_{rms} \approx \frac{1}{3}(0.01\,\text{K}) \approx 0.0033\,\text{K}$$

Since this is a position switched observation, the sensitivity constant, K, is 2. Substitution then yields,

$$\begin{aligned}\Delta\tau_{int} &= \frac{(2(341\,\text{K})/0.0033\,\text{K})^2}{7.7\times10^6\,\text{Hz}}\\ &= 5.547\times10^3\,\text{s}\\ &= 1.54\,\text{h}\end{aligned}$$

6.2 SUPERHETERODYNE RECEIVERS

In THz astronomy, coherent receivers are used to take a piece of spectrum of astrophysical interest at a high frequency and translate it to a lower frequency where it can be amplified and processed. A block diagram of the most basic implementation of a coherent receiver is shown in Figure 6.2. The down conversion is achieved by multiplying the incoming celestial signal with frequency ν_s collected by the antenna/telescope by a much stronger, locally produced tone of frequency ν_{LO}, referred to as a local oscillator (LO). The multiplication process occurs within a nonlinear device, called a mixer. The output of the process is a copy of the astrophysical spectrum at the difference frequency of the celestial signal and the LO. The downconverted copy is referred to as the intermediate frequency (IF) signal and has frequency ν_{IF}. The IF signal then passes through a low-noise amplifier before undergoing further processing. This receiver architecture was devised during World War I by Edwin Amstrong and is referred to as a superheterodyne receiver. Superheterodyne is a contraction for "Supersonic Heterodyne," arising from the fact that more than one signal at frequencies far above audio are employed in the process. This type of receiver is pervasive in our society, and can be found in everything from cellphones to interstellar spacecraft (e.g., Voyager 1). In modern usage, the term is often shortened further to just "heterodyne."

Unlike the 3-port mixer depicted in Figure 6.2, at THz frequencies mixers often have just two ports: an input port through which both the celestial signal and LO tone enters and an output port from which the IF signal emerges. The celestial and LO tones are combined in front of the mixer using a diplexer. Diplexers can range from simple, broadband dielectric beamsplitters (see Appendix 7) to highly tuned Martin–Puplett or Fabry–Perot interferometers (Lesurf, 1990; Goldsmith, 1998). Processing of the IF signal at the mixer output can range from simple square law detection by a diode for total power measurement, to further downconversion–amplification, followed by digitization and Fourier analysis.

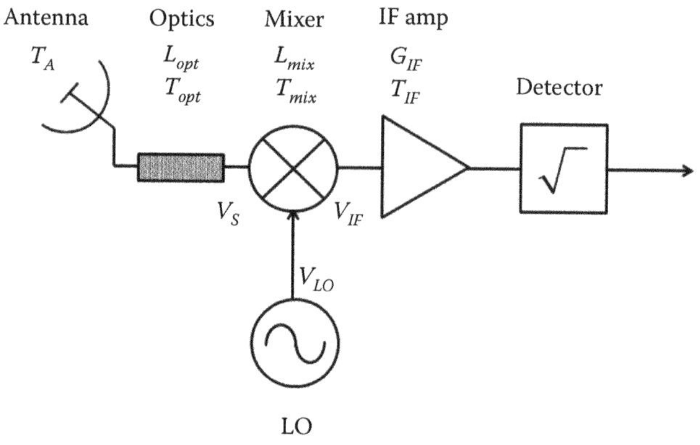

FIGURE 6.2 Basic THz heterodyne receiver block diagram.

In order to gain a better understanding of the downconversion process, let us model the incoming and LO signals as cosine functions, such that

$$V_S(t) = V_S \cos(\omega_S t)$$
$$V_{LO}(t) = V_{LO} \cos(\omega_{LO} t) \tag{6.6}$$

where

$$\omega_S = 2\pi\nu_S,$$
$$\omega_{LO} = 2\pi\nu_{LO},$$

and the mixer itself as a nonlinear device (e.g., a diode) with a current–voltage (I–V) characteristic of the type shown in Figure 6.3. With $V_s(t)$ and $V_{LO}(t)$ applied, the device response can be expressed as a Taylor series expansion, such that

$$I_{mix}(t) = a_o + a_1 V(t) + a_2 V^2(t) + a_3 V^3(t) + \cdots \tag{6.7}$$

where

$$V(t) = V_S(t) + V_{LO}(t) \tag{6.8}$$

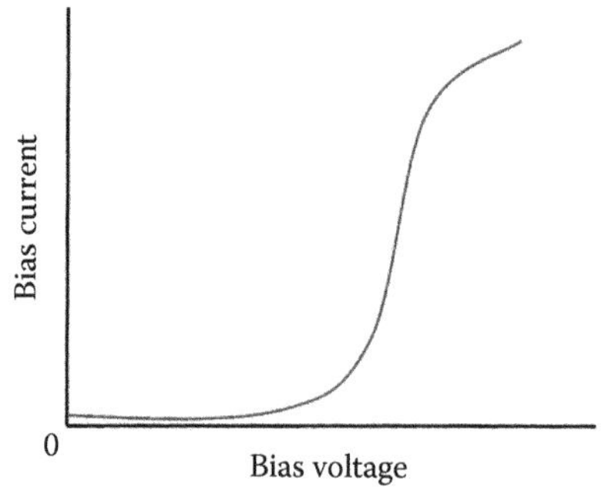

FIGURE 6.3 Nonlinear I–V curve of a mixing device.

The values of the leading coefficients, a_0, a_1, a_2..., are determined by a fit to the mixer's I–V curve. Substitution of Equation 6.8 into Equation 6.7 yields

$$\begin{aligned} I_{mix}(t) = a_0 &+ a_1[V_S\cos(\omega_S t) + V_{LO}\cos(\omega_{LO}t)] \\ &+ a_2[V_S\cos(\omega_S t) + V_{LO}\cos(\omega_{LO}t)]^2 \\ &+ a_3[V_S\cos(\omega_S t) + V_{LO}\cos(\omega_{LO}t)]^3 \\ &+ \cdots \end{aligned} \tag{6.9}$$

In Figure 6.2, the IF output of the mixer feeds a low-noise amplifier (LNA) with a bandwidth $\Delta\nu_{IF}$. Expanding the second order term (see Equation 6.10), we find it contains the product of the signal and LO cosine functions. This product contains sum and difference frequencies, ν_S and $\nu_S + \nu_{LO}$ (see Equation 6.10). It is the signal component at the difference (IF) frequency that conveys the downconverted astronomical signal.

$$\begin{aligned} &a_2[V_S\cos(\omega_S t) + V_{LO}\cos(\omega_{LO}t)]^2 \\ &= a_2\left[(V_S\cos(\omega_S t))^2 + \underbrace{2V_S\cos(\omega_S t)V_{LO}\cos(\omega_{LO}t)}_{V_s[\cos(2\pi t(\nu_S - \nu_{LO})) + \cos(2\pi t(\nu_S + \nu_{LO}))]} + (V_{LO}\cos(\omega_{LO}t))^2\right] \end{aligned} \tag{6.10}$$

In a phenomenological sense, the mixer is a relay, activated by the LO, that modulates the flow of photons through the receiver (see Figure 6.4). The mixer is biased on the knee of its I–V curve so that the voltage excursions of the LO (Equation 6.6) are sufficient to swing the mixing device between a more "ON" state and a more "OFF" state, as depicted in Figure 6.5. The sharper the knee of the I–V curve the better, since less LO and signal power will be needed for the downconversion process, thereby making the receiver more sensitive.

The IF amplifier input bandwidth, $\Delta\nu_{IF}$, defines the spectral window (or band of frequencies) centered on the downconverted signal (IF) frequency that will appear at the receiver's output. Since the difference frequency term falls within the argument of a cosine function, two such frequency bands will be downconverted, one centered at $\nu_{USB} = \nu_S + \nu_{IF}$, and the other at $\nu_{USB} = \nu_S - \nu_{IF}$. The band falling above ν_{LO} is referred to as the receiver's upper sideband (USB), and the one falling below ν_{LO} the receiver's lower sideband (LSB). The sideband containing the desired signal is sometimes referred to as the "signal" sideband,

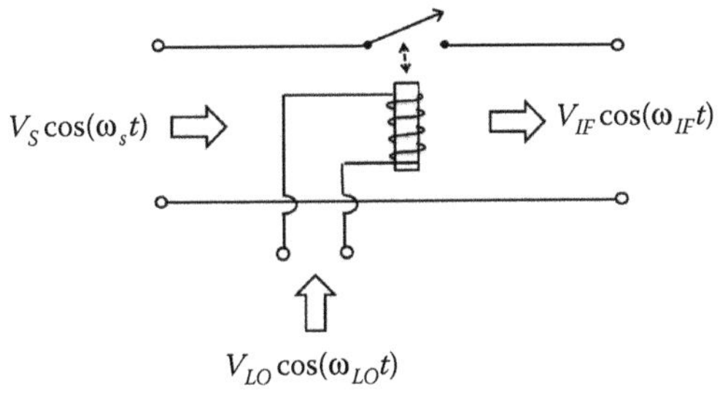

FIGURE 6.4 Mixer relay model. A mixer can be thought of as a relay modulated by the LO.

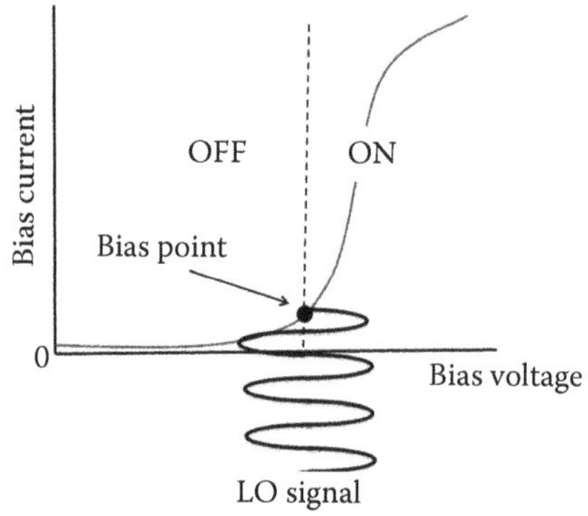

FIGURE 6.5 Mixer I–V/LO interaction. The nonlinear mixing device. (e.g., diode, SIS junction, or HEB) is biased, such that the LO can turn it more "OFF" and more "ON."

and the opposing sideband as the "image" sideband. An illustration of the downconversion process is provided in Figure 6.6.

As we will discuss later, some mixers are designed to separate out the sidebands into two separate IF outputs, while others have only a single IF output where the two sidebands are superimposed on each other. The former type of mixer is referred to as a single sideband (SSB) or sideband separation mixer, while the latter type is referred to as a double sideband (DSB) mixer. Due to its simplicity of construction, the DSB mixer is most commonly used at THz frequencies. The disadvantage of the DSB mixer is that astronomical lines in both the USB and LSB will appear together on the receiver output, potentially making it difficult to identify them. This potential degeneracy can be broken by changing the LO frequency by an amount, $< \Delta\nu_{IF}$, and reobserving the lines. Lines from different sidebands will move in different directions within the IF passband, allowing them to be identified. Groesbeck (1995), Blake et al. (1986), and Sutton et al. (1985) have developed an efficient "CLEAN" algorithm using this approach that can generate an SSB spectrum from a DSB spectrum.

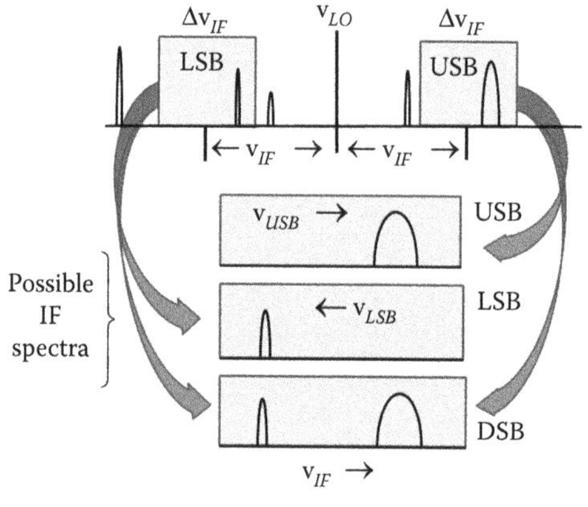

FIGURE 6.6 Mixer downconversion process in which upper and lower sidebands are generated. (Kraus, J., 1966, *Radio Astronomy*, McGraw Hill, New York.)

6.3 RECEIVER NOISE TEMPERATURE

As Equation 6.4 shows, it is the system noise temperature, T_{sys}, that ultimately determines the sensitivity of our receiver system. Depending on the frequency and location, the value of T_{sys} can be dominated by noise from the atmosphere or the receiver itself, T_{RX} (Equation 6.2). At frequencies less than ≈ 50 GHz, T_{sys} of a ground-based observatory is dominated by receiver noise and at higher frequencies by atmospheric absorption, which acts as a diluted black-body absorber. A receiver system can be modeled as a cascade of linear two-port components (see Figure 6.7), each generating a noise power (Kraus, 1966),

$$W_n^{alone} = G_n k T_n \Delta\nu_n \tag{6.11}$$

where

W_n^{alone} = noise power of component operating alone n (watts)
G_n = power gain of component (magnitude)
T_n = equivalent black-body noise temperature of component (K)
k = Boltzmann's constant (1.38×10^{-23} joule K^{-1})
$\Delta\nu_n$ = bandwidth of component (Hz)

When computing the noise power from the *whole* system, W_{sys}, we must take into account that each successive component sees the gain and noise from what came before, such that

$$\begin{aligned} W_{sys} &= W_A + W_{RX} \\ &= G_1 G_2 \ldots G_n k T_{sys} \Delta\nu_{sys} \\ &= G_1 G_2 \ldots G_n k T_A \Delta\nu_{sys} + G_1 G_2 G_3 \ldots G_n k T_1 \Delta\nu_{sys} \\ &\quad + G_2 G_3 \ldots G_n k T_2 \Delta\nu_{sys} + \cdots + G_n k T_n \Delta\nu_{sys} \end{aligned} \tag{6.12}$$

where

W_{sys} = system noise power (watts)
W_A = noise power from the antenna/telescope (watts)
W_{RX} = noise power from receiver (watts)
T_{sys} = system noise temperature (K)
T_A = antenna temperature (K)
T_n = equivalent black-body noise temperature of component (K)
k = Boltzmann's constant (1.38×10^{-23} joule K^{-1})
$\Delta\nu_{sys}$ = system bandwidth, set by narrowest bandwidth component (Hz)

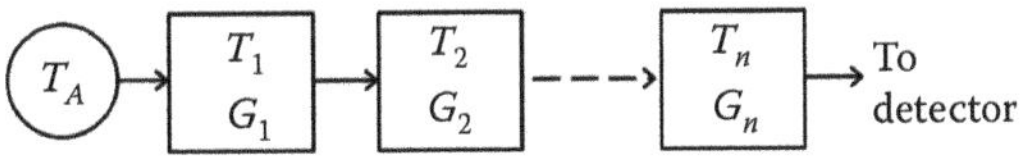

FIGURE 6.7 Linear two-port series model of receiver noise. (After Kraus, J., 1966, *Radio Astronomy*, McGraw Hill, New York.)

We can derive expressions for T_{sys} and the receiver noise temperature, T_{RX}, by dividing both sides of Equation 6.12 by the factor ($k\Delta\nu_{sys} \sum_1^n G_n$).

$$T_{sys} = T_A + T_{RX} \tag{6.13}$$

where

$$T_{RX} = T_1 + \frac{T_2}{G_1} + \frac{T_3}{G_1G_2} + \frac{T_n}{G_1 \ldots G_{n-1}} \tag{6.14}$$

In order to use Equation 6.14 to derive the noise temperature, T_{RX}, of the THz receiver of Figure 6.2, we must first determine the gain and equivalent black-body noise temperature of each component.

6.4 NOISE TEMPERATURE OF THz OPTICAL SYSTEMS

As was discussed in Chapter 5, the purpose of the optical system (e.g., relay mirrors and/or lenses) is to convey the power collected by the telescope (i.e., antenna) to the detection system. Ideally, the optics would do their job perfectly, with an efficiency of $\eta_o = 1$, corresponding to a power gain $G_1 = 1$. However, due to the finite size of our optics, some of the power collected by the antenna will be lost, so that $G_1 = \eta_o < 1$. The optics can also add noise power by way of scattering radiation into the beam, or by generating radiation through Ohmic losses within the optical elements. In essence, to the rest of the receiver, the optics appear to be an emitting/absorbing medium through which the antenna is observed. This situation is analogous to observing an astronomical source through an interstellar cloud with transmission efficiency (or power gain) $\eta = G_1 = e^{-\tau_\nu}$, where τ_ν is the cloud optical depth at frequency ν (Equation 2.10). The temperature, T_{in}^{mix}, "seen" at the input terminals of the mixer in Figure 6.2, is, then,

$$T_{in}^{mix} = T_A G_1 + T_{opt}(1 - G_1) \tag{6.15}$$

where T_{opt} is the physical temperature of the optical system. The total temperature at the output terminals of the receiver, T_{term}^{RX}, is

$$T_{term}^{RX} = T_A G_1 + T_{opt}(1 - G_1) + T_R' \tag{6.16}$$

where T'_R is the receiver noise temperature excluding the optical system.

In order to reference the receiver noise temperature back to the antenna terminals, we simply divide Equation 6.16 by the optical system gain, G_1.

$$T_{sys} = T_A + T_{opt}\left(\frac{1}{G_1} - 1\right) + \left(\frac{1}{G_1}\right)T_R' \tag{6.17}$$

Comparing Equations 6.14 and 6.17, we find the optical system has an effective noise temperature contribution, T_1, of

$$T_1 = T_{opt}\left(\frac{1}{G_1} - 1\right) \tag{6.18}$$

T_1 adds directly into the thermal noise from the antenna, T_A. Since the loss factor, L_{opt}, (i.e., attenuation) through the optical system is equal to $1/G_1$, we can rewrite Equation 6.18 in the form

$$T_1 = T_{opt}(L_{opt} - 1) \tag{6.19}$$

or, more generally,

$$T = T_p(L - 1) \tag{6.20}$$

where

T = noise contribution of component (K)
T_P = physical temperature of component (K)
L = $1/G$ = loss factor, attenuation, or noise figure, F, of component (unitless)

Equation 6.20 is valid for computing the noise temperature contribution from any linear two-port network (e.g., a beamsplitter, attenuator, waveguide, and coax).

EXAMPLE 6.2

Find the equivalent black-body noise temperature, T, of an amplifier with a noise figure, NF, of 0.7 dB and a gain of 13 dB operating at 290 K.

In the technical literature, the noise performance of an amplifier or mixer is sometimes specified in terms of a noise figure, NF, in units of dB. The equivalent noise temperature can be estimated using the relation,

$$T(10^{NF_{dB}/10} - 1)T_0,$$

where

NF_{dB} = noise figure of device (dB)
T_0 = physical temperature of device

The same relation can be used to determine the equivalent noise temperature of any component (e.g., attenuator and optical system) whose loss is specified in dB. In the case of an amplifier or mixer, the noise performance does not always scale directly with temperature, as the expression suggests. The associated product literature should be consulted for accurate performance numbers. For the amplifier being considered here, the equivalent noise temperature is

$$\begin{aligned} T &= (10^{0.7/10} - 1)290\text{ K} \\ &= 50.7\text{ K} \end{aligned}$$

EXAMPLE 6.3

If the output of the amplifier in Example 6.1 is connected to the input of a second amplifier with a noise temperature of $T_2 = 140$ K and 40 dB of gain, what is the system noise temperature? Assume the rest of the receiver has a noise temperature of 800 K.

From Example 6.2, we know,

$$\begin{aligned} T_1 &= 50.7 \text{ K} \\ G_1 &= 13 \text{ dB} \Rightarrow 10^{13/10} = 20 \end{aligned}$$

Here, we have

$$\begin{aligned} T_2 &= 140 \text{ K} \\ G_2 &= 40 \text{ dB} \Rightarrow 10^{40/10} = 10{,}000 \\ T_3 &= 800 \text{ K} \end{aligned}$$

Substitution into Equation 6.14 yields

$$\begin{aligned} T_{RX} &= 50.7 + \frac{140}{20} + \frac{800}{(20)(10{,}000)} \\ &= 50.7 + 7 + 0.004 \\ &\approx 58 \text{ K} \end{aligned}$$

The receiver noise temperature is dominated by the noise of its first component, here an amplifier. In terms of noise, what comes after the second amplifier is of little significance.

EXAMPLE 6.4

Now, instead of the amplifier being the first component in the receiver, let the first component be a mixer with a noise temperature of 274 K, and a conversion loss of 10 dB. What is the new T_{RX}?

A conversion loss of 10 dB corresponds to a below unity gain of

$$G = 10^{-10/10} = 0.1$$

Substitution into Equation 6.14 then becomes

$$\begin{aligned} T_{RX} &= 274 + \frac{50.7}{0.1} + \frac{140}{(0.1)(20)} + \frac{800}{(0.1)(20)(10{,}000)} \\ &= 274 + 507 + 70 + 0.04 \\ &\approx 851 \text{ K} \end{aligned}$$

EXAMPLE 6.5

Now, to make matters worse (and more like Figure 6.2), let us assume the optical system (excluding the antenna itself) that conveys the light from the telescope to the mixer operates at room temperature and has 2 dB of loss (e.g., from atmospheric absorption, scattering, and misalignment). How does this impact the system noise temperature, T_{RX}?

An optical loss of 2 dB corresponds to a below unity gain of

$$G = 10^{-2/10} = 0.63$$

From Equation 6.18, the effective noise temperature of the optical system is

$$\begin{aligned} T &= 290\,\mathrm{K}\left(\frac{1}{0.63} - 1\right) \\ &= 170.3\,\mathrm{K} \end{aligned}$$

Substitution into Equation 6.14 then becomes

$$\begin{aligned} T_{RX} &= 170.3 + \frac{274}{0.63} + \frac{50.7}{(0.63)(0.1)} + \frac{140}{(0.63)(0.1)(20)} + \frac{800}{(0.63)(0.1)(20)(10{,}000)} \\ &= \underbrace{170.3}_{\text{optics}} + \underbrace{435}_{\text{mixer}} + \underbrace{804.8}_{\text{1st IF amp}} + \underbrace{111}_{\text{2nd IF amp}} + \underbrace{0.06}_{\text{all the rest}} \\ &\approx 1521\,\mathrm{K} \end{aligned}$$

Examination of the noise contributions in the above expression is quite informative. We see that just 2 dB of loss in the optical system will effectively double the receiver noise temperature. The degree to which cooling the optics will help, depends on the source of the loss. In a properly designed system below 1 THz, the optical loss is most likely due to optical misalignment, in which case cooling the optics will not help significantly. Above 1 THz, absorption of the light by atmospheric water vapor could be the dominant source of loss. In this situation, one solution is to house the relay optics in an evacuated chamber. For receiver systems that fly on high-altitude balloons or in space where there is little or no atmosphere, the relay optics may only need to be housed in an evacuated chamber for ground-testing.

In a full receiver noise temperature calculation, another significant contributor to the noise is seen to be the first amplifier after the mixer known as the 1st IF amp. This amplifier is usually cooled to ≤20 K to reduce its intrinsic noise, and connected to the mixer by the shortest piece of coax the cryogenic system will tolerate. When cooled, modern IF amplifiers have noise temperatures of between 5 and 10 K, reducing its noise contribution to the system by as much as an order of magnitude. If a 5 K IF amplifier were used in the above system, the value of T_{RX} would drop from 1521 K to 635 K.

6.5 THz MIXER ARCHITECTURES AND NOISE

Outside of the atmosphere, it is the mixer that dominates the noise performance of most THz astronomical receivers. Every mixer is inherently a DSB mixer, although they can be configured to operate in an SSB mode by either suppressing the unwanted sideband using an external, quasi-optical filter (e.g., a Fabry–Perot filter) or by phasing the output of two DSB mixers such that one sideband is suppressed.

6.5.1 DOUBLE SIDEBAND (DSB) MIXER

A DSB mixer noise model is shown in Figure 6.8a. Here, T_{in}^{mix} from Equation 6.15 is the noise temperature of the signal delivered to the mixer input port. T_{in}^{mix} contains all the spectral information that was able to pass through the optical system. The mixer down-converts the USB and LSB components of T_{in}^{mix} to the IF frequency with conversion gains G_{USB} and G_{LSB}. In the downconversion process, the mixer adds in a noise signal, T_{DSB}, to both sidebands. Mathematically, this is equivalent to

$$T_{IF} = \frac{T_{in}^{mix}}{G_{USB}} + \frac{T_{in}^{mix}}{G_{LSB}} + \frac{T_{DSB}}{G_{USB}} + \frac{T_{DSB}}{G_{LSB}} \tag{6.21}$$

or, in terms of conversion loss,

$$\begin{aligned} T_{IF} &= T_{in}^{mix} L_{USB} + T_{in}^{mix} L_{LSB} + T_{DSB} L_{USB} + T_{DSB} L_{LSB} \\ &= (T_{in}^{mix} + T_{DSB})(L_{USB} + L_{LSB}) \end{aligned} \tag{6.22}$$

where

T_{IF} = noise temperature at mixer IF port (°K)
T_{in}^{mix} = noise temperature at mixer input (i.e., RF) port (°K)
$L_{USS} = 1/G_{USB}$ = conversion loss for upper sideband (USB) (unitless)
$L_{LSB} = 1/G_{LSB}$ = conversion loss for lower sideband (LSB) (unitless)

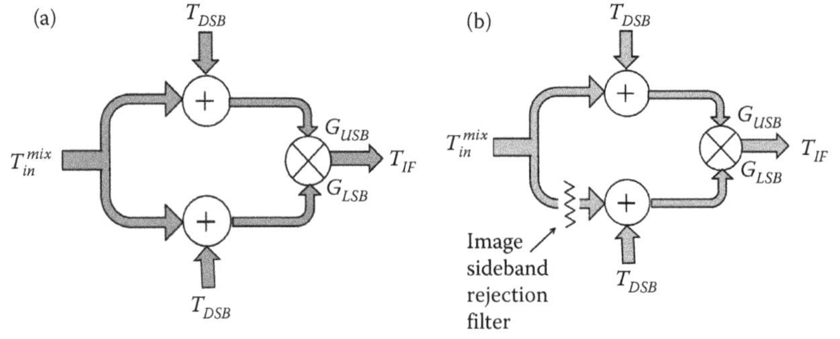

FIGURE 6.8 DSB mixer noise model. (a) Both USB and LSB appear in mixer IF output. (b) LSB input terminated in matched load: IF output includes USB + LSB noise.

Solving for T_{DSB}, we find

$$T_{DSB} = \frac{T_{IF}}{L_{USB} + L_{LSB}} - T_{in}^{mix} \tag{6.23}$$

For such a receiver, the single-sideband noise temperature, T_{SSB}, is given by

$$\begin{aligned} T_{SSB} &= T_{DSB}(1 + C_I^S) \\ C_I^S &= \frac{L_{SIG}}{L_{IM}} \end{aligned} \tag{6.24}$$

where

C_I^S = sideband gain ratio
L_{SIG} = conversion loss within sideband containing desired signal (unitless)
L_{IM} = conversion loss within sideband with no desired signal (unitless)

For THz DSB mixers like that of Figure 6.8a, where $\nu_{IF} = \nu_{THz}$, the assumption that $L_{SIG} \approx L_{IM}$ is generally valid. Substitution into Equation 6.24 then yields

$$T_{SSB} \approx 2T_{DSB} \tag{6.25}$$

Since emission/absorption lines occur in just one sideband, *one should use the higher, SSB noise temperature* when estimating observing times for spectral line observations (Equation 6.4). T_{DSB} noise temperatures are appropriate for use in estimating observing times when a DSB receiver is being used for continuum observations, for example, planets and dust cores.

As mentioned above, one way to operate a DSB mixer in SSB mode is to suppress the unwanted sideband by using a tuned quasi-optical filter at the mixer input (see Figure 6.8b). However, the noise associated with the downconversion of the sideband will still contribute to the mixer's SSB noise temperature, T_{SSB}. Also, the insertion loss associated with the filter itself will add unwanted noise in the signal sideband.

6.5.2 SIDEBAND SEPARATION (2SB) MIXER

When possible, a superior approach is to use a sideband separation mixer (referred to as a 2SB mixer) that provides a separate IF output for each sideband. Since the emission line of interest will appear in only one sideband, this approach reduces the possibility of spectral confusion from lines in the other sideband, and also works to isolate the noise in one sideband from the other, leading to lower noise operation within each.

A block diagram of a 2SB mixer is shown in Figure 6.9. The mixer has four components: two phase quadrature hybrids and two mixers. Each hybrid has two inputs and two outputs, with 1/2 of the power from each input arriving at each output port. Referring to Figure 6.9, we see that the short path through a hybrid adds 0° phase shift to the input signal, while the longer path adds 90° of phase shift. The two mixers are of the DSB type and pumped by the same LO. Let us follow our THz signal through the system.

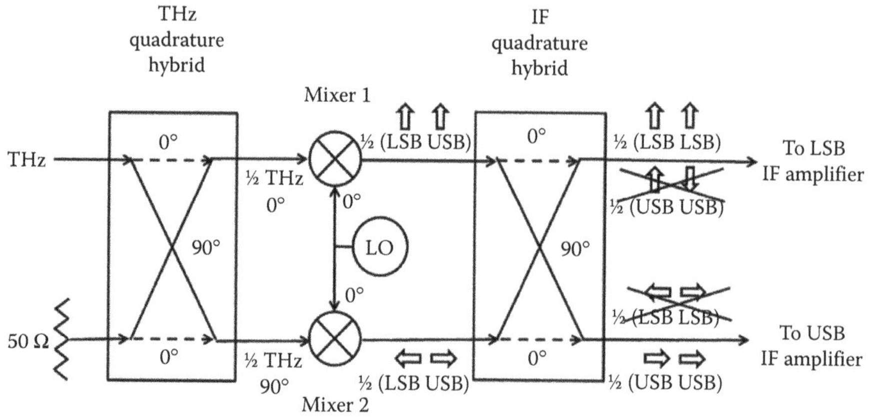

FIGURE 6.9 Single sideband separation mixer. (2SB) Block diagram. Phasor representation of the upper sideband (USB) and lower sideband (LSB) is used to illustrate how sideband suppression is achieved. (Adapted from Billade, B., 2013, Mixers, Multiplier and Passive Components for Low Noise Receivers, PhD dissertation, Chalmers University of Technology; Henderson, B. C. and Cook, J. A., 1985, Image-reject and single-sideband mixers, Watkins-Johnson Tech-Notes, May/June 1985.)

The THz signal arrives at the THz quadrature hybrid where it is split in two, with 1/2 of the THz signal appearing at the input of Mixer 1 with 0° phase shift, and the other half appearing at the input of Mixer 2 with 90° phase shift added to it. The second input port of the THz quadrature hybrid is terminated in a matched load, so only the load's thermal noise (which is usually quite low, since the mixer is cooled to a low temperature) is added to the THz signal outputs. Being DSB, both Mixers 1 and 2 create an upper sideband (USB) and lower sideband (LSB). The USB and LSB coming out of Mixer 1 are in-phase, but due to adding the 90° phase shift to the THz signal, the phase of the USB and LSB signals coming out of Mixer 2 are 180° apart from one another. The outputs from Mixers 1 and 2 then pass through a much lower-frequency IF quadrature hybrid, where both USBs and LSBs from Mixers 1 and 2 appear at each of the outputs. On one output, the LSBs are in-phase and the USBs out-of-phase. Likewise, on the other output port, the USBs are in-phase and the LSBs out-of-phase. The out-of-phase LSBs and USBs cancel each other out, leaving one IF quadrature hybrid port predominantly LSB and the other USB. The LSB and USB signals then go to their own low-noise IF amplifiers.

For this approach to work well, the amplitude and phase response through both mixer chains should be as identical as possible. The metric for characterizing how well a 2SB mixer is able to separate one sideband from the other is the sideband gain/rejection ratio, C_I^S. For example, to achieve a $C_I^S \geq 15$dB (≤3% bleed-through), the amplitude and phase imbalance should be < 3 dB and 20° (Billade, 2013).

6.5.3 BALANCED DSB MIXER

A third type of mixer architecture in growing use at THz frequencies is the balanced DSB mixer. As with the 2SB mixer, it is composed of two hybrids and two DSB mixers (see Figure 6.10). Here, a 90° hybrid is used at the input to inject both the THz signal and LO.

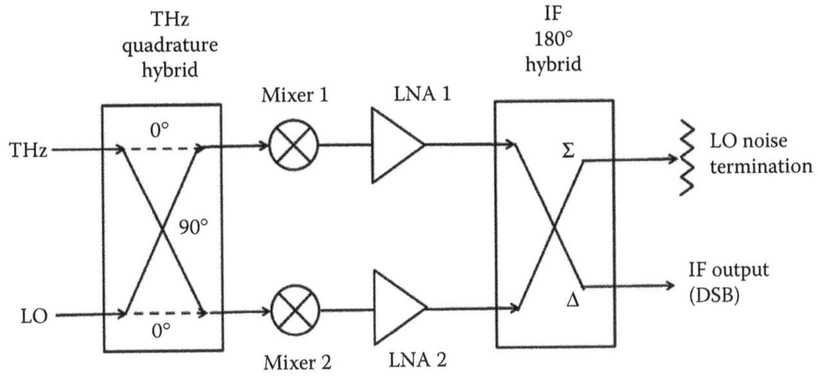

FIGURE 6.10 Balanced DSB mixer block diagram. Though more complex than the single-ended DSB mixer, this architecture can provide low-loss LO injection, increased receiver stability, and lower-noise temperature operation.

The signal and LO power are split, with half of each appearing at the input ports of Mixers 1 and 2. The signal and LO on the input port of Mixer 1 have a 0° and 90° phase shift, respectively. These phase shifts are reversed on the input port of Mixer 2. The IFs from Mixers 1 and 2 then pass through a 180° hybrid, with the downconverted THz signal (LSB and USB) appearing on the difference, Δ, output port and the LO noise on the sum, Σ, port (Kerr and Pan, 1996; Kooi, 2008; Meledin et al., 2008).

The balanced DSB has several advantages over a single-ended DSB mixer. (1) The LO amplitude noise is terminated on a load, increasing overall receiver stability. (This feature is particularly beneficial for THz hot electron bolometer (HEB) mixers, where LO amplitude noise is a major contributor to system instability.) (2) As will be discussed in Section 6.4., LO noise can contribute directly to the system noise temperature. With the balanced mixer, this noise is terminated in a matched load. (3) *Less* LO power is required. With a balanced mixer, all the available LO power can be injected directly into the LO port. For single-ended DSB mixers LO injection is typically performed quasi-optically, either with an interferometer or dielectric beamsplitter—both of which can have significant loss. Even with two mixers being required, the balanced DSB mixer will likely require less LO power. The disadvantages of this approach include a higher parts count (i.e., more to go wrong), more difficult to manufacture, and greater cost.

6.6 IF AMPLIFIERS

The signal power after the mixer in Figure 6.2 is very, very low—typically < 10^{-16} watts/s/GHz for a 10 m telescope aperture. A common unit used in measuring power levels is the dBm, where a 0 dBm power level corresponds to 1 milli-watt (mW) of power. Converting to this unit,

$$\text{Power (dBm)} = 10\log_{10}\left(\frac{\text{Power (watts)}}{1\times10^{-3}}\right) \tag{6.26}$$

we find the output power on the mixer output to be <–130 dBm. A typical detector (e.g., total power diode or the analog-to-digital converter (ADC) of a computer) would like to see of order ~0 dBm of power at its input. These means the IF system of a THz receiver should be capable of providing ~130 dB of gain, a staggering amount! This is usually achieved not by a single amplifier, but by three or four 30–40 dB amplifiers connected in series, as shown in Figure 6.10. Remembering Equation 6.14, we know that when computing the noise contribution of a component, its noise temperature is divided by the gains of the components that came before. For our THz receiver system, this means the burden of low noise performance is greatest on the first IF amplifier in the chain, since the "gain" of the optical system and mixer are <1. The noise performance of the subsequent amplifiers are not nearly so important, since their noise temperatures are cut by factors of a 1000 or more, due to the gain(s) of the proceeding amplifier(s). Before looking into the particulars of noise generation in an IF amplifier, let us look into their basic operating principles.

All electrical amplifiers are based on *Ohm's Law at a point*, which states that current density, J, and electric field, E, in a circuit, are related through the proportionality constant, σ, namely,

$$J = \sigma E \tag{6.27}$$

where

J = current density (A/m²)
E = electric field (V/m)
σ = conductivity (mhos, θ)

Figure 6.11 contains a schematic representation of a vacuum tube, bipolar transistor, and a field effect transistor (FET). Each of these devices contains three electrodes. The two outer electrodes (i.e., cathode and anode for the tube, emitter and collector for the bipolar transistor, and source and drain for the FET) have DC bias voltages applied to them with opposite polarity, thereby setting up an electric field, E, between them. The conductivity between the two outer electrodes is controlled by the center electrode, to which the input signal and a third DC bias voltage is applied. In a vacuum tube, the center electrode is a wire grid, which permits some, but not all electrons emitted by the cathode to flow to the anode (i.e., plate). What faction of the emitted electrons will flow through depends on the magnitude and polarity of the grid voltage. In the case of a bipolar transistor, the exchange of charge is modeled by the passing of "holes" and electrons on a semiconductor (e.g., silicon and germanium) substrate doped with impurities (e.g., gallium and arsenic) to make the collector and emitter have an excess of n-type charge carriers (i.e., electrons) or p-type charge carriers (i.e., holes). The base region between the collector and emitter is doped in the opposite sense. The flow of charge carriers through the base region is controlled by the base current. An FET is also fabricated on a semiconductor substrate with the regions below the source and drain doped with either p- or n-type impurities. The flow of charge in the semiconductor channel between the source and drain is controlled by the electric field applied to the gate.

With proper biasing, small excursions in the voltage (or current) of the input signal, $V_{in}(t)$, can produce an exponential change in the conductivity, $\sigma(t)$, between the electrodes

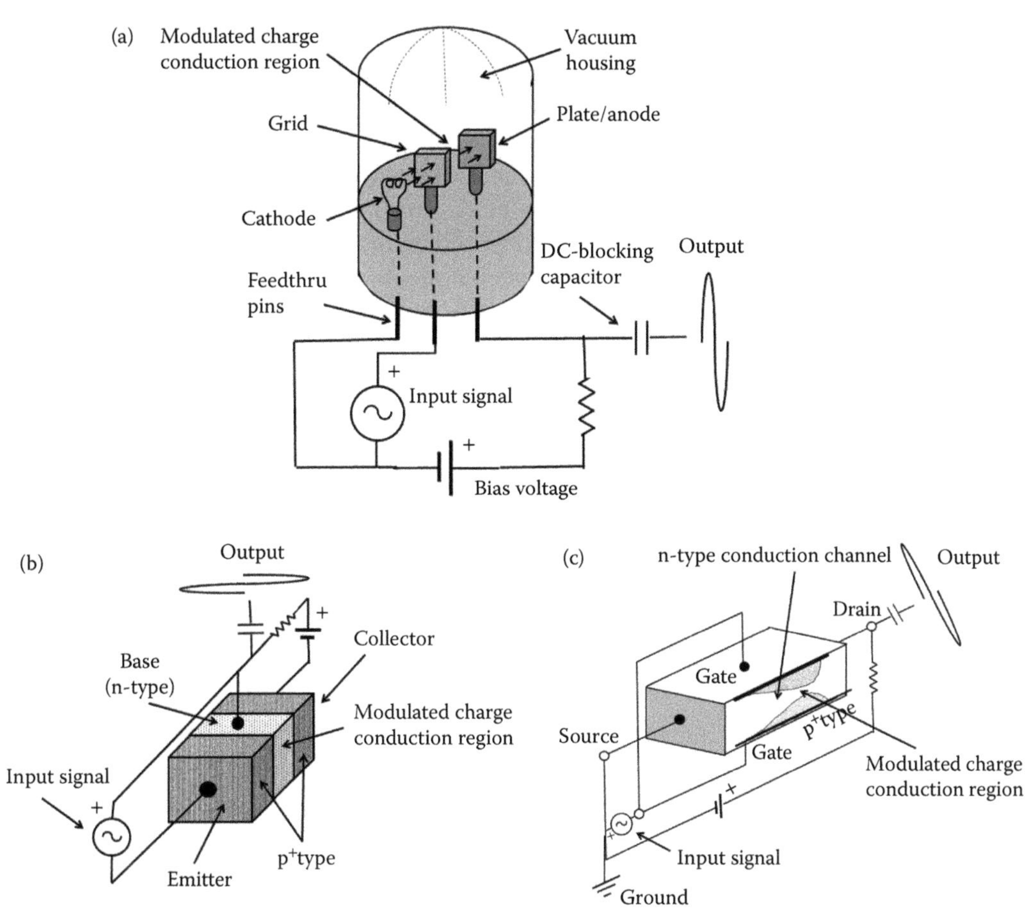

FIGURE 6.11 Amplifying devices. (a) Vacuum tube, (b) bipolar transistor, and (c) field effect transistor. All three amplify by using the weak input signal to modulate the conductivity within a region permeated by a relatively large electric field.

and, thereby, the time-varying output current, $I(t) \propto J(t)$, in the circuit containing the device. With a load resistor across the output, we have $V_{out}(t) \propto I(t)$, yielding a voltage gain, V_{Gain},

$$V_{Gain} = \frac{V_{out}(t)}{V_{in}(t)} \tag{6.28}$$

The detailed physics of how $\sigma(t)$ reacts to the applied signal voltage (or current) will be different for each type of device, but the ability to modulate conductivity, and, thereby, the much larger ambient E-field, is what makes the devices capable of amplification.

The first IF amplifiers were made using triode vacuum tubes, developed by the American inventor Lee de Forest in 1907. The idea behind the solid state field effect transistor was first patented by the Canadian physicist Julius Edgar Lilienfeld in 1925, but was not demonstrated until 1947 by Bardeen, Shockley, and Brattain at Bell Telephone Laboratories, after which its development was rapid. Due to their greater efficiency, lower cost, compactness, and reliability, transistors eventually replaced vacuum tubes as the device of choice in amplifier circuits. Today, FETs are the building blocks of the low-noise IF amplifiers used in THz receivers. Indeed, high-speed versions of FETs, called high electron mobility transistors (HEMTs), have been used in amplifier designs that have been demonstrated as high as ~700 GHz (Samoska, 2011; Tessmann et al., 2012). Ideally, such amplifiers would be used at the signal frequency as a "front-end" amplifier located *before* the mixer in Figure 6.2. However, for now, the relatively high-noise temperatures of these devices limit their use as front-end amplifiers to frequencies ≤115 GHz (the frequency of the astrophysically important CO $J = 1 \rightarrow 0$ transition).

Pospieszalski (1989) developed a small signal equivalent circuit model for noise in a FET. The model (see Figure 6.12) shows the three ports of the FET, the gate (G), drain (D), and source (S). The connection between G and S is modeled by a gate-to-source capacitance, C_{gs}, a gate-to-source resistance, r_{gs}, and a gate-to-source noise voltage source, $\langle e_{gs}^2 \rangle$,

$$\left\langle e_{gs}^2 \right\rangle = 4kT_g r_{gs} \Delta f \tag{6.29}$$

where
T_g = equivalent gate noise temperature ≈ physical temperature of device (K)
Δf = noise bandwidth (Hz)

As part of the model, there is also a noise current source, $\langle i_{ds}^2 \rangle$, on the FET output between D and S.

$$\left\langle i_{ds}^2 \right\rangle = 4kT_d g_{ds} \Delta f \tag{6.30}$$

where
T_d = equivalent drain noise temperature (K)
g_{ds} = drain-to-source conductance (mho)

The two noise sources, $\langle e_{gs}^2 \rangle$ and $\langle i_{ds}^2 \rangle$, are assumed to be uncorrelated. The large-scale, modulated electric field in the FET is modeled by the current source, $g_m V_{gs}$, between D and S, where g_m is the FET transconductance (i.e., the amount of output drain current, i_{ds}, you

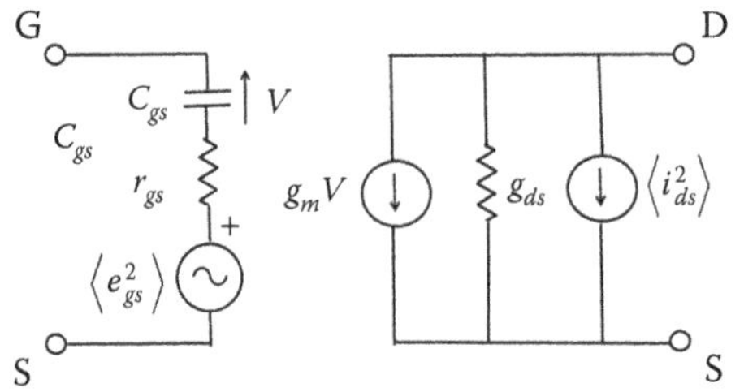

FIGURE 6.12 Simple FET noise model. (Adapted from Pospieszalski, 1989, *IEEE Trans. Microwave Theory Techn.*, 37(9), 1340.)

get for a change in gate voltage, V_{gs}). The minimum noise temperature the FET can provide is, then,

$$T_{\min} \approx \frac{f}{f_{\max}}\sqrt{T_g T_d}$$

$$f_{\max} = f_t\sqrt{\frac{1}{4g_{ds}r_{gs}}} \tag{6.31}$$

where

f = frequency of operation (Hz)

f_t = intrinsic cutoff frequency of FET, frequency where AC-current gain falls to unity (Hz)

The model accurately predicts the noise performance of a wide variety of FET devices, and indicates the noise temperature of the FET will drop as the square root of the device physical temperature. As we will see in later sections, most THz mixers (except for Schottky mixers) require cooling to <10 K. The first IF amplifier is usually cooled as well, typically to ~20 K.

A single FET is capable of providing ~10 dB of gain. In most applications, three or more FETs are packaged together to provide ~30 dB of gain with 5–10 K noise temperature when cooled to ~20 K. For optimum performance, an impedance matching transformer is designed to go on the input and output of each FET. The same basic approach to cascading amplifier stages has been in use since the 1920s, but then with vacuum tubes!

6.7 EFFECTIVE TEMPERATURE MEASUREMENT OF THz COMPONENTS

In the Rayleigh–Jeans limit (when $(h\nu/kT) \ll 1$), a device with bandwidth B and gain G will produce an output power P_d into a noiseless, matched load (T_{load} = 0 K), such that

$$P_d = kT_e GB \tag{6.32}$$

where

T_e = device's effective noise temperature (K)

The most common way in THz astronomy to measure the equivalent noise temperature, T_e, of a device (or an entire receiver system) is by using the Y-factor method. Here, one consecutively places a hot and cold black-body load in front of the device under test (DUT), and measures the corresponding device output power. The ratio of the device's output power, P_H, when looking into a hot load of temperature, T_H, to the device's output power, P_C, when looking into a cold load of temperature, T_C, is the Y-factor.

$$\text{Y-factor} = \frac{P_H}{P_C} \tag{6.33}$$

for,

$$P_H = kGB(T_H + T_e)$$
$$P_C = kGB(T_C + T_e)$$

Solving for T_e we find,

$$T_e = \frac{T_H - Y \cdot T_C}{Y - 1} \quad (6.34)$$

At low temperatures and/or high frequencies, the Raleigh–Jeans approximation leading to Equation 6.34 begins to lose its validity, and the physical temperatures T_H and T_C should be substituted with the corresponding Callen–Welton temperatures, T_H^{CW} and T_C^{CW},

$$\begin{aligned} T_H^{CW} &= \frac{h\nu}{k}(e^{(h\nu/kT_H)} - 1)^{-1} + \frac{h\nu}{2k} \\ T_C^{CW} &= \frac{h\nu}{k}(e^{(h\nu/kT_C)} - 1)^{-1} + \frac{h\nu}{2k} \end{aligned} \quad (6.35)$$

The Callen–Welton temperatures account for zero-point vacuum fluctuations, and include an increase in temperature corresponding to half a photon (Callen and Welton, 1951; Rodriguez-Morales, 2006).

For most astronomical receivers, a room temperature load ($T_H \sim 295$ K) and a load dipped in liquid nitrogen ($T_C \sim 77$ K) provide a large enough contrast in output power to yield a reliable Y-factor measurement. A plot showing the corresponding Callen–Welton temperatures for 295 and 77 K loads from 50 GHz to 10 THz is shown in Figures 6.13 and 6.14 respectively. When performing Y-factor measurements, care should be taken to insure the load has low reflectivity (<~5%) at the frequency of interest and is sufficiently large to intercept the entire beam from the DUT.

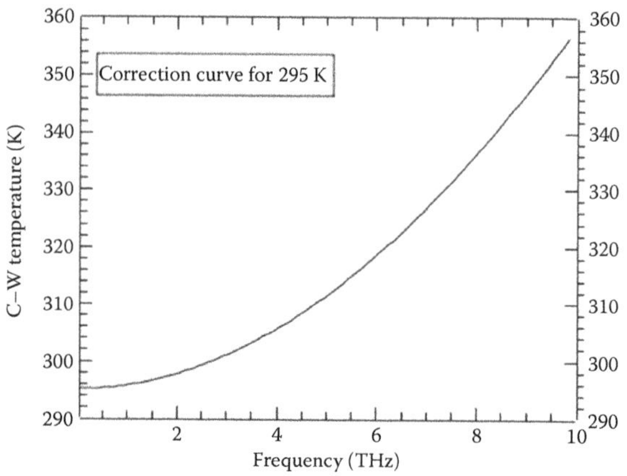

FIGURE 6.13 Callen–Welton black-body correction temperatures corresponding to physical temperatures of 295 K.

EXAMPLE 6.6

A power meter connected to the IF output of a 2 THz receiver reads 10 mW when the receiver is looking at a 290 K load, and 8.3 mW when it is looking at a 77 K load. What is the receiver's noise temperature?

The receiver's Y-factor (Equation 6.33) is

$$Y = \frac{10\ \text{mW}}{8.3\ \text{mW}} = 1.2$$

Using Equation 6.35, we can determine the Callen–Welton corrections to the hot and cold black-body temperatures at 2 THz.

$$T_{290}^{CW} = 292.6\ \text{K}$$
$$T_{77}^{CW} = 86.7\ \text{K}$$

Plugging into Equation 6.34, we find the receiver's noise temperature to be

$$\begin{aligned} T_{RX} &= \frac{292.6\ \text{K} - (1.2)86.7\ \text{K}}{1.2 - 1} \\ &= 943\ \text{K} \end{aligned}$$

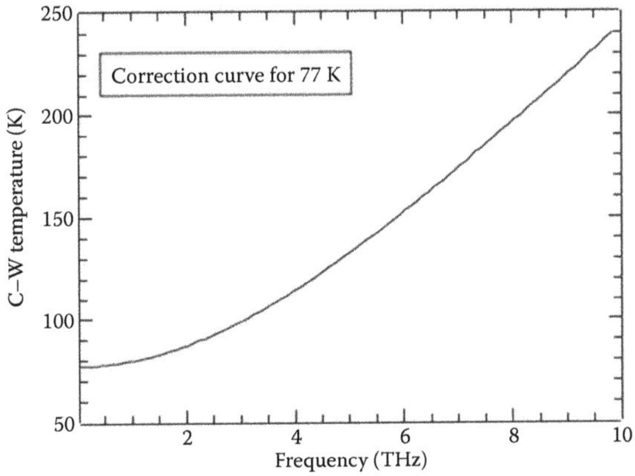

FIGURE 6.14 Callen–Welton black-body correction temperatures corresponding to physical temperatures of 77 K.

6.8 THz MIXERS

THz mixers fall into two broad categories, cryogenic and noncryogenic. Mixers are themselves referred to by the type of mixing device they use. In the cryogenic category, we currently have superconducting–insulator–superconductor (SIS) mixers and hot electron bolometer (HEB) mixers. Both types of devices use the nonlinear properties of superconducting materials to achieve low-noise mixing. In the noncryogenic category, we have Schottky diode mixers, which are based on a semiconductor device and, as such, benefit from cooling, but do not necessarily require it. Considering the weight and/or power associated with cryogenic systems, a Schottky receiver's ability to work uncooled is a huge advantage in space-based or other applications where resources are limited. A plot comparing the DSB noise performance versus frequency of receivers built using these three types of mixers is provided in Figure 6.15.

Unlike incoherent detectors (e.g., bolometers) that only record the intensity of the incoming photons (∝ to number of photons received), heterodyne receivers record both the intensity and phase of the incoming signal. The phase information recovery is made possible by the local oscillator, which effectively "phase stamps" the photons as they arrive. Retention of this phase knowledge is necessary for the downconversion process (Equations 6.9 and 6.10). It also allows one to coherently combine signals from multiple telescopes, either in real time or after the fact (see Chapter 9). However, due to the Heisenberg Uncertainty Principle, retaining knowledge of both amplitude and phase comes at a cost; we cannot measure the incoming energy at any given time to within one photon's worth of energy.

$$E_p = h\nu_p = kT_p \tag{6.36}$$

where

E_p = energy of a single photon (ergs/s)

ν_p = frequency of operation (Hz)

T_p = effective photon temperature (K)

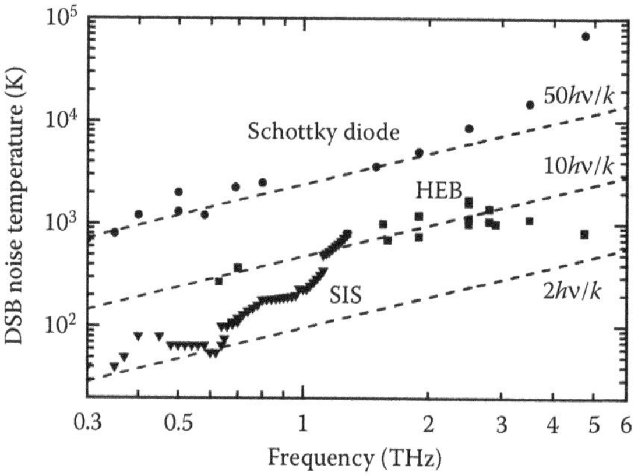

FIGURE 6.15 Measured DSB noise temperatures for Schottky, SIS, and HEB receivers. Curves for 2, 10, and 50 times the quantum noise limit (for SSB operation) are overplotted. (Figure courtesy of H.-W. Hubers. With permission.)

This restriction places a quantum limit on our receiver noise temperature from Equation 6.36 of

$$T_p = \frac{h\nu_p}{k}, \tag{6.37}$$

referenced to one receiver sideband, that is, for SSB operation (Kerr et al., 1996). From Equation 6.37, we see that the quantum noise limit to receiver performance rises steadily with frequency. Equation 6.25 can be used to convert the DSB noise temperatures plotted in Figure 6.15 to SSB noise temperatures. Curves for 2, 10, and 50 times the quantum noise limit are included in the figure. Clearly, there is still room for improvement!

At any specific frequency, Schottky diode mixers have the highest noise and require the greatest amount of LO power (~1 to 3 mW), but can work uncooled, have the widest IF bandwidths, and are very stable. SIS mixers provide the lowest-noise temperatures, can have relatively wide IF bandwidths (~10 GHz), require only modest amounts of LO power (~1 to 3 μW), but have an upper operational frequency limit imposed by the energy gap of their superconducting materials, (e.g., ~1.4 THz for niobium). HEB mixers are relatively low-noise, can work to high frequency, require only small amounts of LO power (~0.1 to 1 μW), but, due to their greater sensitivity to LO power fluctuations, are more difficult to stabilize.

6.8.1 SCHOTTKY DIODE MIXERS

A THz Schottky diode is formed by bringing a small (≤1 μm) diameter metal wire in contact with a semiconductor, as shown in Figure 6.16a. The metal wire is made from gold, platinum, or titanium, while the semiconductor is composed of a GaAs substrate with a few micron thick heavily doped (10^{18} atoms/cm^3) layer, grown on a less doped (10^{17} atom/cm^3) submicron buffer layer (a.k.a. the epilayer). The GaAs substrate and n-type doping is chosen over a silicon substrate and p-type doping because of their greater electron mobility, important for high-frequency operation. When the wire and semiconductor are brought into contact, some of the free electrons from the semiconductor flow to the tip of the wire. This creates a charge depletion region within the semiconductor. The imbalanced charge distribution, in turn, creates an electric field, and, thereby, a diffusion potential, φ, (a.k.a. the built-in voltage/polarity) across the depletion region, which opposes further flow of charge. Electrons attempting to move against the polarity of the depletion region must overcome this energy barrier (a.k.a. Schottky barrier) for conduction to occur. This resistance to the flow of charge in one direction and not the other is what makes the device a diode, capable of rectification. The negative side of the contact (the wire) is called the cathode, and the positive side (the semiconductor) the anode. It is interesting to note that the earliest broadcast radio receivers (ca. 1910–1930) used primitive Schottky diodes. These early devices were often made using a "cat-whisker" cathode and a piece of galena, an abundant, naturally occurring semiconductor, as the anode. With it, one could detect the audio waveform present on top of a low frequency (<~1.6 MHz), amplitude modulated (AM) carrier wave (see Figure 6.16b).

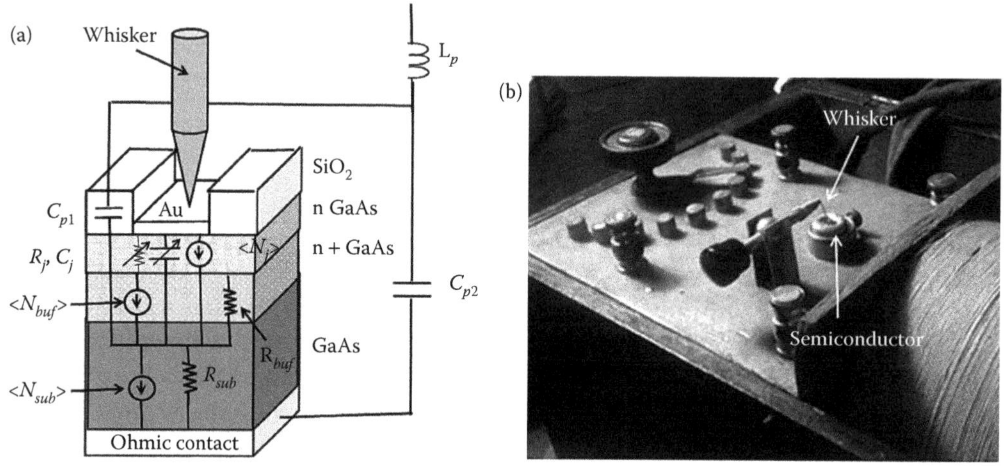

FIGURE 6.16 (a) Diagram and equivalent circuit model of a modern GaAs point contact Schottky diode. (b) Point contact, "cat-whisker" Schottky diode, ca. 1915. The diode was used to rectify the envelope of AM radio broadcast signals (550–1600 kHz). (Photo by author.)

The width, d, of the depletion region can be found from (Maas, 1988)

$$d = \left(\frac{2\phi\varepsilon_s}{qN_d}\right)^{1/2} \tag{6.38}$$

where

d = depletion zone width (m)
ϕ = diffusion voltage across contact (V)
ϵ_s = permittivity of semiconductor (farad/m)
N_d = doping density (atoms/m^3)
q = electron charge (1.6×10^{-19} coulomb)

The width of the depletion region can be modulated by a bias voltage, V_b. When the diode is reversed biased (the plus side of V_b is connected to the cathode), the E field within the depletion region is reinforced (even more electrons flow from the semiconductor to the metal) and the depletion region grows.

$$d = \left(\frac{2(\phi - V)\varepsilon_s}{qN_d}\right)^{1/2} \tag{6.39}$$

When the diode starts to be forward biased, V_b begins working against the junction's intrinsic electric field until it (and the depletion region) collapse, and electrons flood across the diode's junction, constrained only by the diode's series resistance, R_S.

The tip of the metal contact, together with the semiconductor lying beneath, act as plates in a capacitor whose separation is given by d in Equation 6.39. Therefore, the Schottky diode acts as a voltage controlled capacitor (or varactor), with capacitance (Maas, 1988).

$$C(V) = \frac{dQ(V)}{dV} = \frac{C_{jo}}{(1 - (V_b/\phi))^{1/2}} \quad (6.40)$$

where

C_{jo} = zero bias junction capacitance (F)
V_b = bias voltage (V)
ϕ = diffusion voltage across contact (V)

An equivalent small signal circuit model for a Schottky diode is shown in Figure 6.16a.

Charge flow across the Schottky barrier can be through thermionic emission, quantum tunneling, and/or generation–recombination in the vicinity of the barrier. At THz frequencies, conduction appears to occur primarily through the thermionic emission mechanism. For such cases, the I–V characteristic of a Schottky diode takes on an exponential form (Maas, 1988),

$$I(V) = I_{sat}\left[e^{(qV/\eta kT)} - 1\right] \quad (6.41)$$

where

I_{sat} = reverse saturation current (A)

It is the exponential nature of the I–V curve of Equation 6.41 that provides the sharp nonlinearity required for efficient, high sensitivity, mixing (see Figure 6.3).

The upper limit or cutoff frequency, f_c, for Schottky diode operation depends on the junction capacitance and series resistance at the operating bias voltage,

$$f_c = \frac{1}{2\pi R_S C_{jb}} \quad (6.42)$$

where

R_S = total series resistance at operating bias voltage (Ω)
C_{jb} = junction capacitance at the operating bias voltage (V)

Noise in Schottky diode mixers is dominated by noise from the series resistance, R_S, and hot electrons. R_S has contributions from the junction (epitaxial/n–GaAs layer) resistance, R_J, buffer layer (n + GaAs) resistance, R_{buf}, and the substrate (GaAs) resistance, R_{sub}. These resistances and their associated noise sources, $\langle N_J \rangle$, $\langle N_{buf} \rangle$, and $\langle N_{sub} \rangle$ are indicated in Figure 6.16a. Cooling the diode to 20 K can potentially lead to a ~30% improvement in noise temperature (Hubers, 2008). A photograph of a whisker contacted array of Schottky diodes produced at the Jet Propulsion Laboratory or JPL (1996) is shown in Figure 6.17a. More recently, JPL and others (i.e., Virginia diodes, see Figure 6.17b) have used advanced lithographic techniques to produce planar Schottky diodes. These diodes have structures analogous to those of Figure 6.16, but are more robust, compact, and reproducible than their whisker contact forbears.

FIGURE 6.17 Photographs of Schottky diodes. (a) Whisker contact Schottky diode. (Courtesy JPL) and (b) Planar Schottky diode. (Courtesy Virginia Diodes Inc. With permission.)

Whisker contacted Schottky diode mixers in corner cube mounts were the first to be used for heterodyne spectroscopy at THz frequencies. Pioneering observations of both the astrophysically important [CII] and [OI] lines (among others) were first performed on the Kuiper Airborne Observatory using this technology (Boreiko et al., 1988; Boreiko and Betz, 1991). A drawing of a "standard" corner cube mount is shown in Figure 6.18. Here, the whisker is itself bent at a 90° angle with the longest arm serving as a vertical (i.e., monopole) antenna. The horizontal arm attaches to the corner cube apex for support and strain relief. The whisker antenna intercepts photons at an angle θ, both directly and by way of reflection off the walls of the 90° corner cube. The corner cube base serves as the antenna's ground plane. The diode chip itself is mounted to an IF choke/matching network on the end of a coaxial cable. The cable conveys the IF out of the mixer. Brune and Bierschneider (1994) have characterized the Gaussian beam performance of corner cubes with different values for the antenna vertical length, L, and its distance, S, from the cube apex. Their results are provided in Table 6.1. Using their optimized design parameters, the overall power coupling efficiency can be as high as ~82%.

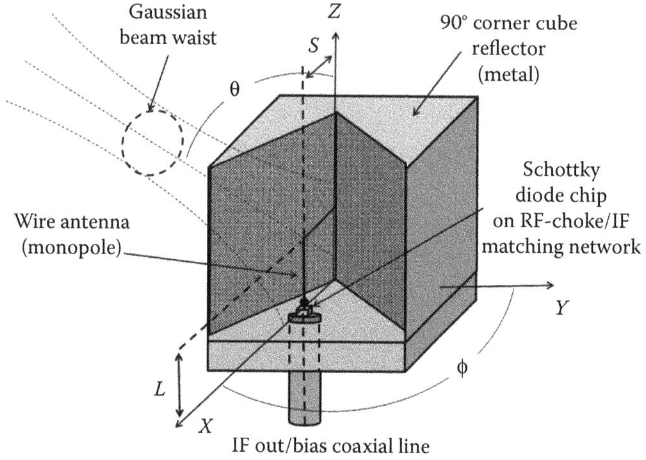

FIGURE 6.18 Schottky corner cube design. A 90° corner cube is used to concentrate an incoming Gaussian beam onto a long wire (1–4 λ) vertical antenna that makes a "whisker" contact to a Schottky diode chip. Here the IF/diode bias cable comes in from the bottom.

TABLE 6.1 Corner Cube Design Parameters

	Design Parameter	Mixer Type: Standard	Mixer Type: Commercial	Mixer Type: Optimum
Antenna parameters	L/λ	4.0	4.0	0.92
	S/λ	1.2	0.93	0.83
Beam parameters	ω_0/λ	1.66	1.46	1.01
	θ (°)	24	26.5	44.5
	ϕ (°)	90	90	90
Optimization results	R_S/Ω	115	91	59
	Quasi-optical coupling (%)	59.5	77.2	88.0
	Overall coupling (%)	59.2	77.0	82.1

Advances in fabrication techniques, computer modeling, and the availability of planar devices, have enabled waveguide and lens/substrate antennas to begin replacing corner cubes as the mount of choice at THz frequencies. These mounts are more robust, compact, and easier to implement into array formats than a corner cube approach. An example of a Schottky waveguide mixer is shown in Figure 6.19. It was designed by Hesler et al. (1997)

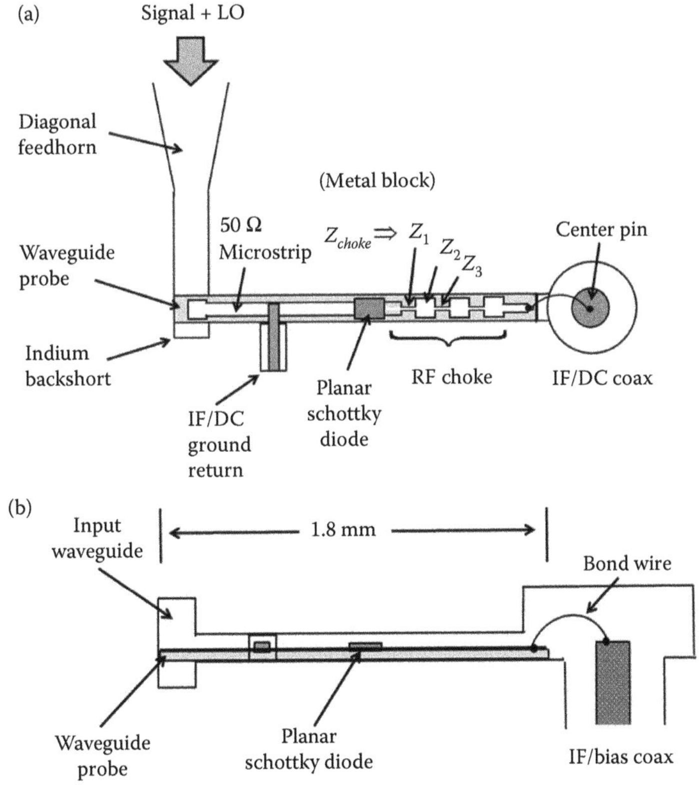

FIGURE 6.19 585–690 GHz Schottky diode waveguide mixer. (a) Side view. (b) Top view. (Adapted from Hesler, J. et al. 1997, *IEEE Trans. Microwave Theory Tech.*, 45(5), 653.)

for operation between 585 and 690 GHz. Before reaching the mixer, the incoming signal and LO (here, an FIR laser) beams are combined quasi-optically on a wire grid using a Martin–Puplett diplexer. Once they arrive at the mixer input, the signal + LO enter a diagonal feedhorn and are funneled into a full-height rectangular (200 × 400 μm) waveguide where they encounter a waveguide probe fabricated using gold microstrip transmission line on a 35 μm thick, 100 μm wide, 1800 μm long quartz substrate. The substrate rests in a form-fitting channel machined into the metal mixer block. From the waveguide probe, the microstrip line extends to the cathode of the planar Schottky diode. The transmission line continues from the diode's anode to a multistage, radio frequency (RF) choke structure composed of alternating high–low impedance sections. The purpose of the structure is to act as a low-pass filter, "choking off" the signal + LO (keeping it from being lost down the microstrip), and letting only the lower frequency (1 to 2 GHz), downconverted IF frequencies from the diode continue on to the coaxial IF output connector.

The choke works by using the impedance transforming properties of a quarter wave transformer (Collins, 1966). The choke first presents a $\lambda_g/4$ long high-impedance (Z_1) section to the diode's output, followed by a $\lambda_g/4$ long low-impedance (Z_2) section. Here, λ_g is the guide wavelength in the microstrip line, given by

$$\lambda_g = \frac{\lambda_0}{n} = \frac{\lambda_0}{\sqrt{\varepsilon_r}} \tag{6.43}$$

where

λ_0 = free-space wavelength
n = index of refraction of substrate material
ϵ_r = relative permittivity of substrate material

Following the low-impedance section, there is another high-impedance (Z_3), $\lambda_g/4$ section. When viewed through Z_1 and Z_2, Z_3 appears to have an even higher impedance (Z_{choke}) given by

$$Z_{choke} = \left(\frac{Z_1}{Z_2}\right)^2 Z_3 \tag{6.44}$$

The greater the value of Z_{choke}, the more of an open circuit the choke structure will appear to be at the output of the diode, and the less leakage of high-frequency photons past the diode there will be. Adding more high–low sections (see Figure 6.19) will further increase the choke structure's efficiency. Seldom are more than three high–low sections required to meet performance specifications.

Waveguide mixers are routinely machined to frequencies of ~2.5 THz using state-of-the-art computer-controlled, numerical mills (Gaidis et al., 2000; Erickson, 2008). Photolithographic and laser micromachining techniques originally developed for the microelectronics industry can be used to fabricate waveguide structures (e.g., feedhorns, couplers, and hybrids) to much higher frequencies, for example, ~5 THz (d'Aubigny et al., 2001; Pütz et al., 2005).

6.8.2 INTRODUCTION TO SUPERCONDUCTIVITY

The most sensitive, lowest-noise mixers at THz frequencies are made from thin superconducting films, usually consisting of niobium or niobium alloys. Superconductivity was discovered experimentally by Kammerlingh Onnes in 1911 (whose lab first liquefied helium in 1908), but was not put on a firm theoretical footing until the work of John Bardeen, Leon Cooper, and Robert Schrieffer (BCS) in 1957. The previous year, Cooper had shown that at very low temperatures, electrons with opposite spin can be bound into pairs by the exchange of a phonon of acoustical energy within a superconductor's crystal lattice. It is the formation of Cooper pairs that make a previously "normal" metal a superconductor. The temperature at which this transition occurs is referred to as the critical temperature, T_c, and is a property of the superconducting material. To see how Cooper pairs form, let us refer to Figure 6.20. Here, we see an elastic lattice of positive ions representing the structure of a superconductor (e.g., niobium). Much lighter, negatively charged, free electrons move about the lattice. As an electron passes by, the positive ions within the lattice are attracted to it and bend towards it, but by that point, the lighter, faster moving electron has already passed by. Therefore, in its acoustic wake, the electron leaves behind a tube of positive charge within the crystal lattice. It is the tube of positive charge that attracts the second electron. In formal superconductivity theory, the formation of these tubes is described as being due to an electron–phonon interaction, where phonons are quantized vibrations within the crystal lattice. The tubes can be orders of magnitude longer than the size of an atom. The second electron also leaves behind it a wake of positive charge. If the two electrons are moving in exactly opposite directions, as is depicted, each electron will be attracted to the positive wake of the other, creating a Cooper pair with binding energy, $2\Delta(0)$, here (Kadin, 1999),

$$2\Delta(0) \approx 3.5\, kT_c \tag{6.45}$$

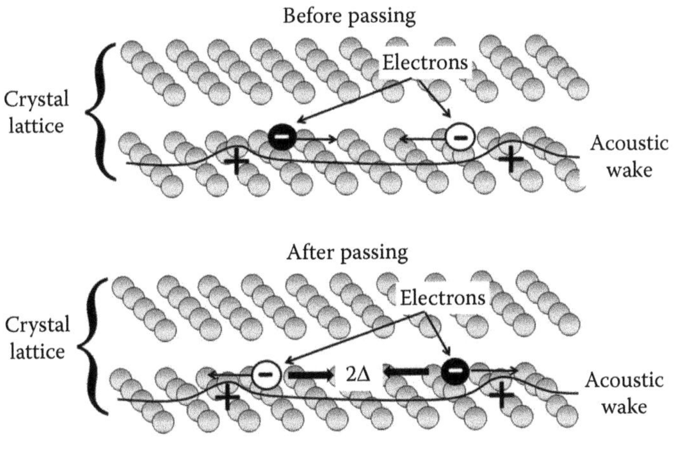

FIGURE 6.20 Cooper pair formation. Cooper pairs form by the attraction between an electron and the positively charge wake left behind in the crystal lattice by a second, passing electron.

where

$2\Delta(0)$ = Copper pair binding energy at 0°K
T_c = critical temperature of superconductor
k = Boltzmann's constant (1.38×10^{-16} erg/K)

As temperatures rise in a superconductor, thermal vibrations in the crystal lattice make it more difficult for Cooper pairs to maintain a strong energy bond, until at temperatures greater than T_c the bond is broken, such that

$$\begin{aligned}\Delta(T) &\approx 1.74\Delta(0)\left(1-\frac{T}{T_c}\right)^{1/2}, \quad T > 0.9T_c \\ &\approx \Delta(0)\left(1-\left(\frac{T}{T_c}\right)^{3.3}\right)^{1/2}, \quad T < T_c \end{aligned} \tag{6.46}$$

Good conducting materials (e.g., silver, gold, and copper) have minimal interaction between free electrons and the lattice at normal temperatures, meaning the lattice appears electrically "soft" to the electrons (due to the constituent atoms readily losing outer shell electrons), and there are relatively few collisions. The material will, therefore, have a low resistance. However, it also means the lattice will be less able to create and propagate the phonons necessary for Cooper pair formation at low temperatures. Consequently, metals that are poor conductors due to electron–lattice collisions at normal temperatures make the best superconductors. The free electrons (sometimes described as an electron cloud) traveling within the superconductor, and the ions forming the crystal lattice, can be thought of as constituting two separate populations of charge particles, each capable of obtaining a different equilibrium temperature, T_e (electron cloud) and T_l (crystal lattice).

The distance over which the electrons in a Cooper pair remain bound (i.e., correlated) is referred to as the coherence length, ξ_o. It is also the characteristic length of the acoustic wake or tube of positive charge left by the passing electrons,

$$\xi_o = \frac{hv_f}{\pi^2\Delta(T)} \tag{6.47}$$

where

ξ_o = coherence length (cm)
h = Planck's constant (6.626×10^{-27} cm² g s⁻¹)
v_f = Cooper pair velocity (cm/s)

The coherence length can vary a great deal from one superconductor to the next (e.g., from 38 nm in niobium to 1600 nm in aluminum).

There are two common misconceptions concerning Cooper pairs (Orlando and Delin, 1991). One is that they are bound together tightly and maintain a monogamous relationship; this is not the case. Let us assume we have a Cooper pair composed of electrons "A" and "B," as depicted in Figure 6.21. Electrons are always moving within a superconductor. Once A and B are separated by more than ξ_o, they are no longer guided by the positive charge of their

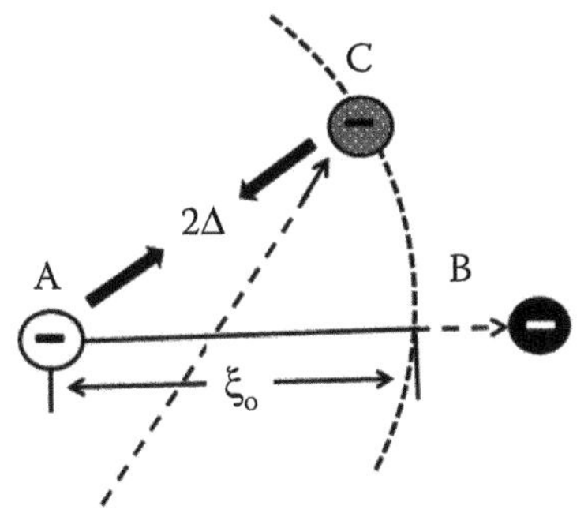

FIGURE 6.21 Cooper pair partner swap. Once the distance between electrons A and B exceeds the coherence length, ξ_o, electron A forms a Cooper pair with electron C. Likewise, electron B will find a new partner.

mutual wake, and become uncorrelated. However, now electron A is attracted by the positive wake left behind by electron C, and a new Cooper pair is formed. Since the coherence length is so large compared to atomic distances, the probability is very high that electron B will have found another partner also. At any given time, we do not know who a particular electron is partnered with, only that it has a partner. Either directly or indirectly, all the Cooper pairs overlap and become correlated, in essence bringing the whole superconductor into the same energy state (i.e., sharing the same eigenstate). This is why superconductors appear to exhibit quantum mechanical properties over macroscopic distances (e.g., levitation). Electrons, protons, and neutrons are classified as spin 1/2 particles, known as fermions. Particles with zero spin, such as photons, are categorized as bosons. Fermions are subject to the Pauli Exclusion Principle, which states that only two fermions, with opposite spins, can be placed in the same energy level. This is why atoms with many electrons form shells of different energies about the nucleus. However, Cooper pairs form a "particle" with a net spin of zero, so they can be treated as bosons. Being bosons, all the Cooper pairs within a superconductor can exist in the same energy state (Kadin, 1999).

This leads to a second misconception—that electrons in a Cooper pair do not undergo scattering, when, in fact, they do. In a normal metal collisional scattering of electrons within the crystal lattice creates disorder (a drag force), which leads to resistance. In a superconductor, the paired electrons will bounce and scatter, but at each scattering event they simply change partners without a loss of energy. Since all Cooper pair partners share the same eigenstate, the changeover occurs in an orderly (correlated) fashion, with the electrons able to carry current without experiencing resistance (Orlando and Delin, 1991).

Magnetism in materials (e.g., iron, nickel, and cobalt) will arise when electrons in the outer orbitals of constituent atoms are unpaired, such that the magnetic field generated by an electron spinning on its axis is not cancelled out by another electron within the same orbital with opposite spin. As discussed above, Cooper pairs form from electrons with opposite spins states; therefore, superconductors are nonmagnetic and will actually work to expel any magnetic field permeating the material when it goes superconducting, that is, when $T \leq T_c$. The magnetic field on the surface of a superconductor does not instantaneously go to zero, but will exponentially decay, such that

$$B(z) = B(0)e^{(-z/\lambda_L)} \quad (6.48)$$

where

$B(z)$ = magnetic field density as a function of depth z into superconductor (webers/m^2 or tesla)

$B(0)$ = magnetic field density at surface (webers/m^2 or tesla)

λ_L = the London penetration depth (m)

The London penetration depth, λ_L, is analogous to the skin depth of an electric field in metals and is given by

$$\lambda_L = \sqrt{\frac{\varepsilon_0 m_e c^2}{ne^2}} \tag{6.49}$$

where

ε_0 = permittivity of free space (8.854×10^{-12} m^{-3} kg^{-1} s^4 A^2)

m_e = electron mass ($9.10938291 \times 10^{-31}$ kg)

c = speed of light (3×10^8 m s^{-2})

e = charge of an electron (1.6×10^{-19} C)

n = superconducting electron density (m^{-3})

For niobium, λ_L has a value of 30 nm.

If an ambient magnetic field is strong enough, it can rip the Cooper pairs within a superconductor apart and cause the material to go normal. The value of magnetic field capable of breaking Cooper pairs is called the critical magnetic field density, B_c. Its value depends upon the binding energy, Δ, of the Cooper pairs, and is, therefore, different for different types of superconductors and changes with temperature, according to

$$B_c(T) \approx B_c(0)\left[1 - \left(\frac{T}{T_c}\right)^2\right] \tag{6.50}$$

where

$B_c(0)$ = critical magnetic field density at 0°K (webers/m^2 or tesla (T))

T = operating temperature (K)

T_c = critical temperature of superconductor

For example, in the case of niobium $B_c(0) = 0.2$ T and $T_c = 9.5$ K, while for NbN and NbTi the corresponding values are 1.5 T and 15.7 K, and 15 T and 10 K, respectively. All three of these materials are classified as Type II superconductors (a.k.a. "hard superconductors"), meaning, they usually exist in a mixed (or vortex) state of normal and superconducting regions. Within these materials, superconducting currents surround regions of normal material in vortices. Type I superconductors are "softer" with lower values of T_c and $B_c(0)$. Notable among Type I superconductors for THz detectors are Pb, Ti, Al, and Ta, with T_c's of 7.19, 0.39, 1.2, and 4.47°K, respectively. At THz frequencies, two types of superconducting mixers are currently in use, ones using HEBs as the mixing device, and those that use an SIS as the mixing device.

6.8.3 SUPERCONDUCTOR-INSULATOR-SUPERCONDUCTOR (SIS) MIXERS

SIS devices are made by fabricating a superconducting "sandwich," with two thin, superconducting films being separated by a thin insulating layer (see Figure 6.22). The geometry of an SIS device is essentially that of a capacitor. If the thickness, t, of the insulating layer is made less than the coherence length, ξ_o, of the superconducting material, then electrons on one side of the insulating barrier can form Cooper pairs with electrons on the other side. Since all Cooper pairs share the same energy (eigenstate), Cooper pairs can quantum mechanically tunnel across the barrier, even with no voltage applied! Another way of stating this is that t is made thin enough so that the quantum wave functions of the two superconductors, Ψ_1 and Ψ_2 overlap, permitting the exchange of Cooper pairs. The tunneling of Cooper pairs at zero bias voltage produces the Josephson supercurrent observed on the I–V curve of SIS devices. The maximum value of the supercurrent is a function of the junction area and the maximum dc current density that can passed through the device at zero voltage, the Josephson critical current density J_c. The value of J_c is a function of the carrier density in the superconductors and the type and thickness of the insulating layer used (Orlando and Delin, 1991).

$$J_c = \frac{eh\sqrt{n_1 n_2}}{4\pi m_e \varsigma \sinh(t/\varsigma)} \tag{6.51}$$

where

e = charge of an electron (1.6×10^{-19} C)
h = Planck's constant (6.63×10^{-34} J · s)
m_e = electron mass (9.1×10^{-31} kg)
n_1 = electron density in superconductors 1 (m^{-3})
n_2 = electron density in superconductors 2 (m^{-3})
t = insulator thickness (m)
ς = decay length of quantum wave function Ψ in insulator (typically a fraction of a nanometer).

On the I–V curve, the Josephson supercurrent should appear as a vertical spike (see Figure 6.23) with height

$$I_c = J_c A_j \tag{6.52}$$

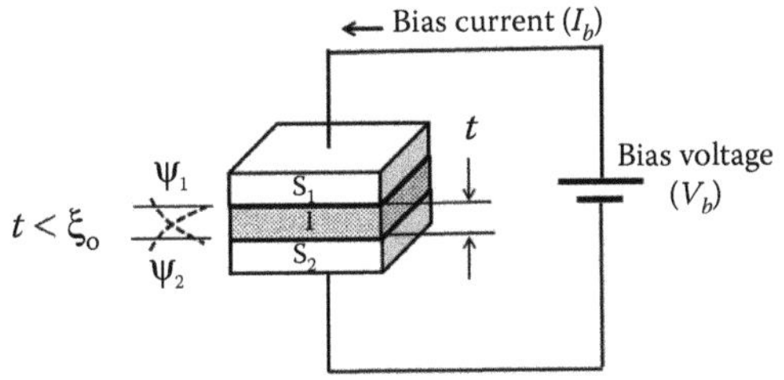

FIGURE 6.22 SIS device geometry. SIS junctions work by placing two superconducting films, S_1 and S_2, within one coherence length, ξ_o, (binding distance of a Cooper pair) of each other. The separation is set by the thickness, t, of an insulating layer, I. By doing so, the wave functions, Ψ_1 and Ψ_2, of the two superconductors overlap, allowing the possibility of quantum mechanical tunneling of Cooper pairs and single electrons to take place between the superconductors.

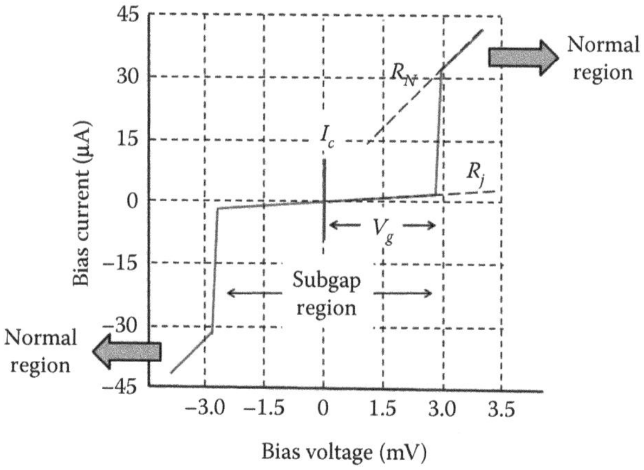

FIGURE 6.23 Unpumped SIS device I–V curve. At zero bias voltage, $V_b = 0$, the Cooper pairs in superconductors S_1 and S_2 share the same energy state. This allows Cooper pairs to travel between the superconductors even when no voltage is applied. This flow of charge is referred to as the Josephson supercurrent, which reaches a value of I_c. As V_b is increased (or decreased), the two superconductors no longer share the same energy state, and only a small leakage current will flow between them; the SIS junction is in an "OFF" state. As the magnitude of V_b is increased, a point is finally reached where the stress on the Cooper pairs is greater than the Cooper pair binding energy, causing them to break; the junction turns "ON." The breaking up of Cooper pairs produces a rush of current at $V_b = V_g$, which continues until all the pairs are broken and the device goes normal.

where

I_c = peak value of Josephson supercurrent (A)
J_c = Josephson current critical density (A/cm^2)
A_j = junction area (cm^2)

If the supercurrent spike on the I–V curve has a slope, this is indicative of a DC contact resistance on the bias lines.

When the bias voltage, V_b, is increased (or decreased) from zero, corresponding to a change in potential energy $P = eV_b$, Cooper pairs on either side of the insulating layer no longer share the same energy state, with the result that Cooper pair tunneling and their associated dc Josephson supercurrent cease; in effect the SIS junction "turns off." As the bias voltage is further increased, a small DC leakage current, due to imperfections in the insulating layer and a small number of thermally excited electrons (with enough excess energy to break their Cooper pairs), is observed through the device until the point is reached where the potential energy associated with the bias voltage equals the binding energy of the Cooper pairs, $P = 2\Delta$. This bias voltage is referred to as the energy gap voltage, V_g,

$$V_g = \frac{2\Delta}{e} \tag{6.53}$$

where

V_g = gap voltage (V)
2Δ = Copper pair binding energy at operating temperature (J)
e = charge of an electron (1.6×10^{-19} C)

Leakage current adds shot-noise to the overall mixer performance and should be minimized. The slope of the subgap current on the I–V curve has the form of a resistance R_j, and serves as a figure of merit for a junction (see Figure 6.23). High values of R_j could indicate a poor quality or damaged insulating layer.

Below V_g the Cooper pairs respond to the applied voltage by oscillating back and forth across the junction at frequency f_J, which increases linearly with V_b.

$$f_J = 483.6 \times 10^{12} V_b (\text{Hz}) \tag{6.54}$$

This is the ac Josephson effect. In the millivolt range common to superconductors within this subgap region, the ac Josephson effect can produce microwave radiation at 100s of GHz. The ac Josephson effect has been used for a variety of purposes, including voltage standards and as LO sources.

For $V_b \geq V_g$, the Cooper pairs start to break and there is a sudden rush of electrons across the insulator, producing a very sharp step in I–V curve. The SIS junction "turns on." In short order, all the Cooper pairs are broken, and the device goes normal, with the total flow of current limited by the transparency (i.e., resistivity, R_N, in Figure 6.23) of the insulating layer. The SIS uses this sharp transition between normal and superconducting conditions to achieve the highly nonlinear I–V curve required for efficient mixing. The I–V curve is symmetric about $V_b = 0$V, with a total voltage swing between the two knees in the I–V curve equal to $V_{max} = 4\Delta/e$. As we will see, this sets the upper frequency limit, f_{max}, of operation of an SIS device to

$$f_{max} \approx \frac{eV_{max}}{h} \tag{6.55}$$

where

f_{max} = upper frequency limit of operation (Hz)
$V_{max} = 2V_g$ = total voltage swing between knees in I–V curve (V)
e = charge of an electron (1.6×10^{-19} C)
h = Planck's constant (6.63×10^{-34} J · s)

One can think of the bias voltage, V_b, as putting a stress on the Cooper pairs. Below the knee in the I–V curve, such that $V_b < V_g$, the stress is not sufficient to break the pairs. However, an incoming photon with an effective voltage,

$$V_p = \frac{h\nu}{e} \geq V_{gap} - V_b \tag{6.56}$$

can provide the additional force necessary to break a Cooper pair and allow tunneling of the newly liberated electron (i.e., quasi-particle) across the insulating barrier. If the absorption of the energy provided by a single photon is not sufficient, the instantaneous absorption of a number, n, of photons can provide the required energy, 2Δ, for pair breaking. For pair breaking to occur, it requires

$$\frac{nh\nu}{e} \geq V_g - V_b \tag{6.57}$$

However, the probability P_n that n photons will be absorbed simultaneously goes down with increasing n, such that

$$P_n \propto J_n^2(\alpha) \tag{6.58}$$

where

$J_n(\alpha)$ = Bessel function of order n
$\alpha = eV_\omega/hf$ = pumping factor
V_ω = peak amplitude of applied signal

We are now in a position to see what happens to the I–V curve of an SIS device in the presence of a continuous LO signal. The applied bias and LO voltages across the junction combine to produce a time-dependent voltage, $V_j(t)$, (Tucker and Feldman, 1985; Billade, 2013),

$$V_j(t) = V_b + V_\omega^{LO}\cos(\omega t) \tag{6.59}$$

where

$V_j(t)$ = total voltage across junction (V)
V_b = DC bias voltage (V)
V_ω^{LO} = peak amplitude of LO voltage (V)
$\omega = 2\pi f$ = LO angular frequency (rad/s)

The resulting I–V curve can be modeled as the superposition of the unmodulated (i.e., no LO) DC bias curve with n versions of itself, each shifted along the voltage axis by an amount nV_p and multiplied by the probability, P_n, such that

$$I_{bias}(V_{bias}, V_p) = \sum_{n=-\infty}^{\infty} [P_n \cdot I_{DC}(V_{bias} + nV_p)] \tag{6.60}$$

The "width" of the I–V curve essentially appears to shrink by an amount hf/e with the addition of each new nV_p. A graphical representation of Equation 6.60 is shown in Figure 6.24. The resultant LO pumped I–V curve is the outer envelope of this family of curves and resembles a staircase. The higher the frequency of the LO, the greater will be V_p and the *wider* each "photon step" will appear. The greater the LO pump power, the greater α will be and the *higher* each photon step will appear. The best mixer performance is generally observed when the SIS device is voltage biased 1/2 photon step below the knee of the I–V curve, where $V_b = V_g - hf/2e$, and is pumped with enough LO power that the resultant LO staircase intercepts the unpumped I–V curve at ≈1/3 the current needed to make the device go normal. Indeed, with sufficiently high LO power, the SIS device will be driven normal and the I–V curve will appear as a diagonal line.

Superimposed on top of the hf/e photon steps caused by single quasi-particle tunneling are a second set of steps with width $hf/2e$. These are called "Shapiro steps" and are due to the frequency modulation of the ac Josephson current by the incident LO signal

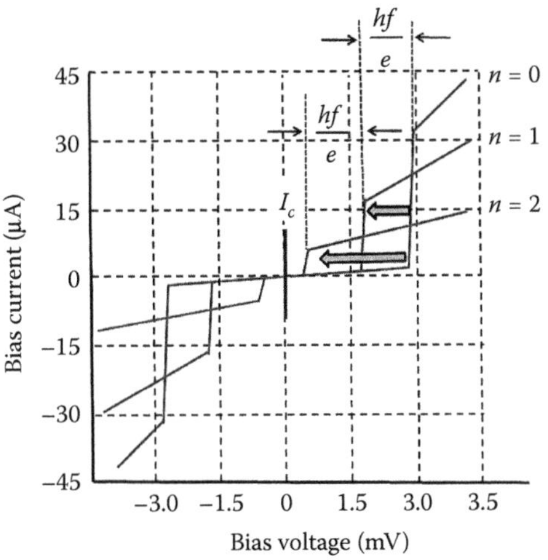

FIGURE 6.24 Formation of pumped SIS device I–V curve. An SIS device can be made to partially turn "ON" at bias voltages less than V_g by irradiating it with photons with a frequency $\leq(eV_g/h)$. These photons effectively shift the normal part of the I–V curve back toward 0V by an amount (nhf/e), where f is the frequency and n is the number of photons "ganging up" to provide the voltage kick at a given instant in time. The more photons required to be in one place at one time to provide a given voltage kick, the less often the event will be, so the corresponding junction current, I_j, will be lower.

(Equation 6.50; Shapiro, 1963). Shapiro steps can be a source of instability in an SIS receiver system and, unfortunately, are located just where you would like to bias your mixer. Mixers optimized to utilize the two-particle Cooper pair tunneling associated with Shapiro steps are called Josephson-effect mixers. These mixers are found to have higher noise and be less stable than their quasi-particle counterparts (Schoelkopf et al., 1995). Fortunately, both Cooper pair tunneling and the associated Shapiro steps can be suppressed by the application of a magnetic field across the SIS junction.

When a magnetic field is applied to an SIS junction, it induces a phase difference, ϕ_o, between the quantum mechanical wave functions of superconductors 1 and 2, Ψ_1 and Ψ_2 (see Figure 6.22), such that the intensity of the Josephson supercurrent, I_c, passing through the junction is modulated much like a beam of light passing through a slit with the same thickness, t, as the insulating barrier (Barone and Paterno, 1982). Namely,

$$I_c(B) = I_c(0)\left|\frac{\sin\pi(\phi/\phi_o)}{\pi(\phi/\phi_o)}\right| \tag{6.61}$$

where

$\phi = BLD$ = magnetic flux passing through the junction (Wb)
$\phi_o = hc/2e$ = the flux quantum = 2.07×10^{-15} Wb = 2.07×10^{-7} gauss cm^2)
L = the length of the junction (m)
$d = (2\lambda_L + t)$ = magnetic penetration assuming both superconductors are the same (m)
B = magnetic flux density in the vicinity of the junction (webers)
$I_c(0)$ = Josephson supercurrent with no applied magnetic field (A)

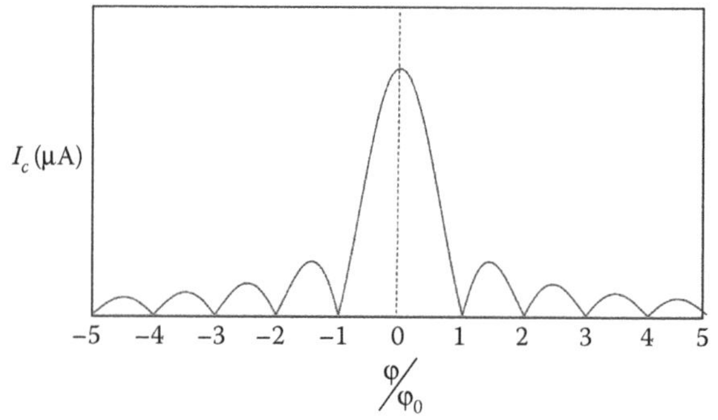

FIGURE 6.25 Magnetic field modulation of Josephson supercurrent. When the SIS junction of Figure 6.22 is immersed in a magnetic.

A plot of $I_c(B)$ in units of φ/φ_0 is shown in Figure 6.25. Examination of the plot reveals a classic (sin x/x) modulation of the supercurrent, $I_c(B)$, with a peak at $\varphi/\varphi_0 = 0$ and minima at $\varphi = n\varphi_0$. For low-noise, stable operation of an SIS receiver at THz frequencies, the SIS junction is typically bathed in a magnetic field capable of placing the junction in an $n = 1$ or 2 supercurrent minimum. In a practical mixer, an electromagnet is often used to generate the required field, which is then conveyed to the SIS device by ferromagnetic (i.e., high magnetic permeability) concentrators. With the proper level of magnetic field, the Josephson supercurrent, Shapiro steps, and associated instabilities can be minimized or eliminated.

EXAMPLE 6.7

The gap voltage of a commonly used Nb/Al-oxide/Nb SIS device is $V_g \approx 2.4$ mV. What is the maximum frequency at which the device can be used to provide low-noise performance?

Substituting into Equation 6.55, we find,

$$f_{\max} \approx \frac{2eV_g}{h} \approx \frac{2(1.6 \times 10^{-19}\,\mathrm{C})(2.4 \times 10^{-3}\,\mathrm{V})}{6.63 \times 10^{-34}\,\mathrm{J \cdot s}}$$

$$\approx 1.16\,\mathrm{THz}$$

EXAMPLE 6.8

A photon step on an SIS I–V curve has a width of ~2 mV. What is the approximate frequency of the LO?

The voltage width of an LO step is given by,

$$V_{step}^{LO} \approx \frac{hf_{LO}}{e}$$

Rearranging for f_{LO}, we have

$$f_{LO} \approx \frac{eV_{step}^{LO}}{h} \approx \frac{(1.6 \times 10^{-19}\,\mathrm{C})(2 \times 10^{-3}\,\mathrm{V})}{6.63 \times 10^{-34}\,\mathrm{J \cdot s}}$$

$$\approx 0.483\,\mathrm{THz}$$

6.8.4 HOT ELECTRON BOLOMETER (HEB) MIXERS

HEBs are thermal (i.e., bolometric) devices, whose resistance depends on temperature. They are formed by fabricating a short (~1 μm), thin (~0.1 μm), superconducting bridge between two normal (e.g., gold) electrodes. The theory of operation for an HEB mixer is shown pictorially in Figure 6.26. The input and local oscillator signals are conveyed to an HEB either quasi-optically or via waveguide, and enter the bridge through contact pads, which form the base of the antenna.

The I–V curve of an HEB is shown in Figure 6.27. Similar to the SIS junction, at zero bias voltage, V_B, the HEB is a short circuit. As the magnitude of V_B is increased from zero (either in the positive or negative direction) Cooper pairs within the bridge begin to break and the device no longer behaves as a pure superconductor. The HEB current, I_B, initially remains constant and then begins to increase as the bridge transitions to being a normal resistor. To operate as a mixer, the HEB is biased so that the combination of DC bias, LO power, and bath temperature place it on a nonlinear transition between a normal and superconducting state. In this transition region, the central part of the bridge is heated to its critical temperature, T_c, and driven normal, while adjacent ends of the bridge remain superconducting. The region of the bridge that is driven normal is referred to as the "hotspot" and has a length L_H in Figure 6.26. The time-varying heating associated with the incident LO and signal, modulate the size of the normal region and, thereby, the bridge's conductivity, at the intermediate frequency (IF). It is this modulation that yields the downconverted signal that is passed on to a low-noise IF amplifier. How high an IF frequency an HEB mixer can support is determined by how fast heat can be transferred out of the bridge, either by electron diffusion through the contact pads at the ends or by electron–phonon coupling to the crystal lattice in the substrate material. Being thermodynamic devices, HEBs do not suffer from the same high-frequency limit imposed by quantum mechanics on an SIS mixer, thereby permitting their use at "super-terahertz" frequencies ($f > 2$ THz).

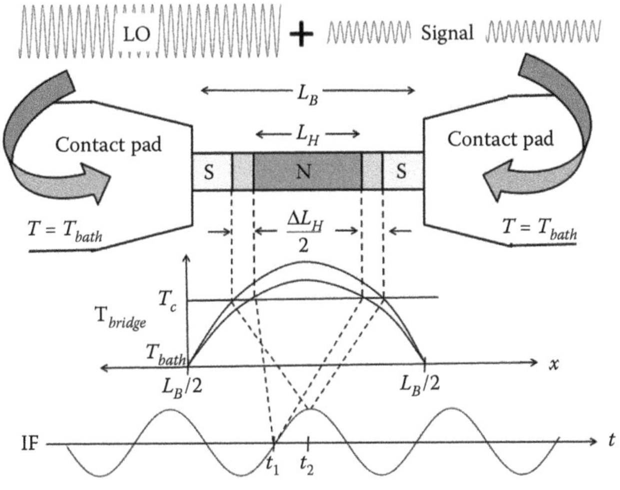

FIGURE 6.26 Hot-spot model of HEB mixing. The signal + LO power is conveyed to the HEB bridge through contact pads on either end. The bridge is biased so that a "hot spot" is formed, over which the bridge goes normal. The incoming signal + LO modulates the size of the "hot spot" at the intermediate frequency (IF).

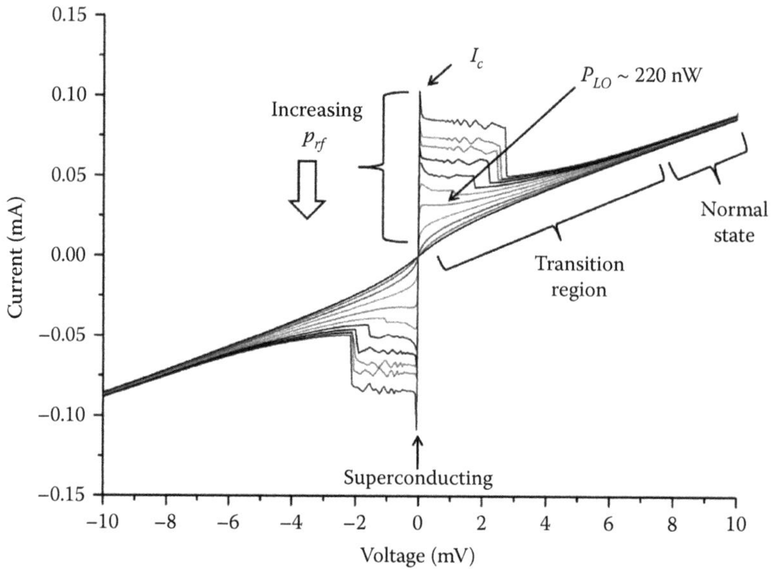

FIGURE 6.27 HEB I–V curve. At zero bias voltage (V_b = 0V), the HEB behaves as a short circuit with a maximum critical current, I_c. As V_b is increased from 0V, the HEB bridge begins to heat up, driving the transition from a superconducting to normal state. Incident RF (i.e., LO) power will hasten the transition, reducing the amount of current that can flow through the device at a given bias voltage.

The heat transfer through a segment, Δx, of an HEB bridge is illustrated in Figure 6.28. Inside the "hot spot," the corresponding heat transfer equation can be written as (Wilms Floet et al., 1999; Wilms Floet, 2001)

$$-K\frac{d^2T}{dx} + \frac{c_e}{\tau_{e-ph}}(T - T_b) = j^2\rho + p_{rf} \tag{6.62}$$

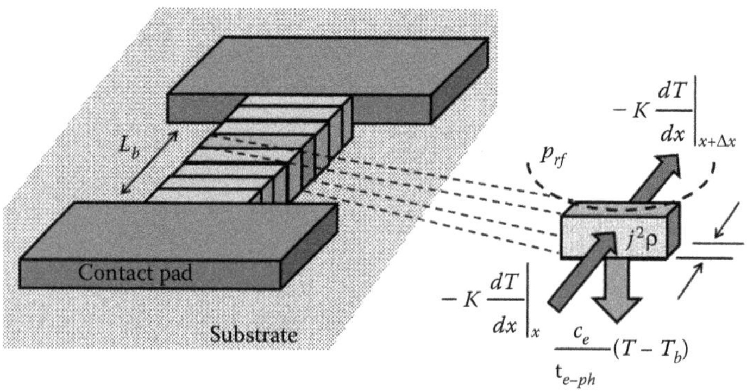

FIGURE 6.28 HEB thermodynamics. The heat flow through the bridge of an HEB is a function of the applied bias voltage (V_b), incident RF power RF (p_{rf}), electron heat capacity (c_e), electron–phonon interaction time (τ_{eph}), and bath temperature (T_b).

and outside the "hot spot" as

$$-K\frac{d^2T}{dx} + \frac{c_e}{\tau_{e-ph}}(T - T_b) = p_{rf} \tag{6.63}$$

where

K = thermal conductivity ($WK^{-1}\ cm^{-1}$)
 $= (\pi^2k^2/3e^2)(T/\rho)$: Wiedemann–Franz Law
ρ = film resistivity (Ω/m)
 $= (Ne^2l_e/m_e v_F) \propto V_b^{-(1/2)}$
c_e = electron heat capacity (JK^{-1})
 $= (N\pi^2k^2/2E_F)T \propto V_b^{-1}$
E_F = Fermi level: potential energy of electrons (J)
 $= (m_e v_F^2/2) = eV_b$
V_b = bias voltage (V)
$\tau_{e\text{-}ph}$ = electron–phonon interaction time (s)
p_{rf} = incident LO power density (Wm^{-3})
N = electron density (m^{-3})
l_e = electron mean free path (m)
T_b = bath temperature (4.3 K for helium at 1 atmosphere)

The distance over which a hot electron will travel in a microbridge before returning its excess energy to the He bath, is called the thermal healing length, λ_{th} (Skocpol et al., 1974), and is given by the ratio between the conduction of heat within the microbridge itself to the heat transfer coefficient to the substrate.

$$\lambda_{th} = \left[\frac{K\tau_{e-ph}}{c_e}\right]^{1/2} = \left[\frac{v_F l_e \tau_{e-ph}}{3}\right]^{1/2} = \left[D\tau_{e-ph}\right]^{1/2} \propto V_b^{1/4} \tag{6.64}$$

where D is the electron diffusion constant ($cm^2\ s^{-1}$).

If the microbridge length is greater than the healing length, $L_b > \lambda_{th}$, electron–phonon interactions will dominate the cooling process and the device is referred to as a phonon-cooled HEB. On the other hand, if $L_b < \lambda_{th}$, diffusion through the end contacts is the dominate cooling mechanism, the device is referred to as a diffusion-cooled HEB. In many instances, the cooling of an HEB may contain elements of both mechanisms.

Under the assumption the contacts are able to keep the ends of the bridge at the helium bath temperature T_b and the temperature at the boundaries of the hot spot is T_c, an expression for the I–V curve of an HEB can be derived from the solution to the heat-transfer equation, Equation 6.54 (Wilms Floet et al., 1999),

$$j(V_b, p_{rf}) = \left(\frac{c_e}{\rho\tau_{e-ph}}\left(A - \frac{A\cosh(B) + (p_{rf}\tau_{e-ph}/\rho)}{\sinh(B)\tanh(L_h/\lambda_{th})}\right)\right)^{1/2} \tag{6.65}$$

where

$$A = T_c - T_b - \frac{p_{rf}\tau_{e-ph}}{\rho}$$

$$B = \frac{2L_h - L_b}{2\lambda_{th}}$$

and j = bridge current density (A cm^{-2}).

Both ρ and c_e (and, therefore, λ_{th}) in Equation 6.65 depend on the bias voltage V_b. The interplay of voltage dependences in Equation 6.65 yields a negative resistance at low values of bias voltage (e.g., $V_b < 0.5$ mV for NbN bridges) and p_{rf}. As p_{rf} and/or V_b are increased, the slope of the I–V curve switches from negative to positive (see Figure 6.27). Optimum mixer performance is typically achieved at a V_b between 0.5 and 1 mV (for NbN).

Maintaining the proper balance between V_b, p_{rf}, and T_b, can be a challenge, and often limits the overall stability of an HEB receiver system and, thereby, how long one may integrate on a signal before needing to recalibrate (i.e., the system's Allan time). For HEB mixers, Allan times can vary from a few seconds to ~30 s, depending on the desired spectral resolution and type of thermal stabilization used. Being bolometric devices, thermal time constants (e.g., τ_{e-p}) determine the mixer's response time and, consequently, the IF bandwidth. The IF bandwidth is important because it determines how much of an astrophysical spectrum can be downconverted at any given time. This is especially true at THz frequencies, where the Doppler effect produced by the rotation of a galaxy (e.g., the Milky Way) can spread the spectral region of interest out over many GHz. Depending on the material and electron cooling mechanism (diffusion or phonon) employed, HEB IF bandwidths can be between 1 and ~6 GHz (see, for example, Figure 6.29). Due to their smaller λ_{th} values, diffusion-cooled HEBs are expected to support higher IF frequencies than their phonon-cooled counterparts.

6.9 THz LOCAL OSCILLATORS

The local oscillator (LO) module of a heterodyne receiver provides a tone by which the incoming signal is multiplied (i.e., modulated) within the mixer. There are four requirements placed on this tone.

1. The LO tone must be at a frequency, ν_{LO}, such that $\nu_{LO} = \nu_S \pm \nu_{IF}$, where ν_S is the frequency of the signal being downconverted, and ν_{IF} is the intermediate frequency (IF) output of the mixer. The higher the IF frequency a mixer can support, the easier it is to find an LO source that can meet this specification. Schottky mixers can support IFs of 10s of GHz, SIS mixers typically up to 10 GHz, and HEBs between 3 and 6 GHz.
2. The LO tone must have sufficient power to swing the mixer's I–V curve back and forth between an "ON" and "OFF" state (see Figure 6.5). For example, Schottky mixers typically require ~0.5 to 1 mW of LO power (Erickson and Goyette, 2009), while SIS and HEB devices often require ~1 to 10 μW (Belitsky, 1999) and ~0.1 to 0.5 μW of LO power (Baselmans et al., 2004), respectively (depending on device characteristics

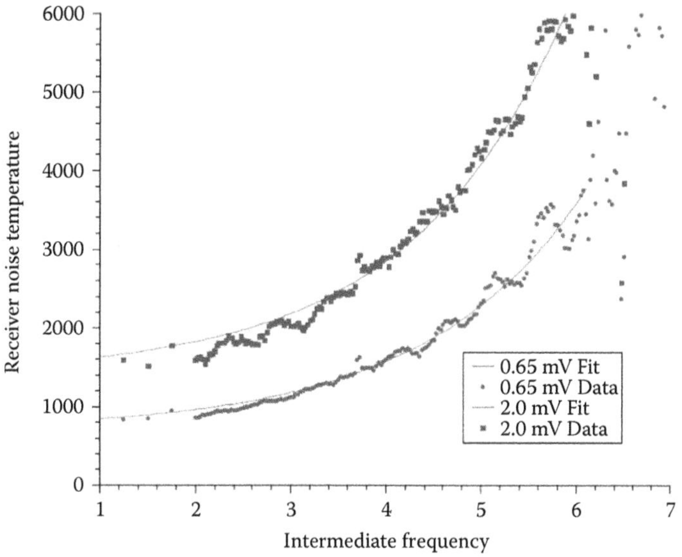

FIGURE 6.29 HEB IF bandwidth. Driven by thermodynamics, the upper intermediate frequency (IF) supported by an HEB, is limited by diffusion and electron–phonon interaction times. The shorter the HEB bridge and the better it is heat-sunk to the substrate, the higher the allowed IF frequency will be. The sensitivity of the HEB decreases with increasing IF because it does not have time to dump all the heat from the absorption of one set of LO and signal photons before the next set comes along. Above are curves showing the IF response of a $2 \times 0.2\ \mu m^2$ NbN (phonon-cooled) HEB at 4.7 THz. (Adapted from Kloosterman, J., 2014, Heterodyne Arrays for TeraHertz Astronomy, PhD dissertation, University of Arizona.)

and frequency). The listed power levels are referenced to the mixer input. When specifying the output power of an LO, one should include any losses associated with the LO injection (i.e., diplexing) scheme, which can be quite high (90% to 99% for a dielectric beam splitter).

3. The LO tone intensity at the mixer should be sufficiently stable that it does not negatively impact the desired Allan time of the receiver. The amount of LO power variation that can be tolerated depends, to a large extent, on the type of mixer being used. The higher LO power requirement of Schottky mixers and the quantum nature of SIS mixers makes them less susceptible to low-level power variations than the thermodynamically balanced HEB mixer.
4. The LO tone must have sufficient spectral purity (frequency and/or phase) for the desired application. For single dish THz spectroscopy, a rule of thumb is that the frequency stability of an LO should be more than ~5 times better than the linewidth to be measured. In interstellar medium (ISM) studies, the linewidth specification is often set by the turbulent velocity dispersion within the object of interest, typically $\Delta \upsilon \sim 1$ km/s for a Milky Way cloud. A velocity dispersion can be converted to a frequency dispersion (i.e., linewidth in Hz) using the relationship,

$$\Delta \upsilon = \frac{\Delta \nu}{\nu_0} c \tag{6.66}$$

where

$\Delta \upsilon$ = linewidth (m/s)

ν_0 = line frequency (Hz)

$\Delta \nu$ = linewidth (Hz)

c = speed of light (m/s)

The demand for spectral purity is seen to decrease with increasing frequency. For example, in the case of the ^{12}CO $J = 3 \rightarrow 2$ line at 345 GHz, we find that, for a $\Delta \upsilon \sim 1$ km/s, the corresponding linewidth is 1.1 MHz, suggesting an LO frequency stability of ~0.2 MHz would be sufficient. While for the [CII] line at 1.9 THz, the corresponding linewidth is 6.3 MHz, suggesting an LO frequency stability of only ~1 MHz is necessary.

There are numerous device technologies capable of providing LO power at THz frequencies. The frequency coverage and output power available from these devices is summarized in Figure 6.30 (Mehdi et al., 2008).

High electron mobility transistor amplifiers have recently provided gain up to ~700 GHz (see Section 6.6). Photomixers and resonant tunneling diodes (RTDs) have been demonstrated at THz frequencies, but low-output powers continue to be a challenge. Dual color, vertical external cavity surface emitting laser (VECSEL) sources hold promise as a tunable, high power source in the 1–2 THz range, but have only recently emerged as a viable technology (Scheller et al., 2011; Paul et al., 2013). Far infrared (FIR) lasers, carcinotrons, and backward-wave oscillators can generate large amounts of power, but their narrow bandwidths, large power requirements, and size, often limit their use in observational astronomy.

The most commonly used THz LO sources are Schottky diode frequency multipliers. Frequency multiplied LO sources have been used extensively in ground, airborne, and space-based millimeter/submillimeter-wave observatories for decades, and have recently been demonstrated up to 2.7 THz (Maestrini et al., 2011). However, their output power decreases exponentially with frequency, making it an ever greater challenge to push them

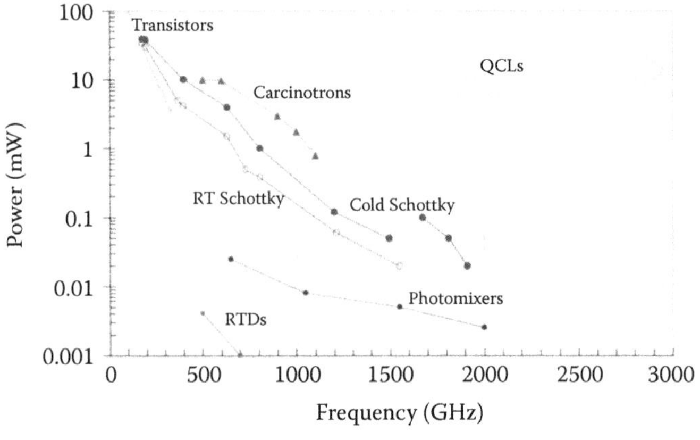

FIGURE 6.30 Available output powers from THz sources as a function of frequency. The terms RT Schottky and cold Schottky refer to room temperature and cold frequency multiplied sources. (From Mehdi, I. et al. 2008, *Proceedings of 19th International Symposium on Space Terahertz Technology*, Groningen, April 28–30, 2008, p. 196. With permission.)

into the "super-THz" range. An alternative technology, the quantum cascade laser (QCL), has been successfully demonstrated in laboratory receivers up to 4.7 THz (Kloosterman et al., 2013), but requires cryogenic cooling to operate. Since THz detectors are often cryogenic in nature, the cooling requirement of QCLs can, in principle, be accommodated without excessive added complexity.

6.9.1 FREQUENCY MULTIPLIED SOURCES

Mixers and frequency multipliers are closely related. The operation of both requires pumping a nonlinear device (at THz frequencies, typically a varactor diode) and a relatively strong tone. This situation is described by Equation 6.9. In the case of the mixer, there are two tones, a signal and LO. In the case of the multiplier, there is only one tone, such that V_S is set to zero in Equation 6.9, leaving the cosine terms containing the frequency component ω_{LO} raised to a power n, that is, $\cos^n(\omega_{LO}t)$, where $n = 1, 2, \ldots$ In a generic multiplier, for each power of n, the multiplier will generate a harmonic signal (among others) of frequency, $n\omega_{LO}$, with a multiplicative amplitude coefficient $1/2^{n-1}$. Therefore, one can think of a frequency multiplier as a harmonic generator with ever decreasing amounts of power, P_n, available in its higher frequency components. In the case of varactor multipliers, which are based on the nonlinear capacitance of a reversed-biased *pn* junction, the amount of harmonic output power that can be produced is governed by the Manley–Rowe relations (Manley and Rowe, 1956). For an n-th harmonic multiplier, the highest possible value of P_n occurs, in theory, when there is only "real" power in the circuit in the harmonic of interest. In such a case, the multiplier is said to be 100% efficient. This condition is met when the multiplier is designed such that the diode's junction is terminated in a pure reactance (no real power) at all but the desired frequency. For a Schottky varactor diode multiplier, the highest efficiency is achieved when the mount is designed such that the device sees a short circuit at all but the desired output frequency (Maas, 1988). In practice, this is not always possible. The "unused" short-circuited, frequency harmonics are called "short-circuit idlers." Since large currents can flow in the idler circuits, it is important to maximize their Q (here, ratio of reactive to real power) so that a minimum amount real power is lost in parasitic resistances.

A block diagram of a modern THz LO chain is shown in Figure 6.31. The chain begins with a computer controlled, microwave synthesizer capable of producing a ~10 to 20 mW continuous-wave (CW) tone at frequency, f_{synth}. Typically, the frequency of the tone at the

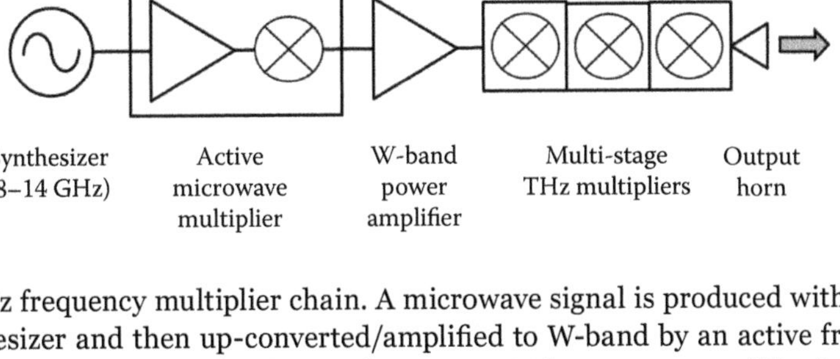

FIGURE 6.31 THz frequency multiplier chain. A microwave signal is produced within a computer-controlled synthesizer and then up-converted/amplified to W-band by an active frequency multiplier. The W-band signal level is then boosted to ~100 mW by a power amplifier before entering a chain of varactor frequency multipliers.

synthesizer output falls within X (8–11 GHz) or K_u band (12–18 GHz). The synthesized tone then passes through one or more active multipliers (a unit consisting of a frequency multiplier followed by a power amplifier) until a frequency, ν_W, of ~100 GHz is reached. This frequency falls within the W-band designation (70–110 GHz). The tone then enters a W-band power amplifier that boosts its power level up to ~100 mW. From this point, the tone's frequency needs to be multiplied by a factor n to reach the desired LO frequency, ν_{LO}.

$$n = \frac{\nu_{LO}}{\nu_W} \tag{6.67}$$

To achieve this, the tone passes through one or more frequency multipliers, with multiplication factors of ×2, ×3, or perhaps ×4. The efficiency of a multiplier drops sharply with its multiplication factor, so most frequency multiplier chains (FMCs) are composed of doublers and/or triplers. A plot of FMC output power and efficiency for different combinations of doublers and triplers utilizing planar Schottky diodes is shown in Figure 6.32. Photographs of a 1.5 THz planar diode multiplier chip and a complete waveguide FMC chain are provided in Figures 6.33a and 6.33b (Mehdi et al., 2008).

An ideal local oscillator would produce a sinusoidal waveform described by

$$V_{LO} = V_o \cos(\omega_{LO} t) \tag{6.68}$$

where

V_{LO} = voltage amplitude of LO waveform (V)
V_o = peak amplitude (V)
$\omega_{LO} = 2\pi f_{LO}$ = LO angular frequency (rad/s)
f_{LO} = LO frequency (Hz)

A real LO chain composed of an oscillator, amplifiers, and multipliers produces a tone that carries with it a white noise component consisting of amplitude modulated (AM) noise, $a_n(t)$, and phase modulated (PM) noise, $\theta_n(t)$ (Grebenkemper, 1981).

$$V_{LO} = [1 + a_n(t)]\cos[\omega_{LO} t + \theta_n(t)] \tag{6.69}$$

This LO noise can potentially add to the noise temperature of the system and produce unwanted broadening of the spectral lines being observed. The white noise component contributed by each amplifier is half $a_n(t)$ and half $\theta_n(t)$. In a balanced mixer, the LO noise is terminated in a load, while it is still present in a sideband separating mixer since the sidebands of the LO are not separated (Bryerton et al., 2007).

As discussed earlier, the output of a mixer has an upper and lower sideband. Along with the incoming signal from the telescope, the mixer will downconvert the amplitude and phase-noise components of the LO into each mixer sideband at the IF frequency. The mixing process leads to the sidebands having opposite phase. In a single ended mixer (like those commonly used at THz frequencies), the two sidebands are superimposed on each other at the mixer's output. Since the phase noise in the sidebands is 180° out of phase, they will, to first order, cancel out, while the amplitude noise components add coherently.

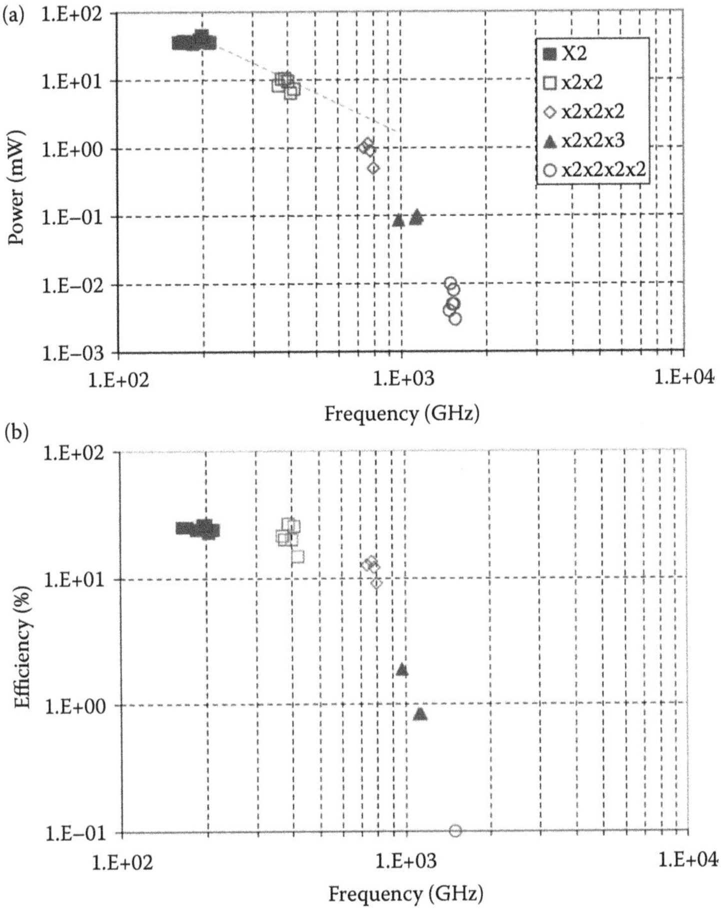

FIGURE 6.32 THz frequency multiplier performance. (a) Frequency multiplier output power vs. frequency. To generate a THz tone with sufficient output power to drive a mixer, several separate multiplier stages can be daisy chained together to provide the required multiplication factor. (b) THz frequency multiplier efficiency vs. frequency. The efficiency of single and multiple frequency multipliers drops exponentially with multiplication factor n. With cooling the I–V curve of a varactor diode becomes sharper and its frequency conversion efficiency (and power therefore output power) improved. The output power of an LO chain can be increased by ~50% with cooling to ~77 K (Adapted from Mehdi et al., 2003, *Proc. SPIE*, 4855, 435.).

A proven way of reducing amplitude noise is to run the amplifiers in the LO chain near saturation, where an amplifier's gain (and susceptibility to noise) is undergoing compression.

The amplitude noise contribution to the signal-to-noise (SNR) level through a properly pumped multiplier will remain constant. However, if the multiplier is underpumped (as is often the case at THz frequencies), the amplitude-noise contribution will increase exponentially. The phase-noise contribution to the SNR of an LO chain increases with the square of the multiplication factor, n. At the end of a THz LO chain, where n can be as high as ~200, phase noise grows quickly. As discussed earlier, on the output of a mixer the phase noise in the two sidebands tends to cancel one another. If any element is added before the mixer with (or if the mixer itself has) unequal amplitude or phase response

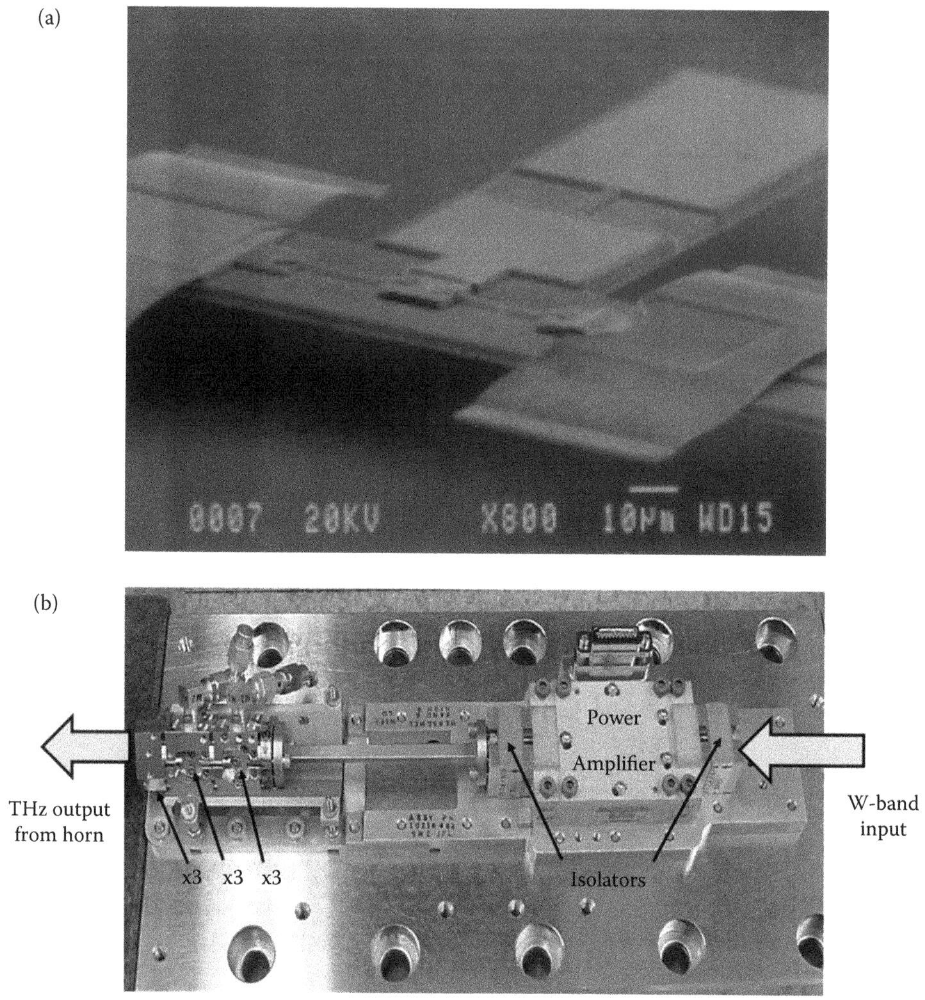

FIGURE 6.33 (a) SEM photograph of a balanced THz doubler chip. (Adapted from Erickson, N. et al., 2002, *13th International Symposium on Space THz Technology*, Harvard, March 2002.). (b) a 1.5 THz LO frequency multiplier chain. (Mehdi, I., private communication, 2012.)

between the two sidebands, some of the phase modulation will be converted to amplitude modulation (or vice versa) and add to the system noise (or to line broadening) (Bryerton et al., 2007). LO noise within a synthesizer driven LO chain has been shown to potentially contribute as much as 30% to the noise in a receiver at 420 GHz (Westig et al., 2012).

6.9.2 QUANTUM CASCADE LASERS (QCLs)

QCLs are a class of semiconductor lasers that can be engineered to operate at frequencies from ~1 to 10 THz. The operating frequency of a typical, bulk semiconductor laser is determined by the energy gap between low-energy electrons in the valence band, where electrons are attached to individual atoms, and higher-energy electrons in the conduction band, where electrons are free to move around the crystal lattice. This energy gap and, therefore, the operating frequency of the laser, are largely fixed for a particular material. A QCL, on the other hand, is created by depositing thin layers of varying material composition on a substrate to create a superlattice with a number of energy subbands. The inversion (i.e., active) layer thickness and material can be engineered to produce a population inversion between subbands that creates laser emission at a desired frequency as one or more electron pass through it. For example, a 3.45 THz QCL has been successfully made, using a 10 μm thick, 50 μm wide, and 1000 μm long layer of GaAs/AlGaAs sandwiched between gold conductive layers. The conductive layers serve both to help confine the laser emission and as electrodes (see Figure 6.34). When a DC voltage is placed across the electrodes, electrons are driven through the inversion layer triggering the emission of THz photons (Cui et al., 2013). To select a single frequency and create a THz beam, a periodic series of slots are etched into the conductive top electrode and active layer at ~λ/2 spacing. As the THz photons propagate through the layer, the slots produce constructive interference that guides the photons and forms THz beams out each end of the layer. In optics, this type of periodic structure is referred to as a Bragg reflector or distributed feedback array (DFB) (Williams et al. 2005). In antenna theory, it is analogous to a Yagi–Uda antenna (Yagi, 1928). Before cable TV, almost every rooftop had one. In the case of the Yagi–Uda antenna, instead of slots, there are metallic rods, with the lengths and spacing tuned so the beam emerges from only one end of the antenna. The same "end-fire" arrangement can

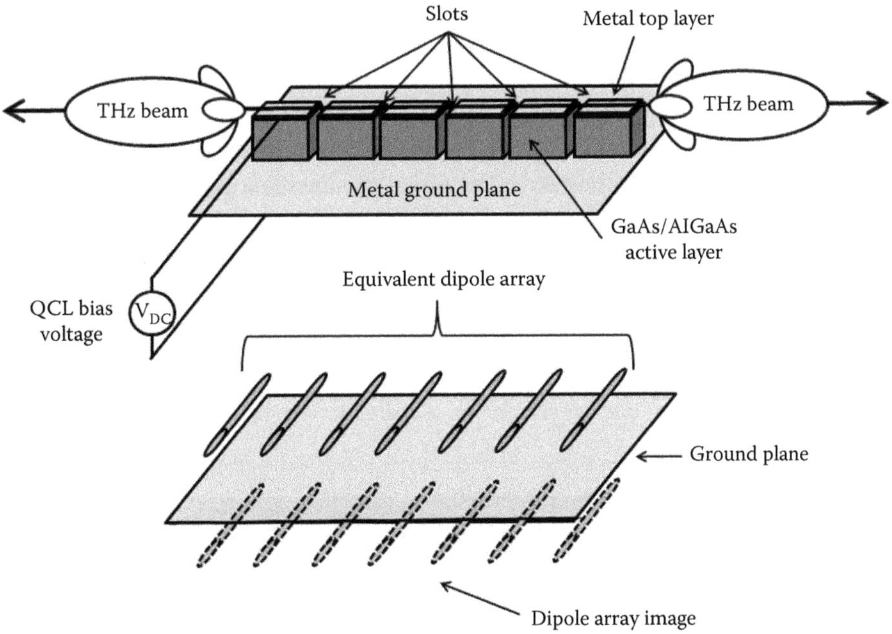

FIGURE 6.34 QCL schematic. (Top) The QCL is formed by creating a 10 × 50 × 1000 μm layer of GaAs/AlGaAs sandwiched between gold conductive layers/electrodes. A DC voltage is applied across the electrodes, driving electrons through the GaAs/AlGaAs inversion layer triggering THz laser emission. Deep slots are etched at ~λ/2 spacing to select the frequency and direct the THz radiation out the ends of the structure. (Bottom) Equivalent dipole model. (Adapted from Cui, M., Hovenier, J., Ren, Y. et al., 2013, *Appl. Phys. Lett.*, 102, 111113.)

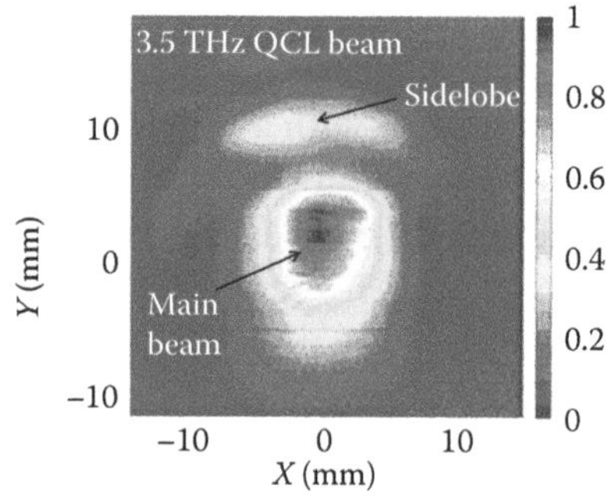

FIGURE 6.35 DFB QCL beam pattern at 3.5 THz. The full-width-half maximum (FWHM) of the main beam is 13° × 17°. The presence of the sidelobe is consistent with theory. (Adapted from Cui, M. et al., 2013, *Appl. Phys. Lett.*, 102, 111113.)

also be realized with a DFB. The slots in the top conductive layer of the QCL see a reflection of themselves in the bottom conductive layer, which acts as a ground plane (Cui et al., 2013). With proper vertical spacing the reflected "image" of the slots makes the structure behave like a stacked Yagi–Uda array, helping to confine the beam in the vertical direction. The resulting beam pattern from such a structure is shown in Figure 6.35.

QCLs can operate in both pulsed and continuous wave (CW) mode. Pulsed mode is good for radar applications, while CW mode is employed in LOs. In CW mode, QCLs have narrow linewidths (e.g., ~100 Hz) and excellent power stability. They can be voltage tuned over ~1.5 GHz (Kloosterman et al., 2013), which allows them to be placed in a frequency or phase lock-loop (Richter et al., 2010; Hayton et al., 2013). Narrow intrinsic linewidths and the ability to frequency (or phase) lock the LO is necessary in spectroscopic studies. LOs that wander too much in frequency have the effect of smearing out the observed line, so that it appears weaker and smoother than it really is.

To date, only cooled QCLs have been successfully used as a THz LO with an HEB. Figure 6.36 is a plot showing optimal HEB pump current versus QCL operating temperature at 4.7 THz. For this HEB/LO combination, optimum performance could be achieved with QCL operating temperatures up to ~60 K (Kloosterman et al., 2013). Work is continuing on increasing the power output of THz QCLs operating at higher temperatures (Razeghi et al., 2013).

6.10 RECEIVER BACK-ENDS

So far, in this chapter, we have focused on the receiver "front-end," the part of the receiver that downconverts the THz signal to microwave frequencies and amplifies it by ~120 dB (i.e., a thousand-billion times) to a level where it can be detected or further processed. In a system used for spectroscopy the signal is processed by a "back-end" spectrometer in order to produce a power spectrum. In other systems, only a measurement of the observed total power is required (e.g., observing continuum emission from planets or dust cores). In these systems, the back-end may be as simple as a total power detector. Even in spectroscopic systems, the ability to measure the total power before the spectrometer is a valuable

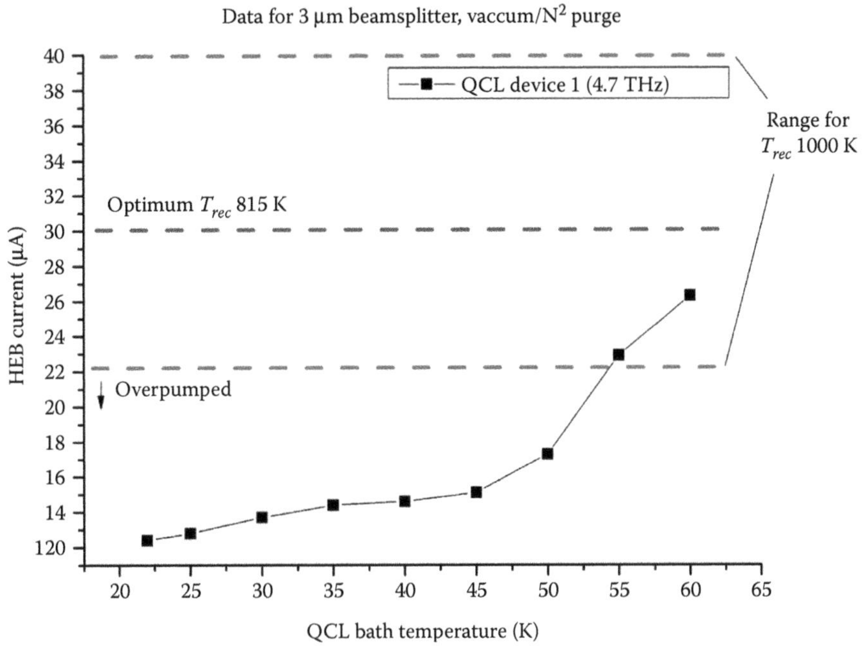

FIGURE 6.36 Temperature dependence of QCL output power at 4.7 THz. The bias current measured on an HEB is inversely related (see Equation 6.57) to the amount of LO power incident upon it. In the above plot a 4.7 THz QCL is being used as the LO to "pump" an HEB mixer. Here, we see that the QCL output power drops as the QCL's operating temperature is increased. (After Kloosterman, J. et al., 2013, *Appl. Phys. Lett.*, 102(1), 011123.)

asset in characterizing the system's performance. For this purpose, a power divider can be used to split the IF signal between a total power detector and spectrometer, so that they can be read simultaneously. If necessary, additional amplifiers can be added to make up for the power loss incurred by the splitter (see Figure 6.37). This is an important advantage of the heterodyne approach to signal detection; once a signal has been downconverted, there is no penalty in sensitivity incurred for making multiple copies of the signal for processing, either for the present or future. This latter point is particularly important for very long baseline interferometry (VLBI) where the data from multiple telescopes may not be processed until days (or years) later.

6.10.1 TOTAL POWER DETECTION

A total power detector consists of a bandpass filter and power detection diode. The bandpass filter is used to restrict the frequency range of the IF signal to a known amount, typically to within 50–500 MHz of the IF center frequency. The diode rectifies the incident IF waveform, meaning that it passes only the positive (if forward biased) or negative (if reversed biased) time-varying voltage associated with the IF signal. Since there is no longer an opposing voltage component to cancel it out, the output voltage across the diode will be a unipolar, DC signal. In this configuration, the diode is said to be a half-wave rectifier. Four diodes, two forward and two reversed biased, can be configured into a full-wave rectifier,

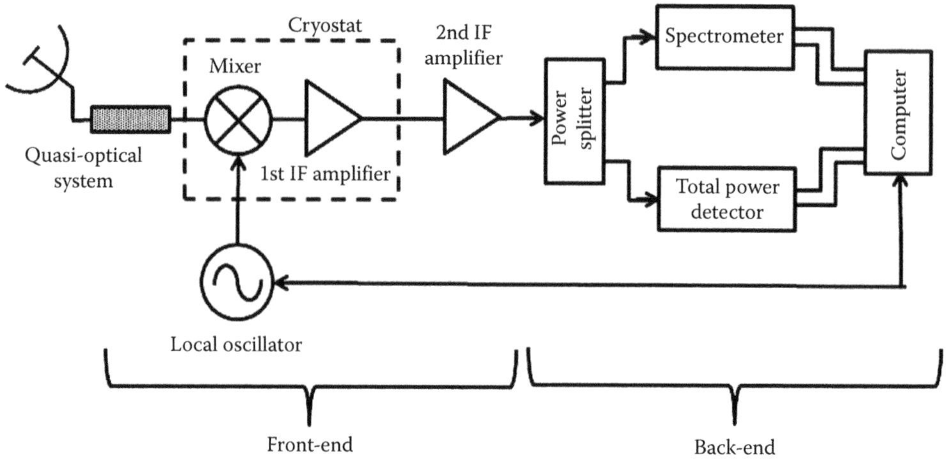

FIGURE 6.37 THz receiver system. A heterodyne receiver system consists of a front-end and back-end. The front-end downconverts the incoming signal from THz to microwave frequencies and then amplifies it to milliwatt power levels. At low THz frequencies (<1 THz), a low-noise amplifier (when available) can be used as a preamplifier ahead of the mixer. THz mixers and amplifiers either require or benefit from being cooled to low temperatures within a cryostat. The back-end typically consists of a spectrometer and/or total power detector read out by a computer. The computer also sets the frequency of the local oscillator.

allowing both halves of the signal to be converted into DC. If a transformer is used, a full-wave rectifier can be realized with just two diodes. The operation of full- and half-wave rectifiers is illustrated in Figure 6.38. Diodes, whether they are used for mixing, multiplying, or rectifying, have nonlinear I–V characteristics of the type shown in Figure 6.3. When the incoming voltages are within the nonlinear portion of the I–V curve, the diode is said to be

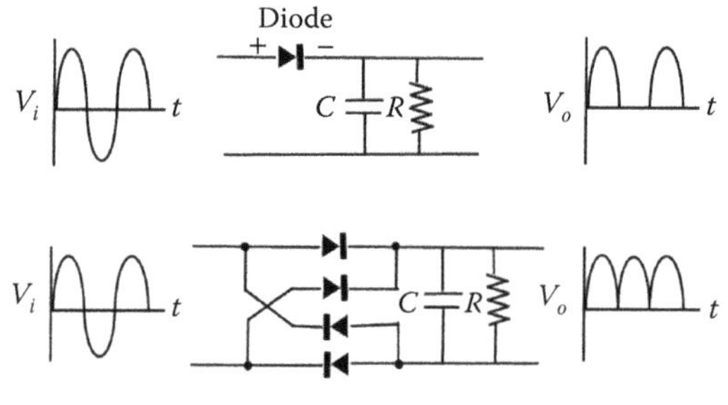

FIGURE 6.38 Total power detector. A total power detector consists of one or more diodes that rectify a signal, i.e., convert it from an AC to DC voltage/current. It makes use of the fact that a diode is polarized and will only pass current in one direction, with the direction depending on how the diode is physically oriented in the circuit. In the above figure, examples of a half-wave (top) and full-wave (bottom) rectifier are shown. In the half-wave rectifier a single diode is used to permit only the positive going part of the incoming signal to pass through. A full-wave rectifier arranges multiple diodes so that the negative going components of the input signal also pass through. An integrating circuit/low-pass filter consisting of a capacitor and resistor can be added (as show here) to smooth the output response of a rectifier.

operating in the "square-law" or small signal region. Here the diode's output voltage, V_o, is proportional to the IF power, P_i^{IF}, and to the square of the input IF voltage, V_i.

$$V_o = \alpha V_i^2 = \alpha P_i^{IF} \tag{6.70}$$

where α is the proportionality constant.

To filter/smooth out the high frequency component of the rectified signal an integrator, composed of a capacitor and resistor, can be attached across the rectifier's output. The integrator will have a time constant, τ_{int}, such that

$$\tau_{int} = \sqrt{RC} \tag{6.71}$$

where

τ_{int} = time constant (s)
C = capacitance (F)
R = resistance (Ω)

Once filtered, the IF output voltage can be recorded on a stripchart and/or put through an analog-to-digital converter (ADC) for storage on a computer.

6.10.2 SPECTROMETERS

The signal provided to the spectrometer by the receiver front-end can be thought of as a superposition of independent sinusoid waves, each with its own frequency, amplitude, and phase (see Figure 6.39). The purpose of the spectrometer is to sort out these waveforms and plot the power in them as a function of frequency, that is, produce a power spectrum. Other than the frequency stability of the local oscillator (which can be good to 1 part in a billion), it is the frequency resolution, $\Delta\nu$, provided by the spectrometer that determines the ability of the receiver to resolve the structure of spectroscopic line profiles.

The range of frequencies a spectrometer can process at any one time is set by the instantaneous bandwidth of the THz mixer/IF system, type of spectrometer technology

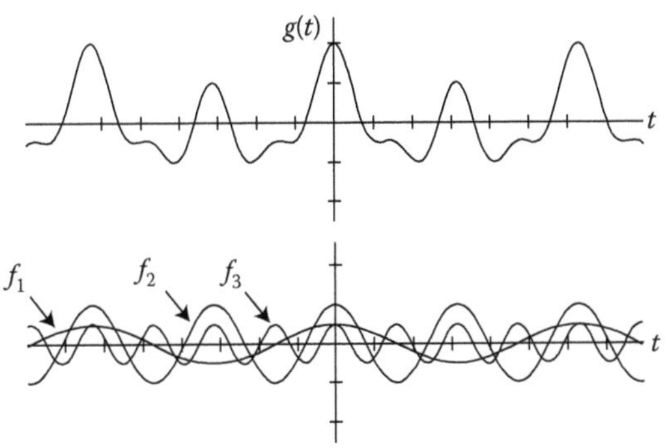

FIGURE 6.39 Sample IF signal. (Top) The voltage waveform of the IF signal $g(t)$. (Bottom) The same signal, but modeled as a superposition of sinusoidal waveforms.

used, and/or cost. For example, some of the earliest THz (460–492 GHz) receivers utilized indium antimonide (InSb) hot electron bolometers (HEB) mixers. These mixers had low noise (<1000 K), but their instantaneous bandwidth was only a few MHz (Schulz et al., 1987; Walker et al., 1988). The backend spectrometer consisted of a single, 1.4 MHz wide bandpass filter and total power detector (similar to that described in Section 6.11.1). In order to produce a spectrum, the frequency of the LO (and therefore the receiver's observational frequency) was scanned in 1 MHz steps, with a stop at each frequency point to integrate for 20–50 s. Multiple scans were made and averaged to yield the desired signal-to-noise ratio.

6.10.2.1 FILTERBANKS

The first spectrometers capable of supporting an IF bandwidth of 100's of MHz were filterbank spectrometers, like that pictured in Figure 6.40. Here, a power splitter is employed to divide up the IF signal into N_{ch} copies of itself. Each copy then passes

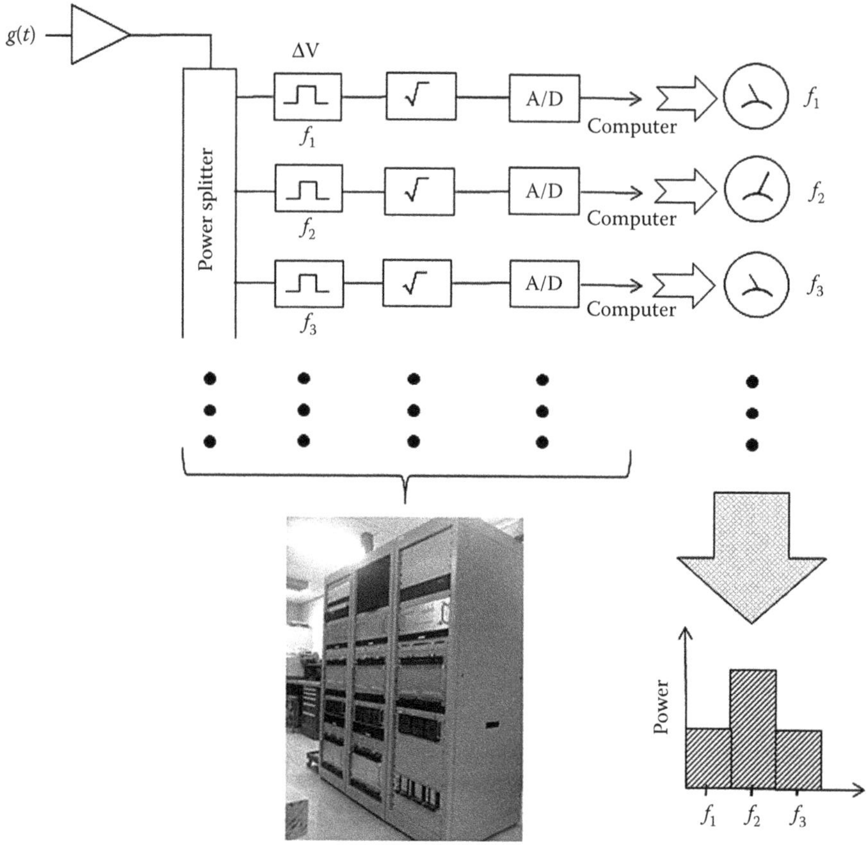

FIGURE 6.40 Filterbank spectrometer. A power splitter is used to generate multiple copies of the IF signal, $g(t)$. Each copy passes through a bandpass filter of width $\Delta\nu$ with a center frequency adjacent to that of its neighbors. Once filtered, each (now independent) IF channel goes through a power detector and A/D converter whose output is read by a computer. The cartoon meters indicate the magnitude of the power in each spectrometer channel. A power spectrum is made by plotting these powers versus frequency. The accompanying photograph is of the Arizona Radio Observatory 1 MHz × 2048 and 250 kHz × 512 channel filterbank spectrometers. (Photo courtesy Forbes, D., 2012, *Arizona Radio Observatory*, University of Arizona.)

through a bandpass filter of width $\Delta\nu_{ch}$. The center frequency of each filter, ν_n^{center} is tuned to be offset by ~$\Delta\nu$ from its neighbor. (The exact values of ν_n^{center} is subject to how sharp the "skirt" of the bandpass filter turns out to be.) The output of each filter is passed to a power detecting diode whose output voltage is digitized by an A/D converter and read by a computer. Each output channel of the filterbank corresponds to one frequency channel of the receiver's power spectrum. The total instantaneous bandwidth of the filterbank spectrometer is

$$BW_{FB} = N_{ch}\Delta\nu_{ch} \tag{6.72}$$

Filterbanks are simple in principle, but are hardware intensive, somewhat bulky per unit bandwidth, and have an inflexible architecture. Today's HEB, SIS, and Schottky mixers/IF systems all have several GHz of instantaneous bandwidth, making the realization of filterbanks to support them increasingly difficult.

Alternative spectrometer architectures include the chirp transform spectrometer or CTS (Hartogh, 1997), the acousto–optical spectrometer or AOS (Schieder et al., 1998), the autocorrelator or AC (Weinreb, 1961), and the fast-Fourier transform spectrometer or FFTS (Klein et al., 2006). The CTS can provide a few thousand frequency channels over a bandwidth of ~200 MHz. An AOS can provide approximately one thousand channels over ~1 GHz of bandwidth. At THz frequencies the Doppler Effect (see Equation 2.51) often dictates an IF bandwidth requirement >1 GHz. For this reason, together with their operational flexibility, the FFTS and autocorrelator spectrometer (ACS) are the two most commonly used spectrometers in today's wideband THz receiver systems.

6.10.2.2 FAST FOURIER TRANSFORM (FFT) SPECTROMETER

The ability to produce a power spectrum of a signal by performing a Fourier transform is an extension to the idea of the Fourier series. In a Fourier series, a time-varying signal (like our IF in Figure 6.39) can be decomposed into a sum of sinusoidal functions, that is, sines and cosines. The power spectrum of the signal is simply how much of the received power is within each of these components at a given frequency in a given amount of time.

Formally, the Fourier transform of a continuous function $g(t)$ can be written as

$$\begin{aligned} \hat{g}_\nu &= \int_{t_1}^{t_2} g(t)e^{-i2\pi\nu t}dt \\ P_\nu &= \alpha\,|\hat{g}_\nu|^2 \end{aligned} \tag{6.73}$$

where

P_ν = power spectral density (W/Hz)
$g(t)$ = time-varying input signal (V)
$\hat{g}_\nu$ = Fourier transform of $g(t)$ at ν
α = proportionality constant
ν = frequency of observation (Hz)
$\omega = 2\pi\nu$ = angular frequency (rad/s)
t_1 = start of integration (s)
t_2 = end of integration (s)

Examination of Equation 6.73 reveals how the transform works. For a given frequency, ν, the input function, $g(t)$, is first multiplied by sinusoidal functions (through Euler's identity) in the complex plane. The product is, in effect, a phase-sensitive, sinusoidal representation of $g(t)$. The product is then integrated over time, $\Delta t = t_2 - t_1$, which is equivalent to summing up all the voltages within the representative sinusoidal function at frequency, ν, in the designated time interval. To get a power spectrum, the expression for $\hat{g}_\nu$ is evaluated at each frequency of interest and squared to get P_ν. If one wants, for example, a power spectrum with k points, then Equation 6.73 must be evaluated k times within each Δt.

In most radio astronomy applications $g(t)$ is sampled at discrete points in time, $t = nT$, where T is the sampling interval in seconds. The corresponding sampling frequency is then $f_S = 1/T$ in Hz. The Fourier transform of Equation 6.73 then takes on its discrete form (DFT),

$$\hat{g}_k = \frac{1}{N}\sum_{n=0}^{N-1} g_n e^{-i2\pi\nu(n/N)} \quad \text{for } k = 0,\ldots,N-1 \tag{6.74}$$

where $N = \Delta t/T$ = number of sample points.

Evaluating this expression for say, $N = 1000$ is quite onerous, even for a fast microprocessor. There are N^2 complex multiplications and $N(N-1)$ complex additions. In order to reduce this processing burden, FFT techniques were developed. Most utilize the recursive Cooley–Tukey algorithm (Cooley and Tukey, 1965), which breaks the DFT into smaller DFTs that can be handled more efficiently, reducing the number of operations from N^2 to $N \log N$, where N is a power of 2.

In an FFTS, the analog input IF signal (here, $g(t)$) is sampled by an analog to digital converter (ADC) at a clock frequency f_S. The time-varying binary equivalent of the IF signal is sent to a field programmable gate array (FPGA) where the FFT is performed over the designated integration time, Δt. The Nyquist sampling theorem limits the overall bandwidth of an FFTS to $BW_{FFTS} \le f_S/2$. The resulting power spectrum from a single integration is often referred to as a subscan. Multiple subscans are sent to an instrument computer where they are stored and/or averaged. A block diagram and photograph of an FFTS is shown in Figure 6.41.

6.10.2.3 AUTOCORRELATOR SPECTROMETER

When digital spectrometers for radio astronomy were first developed, the processing horsepower to perform the required DFT or FFT to produce a power spectrum from an analog signal did not yet exist. The solution was to use simple digital logic circuits to generate an autocorrelation function (ACF) of the input signal in the time domain and then use a much simplified Fourier transform algorithm (via the Wiener–Khinchin Theorem) to produce a power spectrum in the frequency domain. This approach, pioneered by Weinreb (1961), has the added benefit of allowing all the information needed to produce a spectrum to be collected and, if need be, recorded before a single Fourier transform is made.

A block diagram of a basic autocorrelation system is shown in Figure 6.42. In an autocorrelator system, the input signal is clipped and sampled so that it appears to be a stream of logical "high" and "low" signals. The "digitized" signal stream then follows two paths. One path leads it through a series of digital delay lines, each delaying the signal propagation by

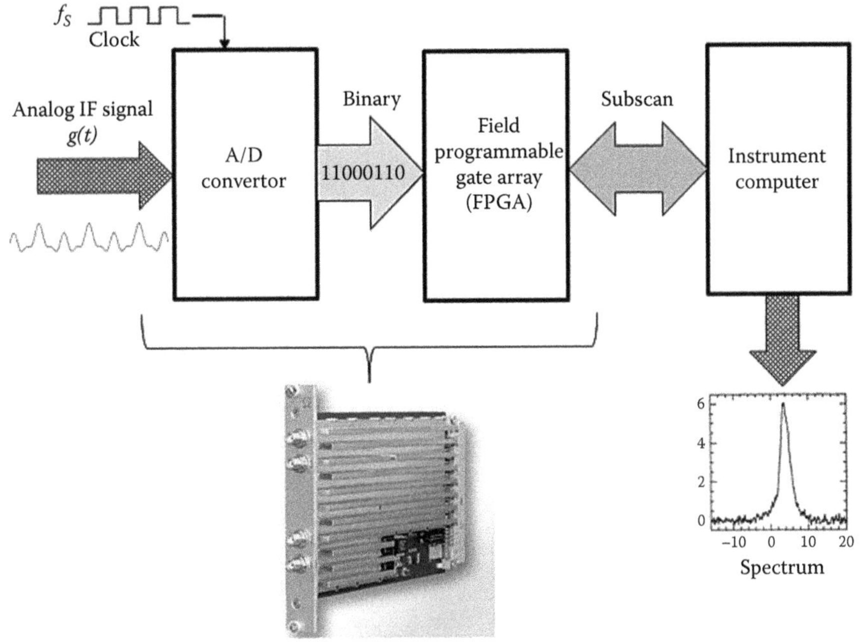

FIGURE 6.41 FFTS block diagram. The input IF signal $g(t)$ is sampled and digitized at a frequency, f_S. The equivalent binary signal is conveyed to a field programmable gate array (FPGA) where the FFT is performed. The resulting power spectra are read out and accumulated by an instrument computer. The accompanying photograph is of an FFT module housing four, 8-bit ADCs capable of providing up to 2 GHz of bandwidth. (Photo courtesy Omnisys Inc. With permission.) A Virtex-4 FPGA with a pipelined FFT core processes the input data and generates a power spectrum. A locally generated, interlaced, sampling clock of 1.1 GHz is fed to all ADCs generating an effective sampling rate of 2.2 GS/s.

an amount, $\Delta\tau$. After each delay, the signal goes into a "AND" logic gate which compares it with the polarity (high or low) of the real time signal. If they are the same, the AND gate produces a logical 1 on its output and increments a binary counter by 1. When logical ones are generated early in the series of $\Delta\tau$s, it means the input signal contains significant power in waveforms with a short period, that is, high frequency. When logical ones are generated on the output of AND gates further down the chain, it means that the incoming signal has significant power in waveforms with a longer period, that is, low frequency. The output of the binary counters at a time, t, is the autocorrelation function $R'(n\nabla\tau)$ of the input signal, $g(t)$ (Kraus, 1988).

$$
\begin{aligned}
R'(n\Delta\tau) &= \frac{1}{N}\sum_{k=0}^{N-1} g(k\Delta t)g(k\Delta t + n\Delta\tau), \quad n = 0,1,2,\ldots N-1 \\
R(n\Delta\tau) &= \sin\left[\frac{\pi}{2}R'(n\Delta\tau)\right]
\end{aligned}
\tag{6.75}
$$

where

$R'(n\Delta\tau)$ = autocorrelation function (V/s)

$R(n\Delta\tau)$ = normalized autocorrelation function for a Gaussian noise distribution

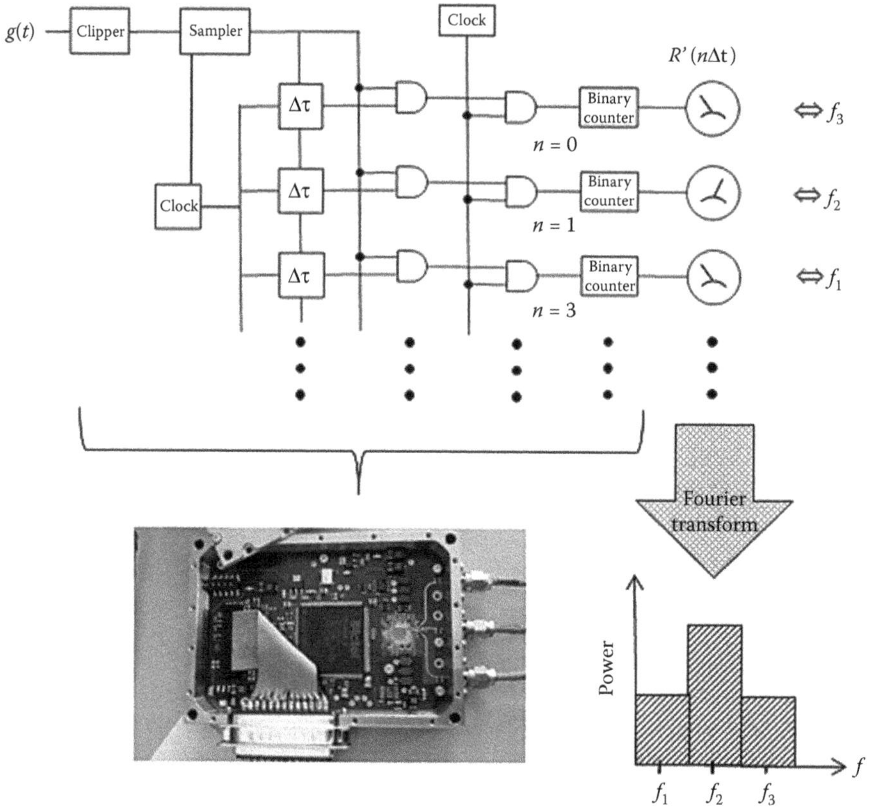

FIGURE 6.42 ACS block diagram. The input IF signal $g(t)$ is clipped and sampled at a frequency, f_S. The equivalent binary "high" and "low" representation of the signal is then compared with an ever increasingly time-delayed version of itself. The power present in the higher-frequency components of the input signal are registered when the delays are short. The lower-frequency components of the signal register after the delay time has built-up. The output of the correlator is the correlation function $R'(n\Delta\tau)$ in the time domain. A power spectrum is generated by taking the FFT of $R'(n\Delta\tau)$. The accompanying photograph is of an ACS module with a 1.5 bit digitizer that can support clock speeds (i.e., f_S) up to 14 GHz. (Photo courtesy of Omnisys Inc. With permission.)

$g(t)$ = time-varying input signal (V)
N = total number of samples
n = number of correlator channels

The power spectrum is obtained by performing an FFT of $R(n\Delta\tau)$. As in the case of the FFTS, the bandwidth of an ACS is limited by the sampling frequency, f_s, to be $BW_{ACS} \leq f_s/2 - 1/2\Delta t$. Compared to the ADC used in an FFTS that can be 8 bits or more, the digitizer in an ACS is coarse, with only 1 or 2 bits. But even with just 2-bit resolution, the ACS is 88% efficient. Table 6.2 compares the performance of a 4 GHz bandwidth ACS on a custom ASIC (application specific integrated circuit) to that of a 4 GHz bandwidth FFTS that uses an FPGA. Even though it is ~10% less efficient, the ACS requires *five times* less power. This power savings makes the ACS an excellent choice for balloon or space-based instrumentation, where both power and heat rejection is at a premium.

TABLE 6.2 Comparison of Digital Spectrometer Performance

Spectrometer Type	BW (GHz)	Efficiency (%)	Power (W)
ACS on ASIC	4	88	300
FFTS on FPGA	4	99	1500

Both the ACS and FFTS are designed to support a specific number of channels (spectral resolution elements) within their operational bandwidth, independent of clock speed. Therefore, by slowing the clock speed, it is possible to trade increased spectral resolution for reduced overall bandwidth.

EXAMPLE 6.9

The rotation speed of the Milky Way is ~250 km/s. In order to observe all the velocity components of the Milky Way along any given line of sight, what is the required spectrometer bandwidth at 1.9 THz?

The required bandwidth, $\Delta\nu_{total}$, can be found from the Doppler relation,

$$\Delta\nu_{total} = \frac{\nu_0 \Delta V_{total}}{c}$$

where

$V_{total} \approx V_{rot} \approx 250$ km/s = galaxy rotation speed

$\nu_0 = 1.9$ THz

Substitution yields,

$$\Delta\nu_{total} \approx \frac{(1.9 \times 10^{12}\ \text{Hz})(250 \times 10^5\ \text{cm/s})}{3 \times 10^{10}\ \text{cm/s}}$$
$$\approx 1.58\ \text{GHz}$$

EXAMPLE 6.10

Due to turbulence, many molecular clouds have atomic/molecular linewidths of 2–3 km/s. What spectrometer channel width, $\Delta\nu_{chan}$, is required to resolve these lines with a velocity resolution of 1 km/s? Assume an observational frequency of 1.9 THz (i.e., the [CII] line).

Substituting into the Doppler expression in Example 6.10 with a $\Delta V_{chan} = 1$ km/s, we find,

$$\Delta\nu_{chan} = \frac{1.9 \times 10^{12}\ \text{Hz}(1 \times 10^5\ \text{cm/s})}{3 \times 10^{10}\ \text{cm/s}}$$
$$= 6.3\ \text{MHz}$$

6.11 RECEIVER STABILITY AND ALLAN TIME

To be useful, it is not enough for a receiver to have a low noise temperature, it must also be stable. Indeed, the *more* sensitive the receiver, the *less* stable it will tend to be. Many things can affect the stability of a THz receiver. For ground-based observations it could be short-term variability in the atmospheric transmission due to weather passing through the telescope beam. It could be reflections or motion (e.g., due to microphonics) between various components in the quasi-optical system, variations in LO power, an unstable IF amplifier, oscillations in the bias system, changes in the mixer's physical temperature, or some combination of the above. For the THz observer what it all boils down to is the receiver's Allan time. The Allan time is how long you can integrate before the noise integrates down two standard deviations slower than expected from the radiometer equation (Equation 6.4) (Ossenkopf, 2008). Over time, an increasing fraction of the system noise becomes correlated, such that it does not average down by the square root of the integration time. Beyond this point, the observing efficiency significantly decreases or may even go negative, *resulting in an increase of the noise level* with time. Once the Allan time is reached (typically between 10 and 30 s), the telescope/receiver system should be switched briefly (~5 s) to a calibration load or sky position (or in the case of frequency switching, off-line) before a new scan of the object commences. The longer the Allan time of a receiver system, the longer one can go between calibrations and the greater will be the percentage of time spent on the target source.

A receiver's Allan time is determined from a plot of Allan variance versus integration time (see Figure 6.43). The Allan variance can be computed for the receiver's total power or, more appropriately here, spectroscopically. The spectroscopic Allan time measures deviations over the channels of interest in a spectrometer (i.e., the channels over which you expect the target spectral line to occur) to the continuum level fluctuations. Let $d_k(n)$ represent the power in channel k at a sample time n when an input signal is present and z_k represent the "zero" power level in the channel, that is, the measured power in the same spectrometer channel with no input signal. Since there is no signal, z_k should be constant with time. The signal can be spectroscopically normalized by dividing the difference between $d_k(n)$ and z_k by the temporal average of each channel (Kooi, 2008; Ossenkopf, 2008).

$$\begin{aligned} s_k'(n) &= \frac{d_k(n) - z_k}{\langle d_k(n) - z_k \rangle_n}, \quad \text{where} \\ \langle d_k(n) - z_k \rangle_n &= \frac{1}{N}\sum_{n=1}^{N}[d_k(n) - z_k] \\ z_k &= \frac{1}{N}\sum_{n=1}^{N} z_k(n) \end{aligned} \tag{6.76}$$

where

n = sample number

N = total number of samples

Δt = time interval between samples (s)

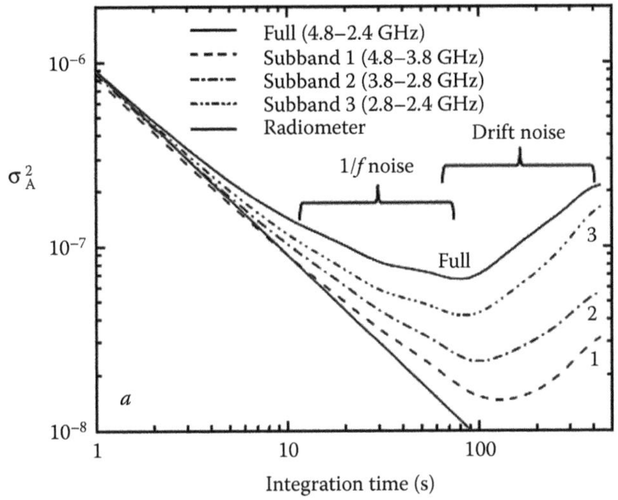

FIGURE 6.43 Spectroscopic Allan variance plot for the Band 6a HEB mixer flown on the *Herschel Space Observatory*. Each curve represents a different section (or subband) of the spectrometer being used. The diagonal line shows the ideal case where "white," uncorrelated noise integrates down according to the radiometer equation. The Allan time is where σ^2_A departs by 2 standard deviations from the radiometeric (white) noise curve; 14.9 s (full), 30.7 s (subband 3), 50.5 s (subband 2), 111 s (subband 1). Departures from this line indicate the presence of unwanted $1/f$ and/or drift noise in the system. (Adapted from Kooi, J., 2008, Advanced Receivers for Submillimeter and Far Infrared Astronomy, PhD dissertation, California Institute of Technology.)

k = channel number
$d_k(n)$ = signal data point n in channel k (V)
$z_k(n)$ = zero power data point n in channel k (V)
$s'_k(n)$ = spectroscopic normalized difference for channel k for data points at n (V)

The next step is to subtract the normalized mean from each spectrometer channel of interest.

$$s_k(n) = s'_k(n) - \langle s'_k \rangle, \quad \langle s'_k \rangle = \frac{1}{N}\sum_{n=1}^{N} s'_k(n) \tag{6.77}$$

The Allan variance for channel k is, then,

$$\sigma^2_{A,k}(T) = \frac{1}{2(N-1)}\sum_{n=1}^{N}(s'_k(n) - \langle s'_k \rangle)^2 \tag{6.78}$$

where
$T = n\Delta t$ = integration time (s)
$\sigma^2_{A,k}(T)$ = spectroscopic Allan variance of channel k for specified T

The average spectroscopic Allan variance is obtained by summing the individual channel Allan variance over the channels where the line will occur.

$$\sigma_A^2(T) = \frac{1}{K}\sum_{k=1}^{K}\sigma_{A,k}^2(T) \tag{6.79}$$

Using the measured time series of $d_k(n)$, Equation 6.79 can be evaluated for increasing values of T and displayed on a log–log plot, as shown in Figure 6.41. The plot shows the system level Allan time performance of the HEB receiver system flown on the Herschel Space Observatory. On the plot radiometric, uncorrelated, "white" noise appears as a straight line with slope T^{-1}. The minimum in the curves gives an indication of the Allan time. Beyond the Allan time, $1/f$ and then drift noise begin to dominate. These types of noise do not integrate down well, if at all. The Allan time is the lowest when the full spectrometer is used.

The Allan times in THz receivers are often dominated by microphonics that lead to time-varying mismatches in the system's quasi-optical beams. This is particularly true in the case of HEB receivers, where even small motions in the LO beam can significantly alter the HEB's delicate thermodynamic balance. Various techniques have been developed to mitigate this problem. Most involve adding an automatic gain control (i.e., AGC) loop that holds the HEB bias current constant by either mechanically or electronically adjusting the incident LO power. Using this approach can lead to more than an order of magnitude improvement in Allan time (see Figure 6.44 and Hayton et al., 2012).

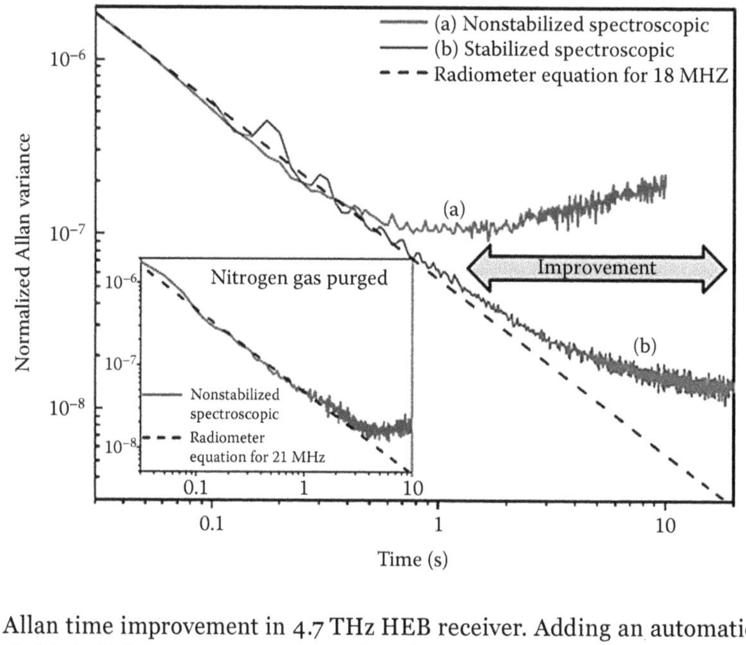

FIGURE 6.44 Allan time improvement in 4.7 THz HEB receiver. Adding an automatic gain control (AGC) loop to hold HEB bias current constant by adjusting LO power can lead to more than an order of magnitude improvement in system Allan time. The measurements were made over 18 and 21 MHz of bandwidth in the main and inset figure, respectively. (Adapted from Kloosterman, J. et al., 2013, *Appl. Phys. Lett.*, 102(1), 011123.)

6.12 HETERODYNE ARRAY CONSIDERATIONS

Large-format THz arrays have been contemplated for over 30 years (Gillespie and Phillips, 1979). However, only recently, through a confluence of technological advances, has it become possible to construct such arrays. The key areas of advancement have been in

1. Mixer technology
2. Local oscillator (LO) technology
3. Intermediate frequency (IF) amplifiers
4. Micromachining
5. Spectrometers
6. Experience with array packaging

The advancements in THz mixer, LO, and IF amplifier technologies were driven largely by the instrument requirements of the *Herschel Space Observatory*. These efforts in mixer development are reflected in Figure 6.15 where receiver double sideband noise temperatures are plotted versus frequency. Below ~1 THz receivers based on SIS mixer technology have the best noise performance. Above ~1 THz receivers HEB mixers provide the best performance. Even at frequencies approaching 5 THz, single pixel receivers with low noise have been demonstrated (Kloosterman et al., 2013). In a parallel effort, solid state LOs with output powers of >100 μW, more than sufficient to drive modest HEB arrays at 1.9 THz, have been constructed (Mehdi, I., private communication, 2014). Above this frequency, quantum cascade lasers (QCLs) can be used to generate output powers of 100s of microwatts. Intermediate frequency (IF) amplifiers with low noise (<5 K), high gain (>30 dB), and low power dissipation (<10 mW) are now available in large numbers and low cost (Groppi et al., 2008).

The need for large-format, heterodyne arrays to conduct large-scale galactic plane surveys has driven the development of micromachining, spectrometers, and array packaging technologies. The first generation THz arrays (e.g., PoleSTAR (Walker et al., 2001), Desert STAR (Groppi et al., 2003)) were constructed using discrete components and stacking individual mixer blocks in a common cryostat. Quasi-optical power dividers and diplexers were used to inject the LO into the signal path to the mixer array. With this approach, array sizes up to 16 pixels have been made (e.g., HARP (Buckle et al., 2009)). For larger arrays, a greater degree of integration is required. SuperCam is a 64 pixel, 345 GHz array constructed in a multi-institutional effort led by the University of Arizona (Pütz et al., 2006; Groppi et al., 2008; Kloosterman, 2014). The focal plane unit is composed of eight mixer modules. Each module contains a 1×8 array of mixers machined into a single split block (see Figures 5.38 and 6.45). The SIS devices are made on 3 μm SOI membranes with selfaligning beam leads. Low-noise, low-power monolithic microwave integrated circuit (MMIC)-based IF amplifiers are integrated in the block and bonded directly to the IF output of the SIS devices. The mixer modules slide into cryogenic "U-mounts," through which all electrical and RF connections are made through blind-mate connectors. The 1-dimensional level of integration employed in SuperCam is what made the construction of this size array practical, both in terms of packaging and cost. Large-format arrays can dramatically increase the scientific throughput of a telescope and, in the process, significantly reduce the manpower and operating costs associated with large-scale survey projects.

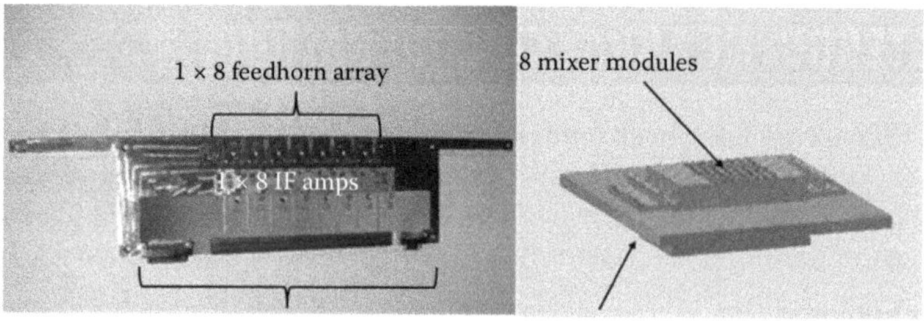

FIGURE 6.45 (Left) 1 × 8 SuperCam mixer module. (Right) Eight mixer modules installed in cryostat cold plate.

The increase in observing efficiency, ϵ_{AY}, of a heterodyne array over a single pixel receiver can be defined in the simplest terms as

$$\varepsilon_{AY} = N_{pix}\left(\frac{\eta_{SP}}{\eta_{AY}}\frac{T_{sys}^{SP}}{T_{sys}^{AY}}\right)^2 \tag{6.80}$$

where

ϵ_{AY} = increase in observing efficiency of array over a single-pixel receiver
N_{pix} = number of array pixels
η_{SP} = optical coupling efficiency of a single pixel
η_{AY} = optical coupling efficiency of an array pixel
T_{sys}^{SP} = system noise temperature of a single pixel receiver (K)
T_{sys}^{AY} = system noise temperature of an array pixel (K)

Examination of Equation 6.72 reveals that mapping speed of an array compared to a single-pixel receiver will increase linearly with N_{pix} only if the noise temperature of an array pixel is comparable to that of an optimized single-pixel receiver (Groppi and Kawamura, 2011). If an array pixel has twice the noise of a single-pixel receiver, it will take four array pixels to match the mapping speed of the single-pixel system.

For future large submillimeter-wave telescopes, such as the 25 m Cornell–Caltech Atacama Telescope (CCAT), large-format arrays are key to all but the most targeted science programs. Heterodyne arrays for as many as 1000 pixels Kilo-pixel Camera (KCAMs) are now being considered for CCAT. Figure 6.46 shows CCAT beam footprints for 64, 256, and 1024 pixel arrays on a familiar object at 0.65 THz. Even for the largest array, the field of view is under 5 arcminutes.

As is well known from the computer industry, a higher level of device integration often leads to leaps in performance at a reduced price. This rule also appears to hold for heterodyne array receivers. Figure 6.47 is a plot of the cost/pixel for receivers built at the University of Arizona versus number of pixels. The costs plotted for a single pixel receiver, PoleSTAR, and SuperCAM are based on actual dollars spent. The costs for larger-array receivers are based on grassroots budgets. The plot shows that the cost per pixel for an array with n pixels is $\propto 1/n^{1/2}$. By continually increasing the level of receiver integration

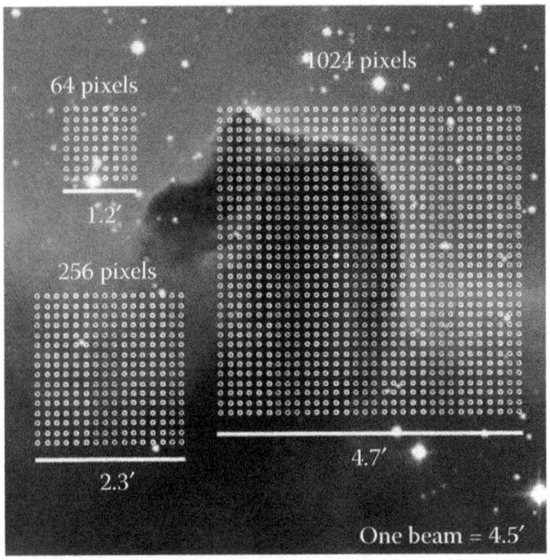

FIGURE 6.46 CCAT beam footprint at 650 GHz for a variety of array sizes. The beam spacing on the sky is ~$2f\#\lambda$ (every other θ_{FWHM} beam). (Adapted from Walker et al., 2008, *Millimeter and Submillimeter Detectors and Instrumentation for Astronomy IV, Proc. SPIE*, 7020(702014), 1–8.)

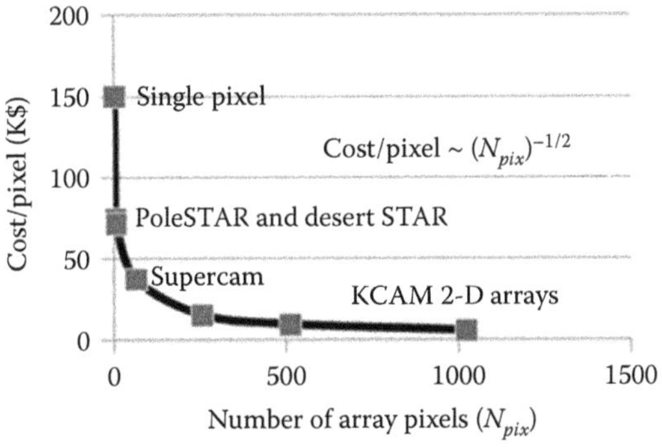

FIGURE 6.47 Cost/pixel of heterodyne arrays versus number of pixels. As long as higher levels of component integration occur, the cost/pixel continues to drop. (Adapted from Walker et al., 2008, *Millimeter and Submillimeter Detectors and Instrumentation for Astronomy IV, Proc. SPIE*, 7020(702014), 1–8.)

from single mixer to multiple mixers sharing a cryostat, to 1-D integration, and ultimately to 2-D integration—the cost per pixel keeps dropping.

SUMMARY

In this chapter, we have introduced the basic principles behind coherent detection and the theory of operation of the components used to create such systems (i.e., mixers, amplifiers, local oscillators, and backend spectrometers). Expressions were derived for characterizing the noise and stability of coherent receivers. Typical system performance parameters were described in the context of THz spectral line observations. In the next chapter, a similar analysis will be performed for incoherent detection systems. A metric will be introduced for determining which type of detector system is best suited for a particular observation.

PROBLEMS

1. Jupiter has an antenna temperature, T_A, of 174.3 K at 0.337 THz. What is the corresponding flux in Janskys that would be observed with a 10 m diameter telescope with an aperture efficiency of 80%?

2. You want to observe the CO $J = 3 \rightarrow 2$ line toward an external galaxy. The typical single-sideband value of T_{sys} at that frequency is 500 K. Using absolute position switching , how much observing time should you request to achieve a $T_{\mathrm{rms}} \approx 1$ mK with 2 MHz spectral resolution (i.e., postdetection bandwidth)?

3. Neutral atomic carbon has an emission line, [CI], at 492.2 GHz. The DSB receiver you are using has an IF frequency of 1.5 GHz. There is a nasty atmospheric absorption feature just below the frequency of the line. What LO frequency should you use to put the line in the receiver's lower sideband (LSB), thereby minimizing the effect of the absorption feature?

4. Due to misalignment the optics between a telescope and receiver has 1 dB of loss.

    a. What is the equivalent noise temperature of the optics if they are at room temperature (290 K)?

    b. If the rest of the receiver system has a noise temperature, T_R, of 100 K, what will the overall instrument noise temperature be with the added loss?

5. Using hot and cold loads, you measure a receiver Y-factor of 1.2 at 4.7 THz. What is the receiver's equivalent double-sideband black-body noise temperature?

6. What type of amplifier is commonly used as the first amplifier in a modern radio astronomy receiver? Briefly describe its principle of operation.

7. You would like to design a receiver system to observe the [CII] emission line at 1.9 THz. Which type of mixer and LO will provide the best performance? Why?

8. You are designing an orbital observatory for high spectral resolution spectroscopy and require a back-end spectrometer capable of processing 4 GHz of instantaneous bandwidth. Which type of spectrometer should you use? Why?

9. You plan to observe the 557 GHz line of water from an orbital observatory toward a protostellar source. What should the spectrometer channel width be in MHz to achieve a 0.1 km/s velocity resolution?

10. What is the quantum noise temperature limit for a receiver operating at 1.9 THz?

11. What is the minimum magnetic flux density (in Gauss) required across a 0.5×0.5 micron SIS junction to suppress the Josephson supercurrent?

12. What limits the Allan time in an HEB mixer? Why is the Allan time important?

13. If an optimized single-pixel receiver has a noise temperature of 50 K and a typical receiver in an array has a noise temperature of 80 K, how many pixels does the array receiver need to have to provide an order of magnitude increase in the mapping speed over the single pixel system? Assume the coupling efficiency to the telescope is the same for both systems.

REFERENCES

Barone, A. and Paterno, G., (eds.), 1982, *Physics and Applications of the Josephson Effect*, John Wiley & Sons, New York.

Baselmans, J. J. A., Hajenius, M., Baryshev, A. M. et al., 2004, Hot electron bolometer mixers with improved interfaces: Sensitivity, LO power and stability, *Proceedings of the 15th International Symposium on Space TeraHertz Technology*, Northampton, MA, NRAO, pp. 1–8.

Belitsky, V., 1999, *Local Oscillator Power Requirements for ALMA SIS Mixers*, MMA Memo #264, NRAO.

Billade, B., 2013, Mixers, Multiplier and Passive Components for Low Noise Receivers, PhD dissertation, Chalmers University of Technology.

Blake, G., Sutton, E., Masson, C., and Phillips, T. G., 1986, The rotational emission line spectrum of Orion a between 247–263 GHz, *Ap. J. (Suppl)*, 60, 357.

Boreiko, R., Betz, A., and Zmuidzinas, J., 1988, Heterodyne spectroscopy of the 158 micron CII line in M42, *Astrophys. J.*, 325, L47–L51.

Boreiko, R. T. and Betz, A. L, 1991, Ionized carbon in the large magellanic cloud, *Ap. J.*, 380, 27.

Brune, J. and Bierschneider, P., 1994, Quasi-optical optimization and modal analysis of corner-cube Schottky-diode mixers, *5th Symposium on Space TeraHertz Technology*, University of Michigan, Ann Arbor, Michigan, p. 682.

Bryerton, E., Morgan, M., Thacker, D., and Saini, K., 2007, Maximizing SNR in LO chains for ALMA single-ended mixers, *Proceedings of 18th International Symposium on Space Terahertz Technology*, Pasadena, CA, March 21–23, 2007, p. 68.

Buckle, J. V., Hills, R. E., Smith, H. et al., and 28 coauthors, 2009, HARP/ACSIS: A submillimetre spectral imaging system on the James Clerk Maxwell Telescope, *MNRAS*, 399, 1026.

Callen, H. and Welton, T., 1951, Irreversibility and generalized noise, *Phys. Rev.*, 83(1), 34.

Collin R. E., 1966, *Foundations for Microwave Engineering*, McGraw-Hill Book Company, New York, p. 260.

Cooley, J. and Tukey, J., 1965, An algorithm for the machine calculation of complex Fourier series, *Math. Comp.*, 19, 297–301.

Cui, M., Hovenier, J., Ren, Y. et al., 2013, Beam and phase distributions of a terahertz quantum cascade wire laser, *Appl. Phys. Lett.*, 102, 111113.

Erickson, N., 2008, A Schottky-diode balanced mixer for 1.5 THz, *19th Symposium on Space TeraHertz Technology*, University of Groningen, the Netherlands, p. 221.

Erickson, N. and Goyette, T., 2009, TeraHertz Schottky-diode balanced mixers, *Proc. SPIE Hol.*, 7215, 721508.

Erickson, N., Maestrini, A., Schlecht, E., Chattopadhyay, G., Gill, J., and Mehdi, I., 2002, 1.5 THz all-planar multiplied source, *13th International Symposium on Space THz Technology*, Harvard, March 2002.

d'Aubigny, C., Walker, C., and Jone, B., 2001, Laser microchemical etching of waveguides and quasi-optical components, *Proc. SPIE*, J. M. Karam and J. Yasaitis (eds.), 4557, 101.

Forbes, D., 2012, *Arizona Radio Observatory*, University of Arizona.

Gaidis, M., Pickett, H., Siegel, P., Smith, C., Smith, R., and Martin, S., 2000, A 2.5 THz receiver front-end for spaceborne applications, *Microwave Theory Tech.*, 48, 733–739.

Gillespie, A. and Phillips, T., 1979, Array detectors for millimeter line astronomy, *A&A*, 73, 14

Groesbeck, T., 1995, The Contribution of Molecular Line Emission to Broadband Flux Measurements at Millimeter and Submillimeter Wavelengths, PhD dissertation, California Institute of Technology.

Groppi, C. E., Walker, C. K., Kulesa, C. et al., 2008, SuperCam: A 64 pixel heterodyne imaging spectrometer, millimeter and submillimeter detectors and instrumentation for astronomy IV, *Proc. SPIE*, W. Duncan, W. Holland, S. Withington, and J. Zmuidzinas (eds.), 7020(702011), 1–8.

Groppi, C.E., Kawamura, J.H., 2011, Coherent detector arrays for terahertz astrophysics applications, *IEEE Trans. on Terahertz Sci. and Technol.*, 1(1), pp. 85–96.

Hartogh, P., 1997, Present and future chirp transform spectrometers for microwave remote sensing, *Proc. SPIE 3221, Sensors, Systems, and Next-Generation Satellites*, 3221, 328.

Hayton, D., Gao, J., Kooi, J., Ren, Y., Zhang, W., and de Lange, G., 2012, Stabilized hot electron bolometer heterodyne receiver at 2.5 THz, *Appl. Phys. Lett.*, 100, 08110.

Hayton, D. J., Khudchencko, A., Pavelyev, D. G. et al., 2013, Phase locking of a 3.4 THz third-order distributed feedback quantum cascade laser using a room-temperature superlattice harmonic mixer, *Appl. Phys. Lett.*, 103, 051115.

Henderson, B. C. and Cook, J. A., 1985, Image-reject and single-sideband mixers, Watkins-Johnson Tech-Notes, May/June 1985.

Hesler, J., Hall, W., Crowe, T., Weikle, R., Deaver, B., Bradley, R., and Pan, S., 1997, Fixed-tuned submillimeter wavelength waveguide mixers using planar Schottky-Barrier diodes, *IEEE Trans. Microwave Theory Tech.*, 45(5), 653.

Hubers, H.-W., 2008, Terahertz heterodyne receivers, *IEEE J. Select. Top. Quant. Electron.*, 14(2), 378.

Kadin, A., 1999, *Introduction to SuperConducting Circuits*, John Wiley & Sons, New York.

Kerr, A. and Pan, S.-K., 1996, Design of planar image separating and balanced SIS mixers, *Seventh International Symposium on Space Terahertz Technology*, Charlottesville, March 1996.

Kerr A. R., Feldman M. J., and Pan, S. K., 1996, Receiver noise temperature, the quantum limit, and the role of the zero-point fluctuations, Electronics Division Internal Report. No. 304 (also distributed as MMA Memo No. 161).

Klein, B., Philipp, S., Kramer, I., Kaseman, C., Gusten, R., and Menten, K., 2006, The APEX Digital Fast Fourier Transform Spectrometer, A&A, Vol. 454, L29.

Kloosterman, J., 2014, Heterodyne Arrays for TeraHertz Astronomy, PhD dissertation, University of Arizona.

Kloosterman, J., Hayton, D., Ren, Y. et al., 2013, Hot electron bolometer heterodyne receiver with a 4.7-THz quantum cascade laser as a local oscillator, *Appl. Phys. Lett.*, 102(1), 011123.

Kooi, J., 2008, Advanced Receivers for Submillimeter and Far Infrared Astronomy, PhD dissertation, California Institute of Technology.

Kraus, J., 1966, *Radio Astronomy*, McGraw Hill, New York.

Kraus J. D., 1988, *Radio Astronomy*, 2nd edition, Powell, Ohio, Cygnus-Quasar.

Maas, S., 1988, *Nonlinear Microwave Circuits*, Artech House, Norwood, Mass.

Maestrini, A., Mehdi, I., Lin, R. et al., 2011, A 2.5–2.7 THz room temperature electronic source, *Proceedings of 19th International Symposium on Space Terahertz Technology*, Tucson, AZ, April 25–28, p. 1.

Manley J. M. and Rowe, H. E., Some general properties of nonlinear elements—Part I: General energy relations, *Proceedings of the IRE*, July 1956, p. 904–913.

Mehdi, I., Sclecht, E., Chattopadhyay, P., and Siegel, P., 2003, THz local oscillator sources: Performance and capabilities, *Proc. SPIE*, 4855, 435.

Mehdi, I., Ward, J., Maestrini, A., Chattopadhyay, G., Schecht, E., and Gill, J., 2008, Pushing the limits of multiplier-based local oscillator chains, *Proceedings of 19th International Symposium on Space Terahertz Technology*, Groningen, April 28–30, 2008, p. 196.

Meledin, D., Desmaris, V., Ferm, S.-E. et al., 2008, APEX band T2: A 1.25–1.39 THz waveguide balanced HEB receiver, *19th International Symposium on Space Terahertz Technology*, Groningen, April 28–30, 2008.

Orlando, T. and Delin, K., 1991, *Foundations of Applied Superconductivity*, Addison-Wesley Publishing Company, Reading, Mass.

Ossenkopf, V., 2008, The stability of spectroscopic instruments: A unified Allan variance computation scheme, *A&A*, 479, 915–926.

Paul, J., Scheller, M., Laurain, A., Young, A., Koch, S., and Moloney, J., 2013, Narrow linewidth single-frequency terahertz source based on difference frequency generation of vertical-external-cavity source-emitting lasers in an external resonance cavity, *Opt. Lett.*, 38(18), 3654.

Pospieszalski, M. W., 1989, Modeling of noise parameters of MESFET's and MODFET's and their frequency and temperature dependence, *IEEE Trans. Microwave Theory Tech.*, 37(9), 1340.

Pütz, P., Hedden, A., Gensheimer, P. et al., 2006, 35 GHZ prototype SIS mixer with integrated MMIC LNA, *Int. J. Infrared Milli. Waves*, 27, 1365.

Pütz, P., Tils, T., Jacobs, K., and Honongh, C., 2005, Terahertz waveguide mixer development with micromachining and DRIE, *16th Symposium on Space TeraHertz Technology*, Chalmers University, Goteborg, Sweden, p. 338.

Razeghi, M., Lu, Q., Bandyopadhyay, N., Livken, S., and Bai, Y., 2013, Room temperature compact THz sources based on quantum cascade laser technology, *Proc. SPIE*, SPIE Digital Library, 8846, 884602.

Richter, H., Pavlov, S. G., Semenov, A. D. et al., 2010, Submegahertz frequency stabilization of a terahertz quantum cascade laser to a molecular absorption line, *Appl. Phys. Lett.*, 96, 071112.

Rodriguez-Morales, F., 2006, Modeling and Measured Performance of Integrated Terahertz HEB Receivers and Focal Plane Arrays, PhD dissertation, Dept. of Electrical and Computer Engineering, University of Massachusetts-Amherst.

Samoska, L., 2011, An overview of solid-state integrated circuit amplifiers in the submillimeter-wave and THz regime, *IEEE Trans. Terahertz Sci. Technol.*, 1(1), 9.

Scheller, M., Yarborough, J., Young, A. et al., 2011, High power room temperature, compact, narrow line THz source as a local oscillator for THz receivers, *Proceedings of 19th International Symposium on Space Terahertz Technology*, Tucson, AZ, April 25–28, 2011, p. 208.

Schieder, R., Horn, J., Siebertz, O., Mockel, C., Schloder, F., Macke, C., and Schmulling, F., 1998, Design of large bandwidth acousto-optical spectrometers, *Proc. SPIE*, 3357, 359.

Schoelkopf, R., Zimuidzinas, J., Phillips, T., LeDuc, H., and Stern, J., 1995, Measurements of noise in Josephson-effect mixers, *IEEE Trans. Microwave Theory Tech.*, 43, 977.

Schulz, A., Gillespie, A., and Krugel, E., 1987, CO ($J = 4–3$) spectroscopy of M17SW, *Astron. Astrophys.*, 171, 297–304.

Shapiro, S., 1963, Josephson currents in superconducting tunnelling: The effect of microwaves and other observations, *Phys. Rev. Lett.*, 11(2), 80.

Sutton, E., Blake, G., Masson, C., and Phillips, T. G., 1985, Molecular line survey of Orion A from 215 and 247 GHz, *Ap. J. (Suppl)*, 58, 341.

Skocpol, W., Beasley, M., and Tinkham, M., 1974, Self-heating hotspots in superconducting thin-film microbridges, *J. Appl. Phys. Lett.*, 45, 4054.

Tessmann, A., Leuther, A., Massler, H., and Seelmann-Eggebert, M., 2012, A high gain 600 GHz amplifier TMIC using 35 nm metamorphic HEMT technology, in *IEEE Compound Semiconductor Integrated Circuit Symposium (CSIC)*, La Jolla, CA.

Tucker, J. R. and Feldman, M. J., 1985, Quantum detection at millimeter wavelengths, *Rev. Mod. Phys.*, 57, 1055.

Walker, C. K., Groppi, C., d'Aubigny, D. et al., 2001, PoleSTAR: A 4-pixel 810 GHz array receiver for AST/RO', *Proceedings of the 12th International Symposium on Space TeraHertz Technology*, San Diego, CA, eds. Mehdi and McGrath, JPL.

Walker, C. K., Kulesa, C. A., Groppi, C. E., and Golish, D., 2008, Future prospects for THz spectroscopy, in *Millimeter and Submillimeter Detectors and Instrumentation for Astronomy IV*, Duncan W., Holland W., Withington, S., and Zmuidzinas, J., (eds.), *Proc. SPIE*, 7020(702014), 1–8.

Walker, C., Schulz, A., Krugel, E., and Gillespie, A., 1988, CO ($J = 4–3$) spectroscopy in warm molecular clouds, *Astron. Astrophys.*, 205, 243–247.

Weinreb, S., 1961, Digital radiometer, *Proc. IEEE*, 49(6), 1099.

Westig, M., Justen, M., Jacobs, K., Stutzki, J., Schultz, M., Schomacker, F., and Honingh, N., 2012, A 490 GHz planar circuit balanced Nb-Al2O3-Nb quasiparticle mixer for radio astronomy: Application to quantitative local oscillator noise determination, *J. Appl. Phys.*, 112, 093919.

Williams, B., Kumar, S., and Hu, Q., 2005, Distributed-feedback terahertz quantum-cascade lasers with laterally corrugated metal waveguides, *Opt. Lett.*, 30(21), 2909.

Wilms Floet, D., 2001, Hotspot Mixing in THz Niobium Superconducting Hot Electron Bolometer Mixers, PhD dissertation, Technische Universiteit Delft.
Wilms Floet, D., Miedema, E., Klapwijk, T. M., and Gao, J. R. 1999, Hotspot mixing: A framework for heterodyne mixing in superconducting hot-electron bolometers, *Appl. Phys. Lett.*, 74, 433.
Yagi, H., 1928, Beam transmission of ultra-shortwaves, *Proceedings of the IRE*, 16, 715–740.

7

INCOHERENT DETECTORS

PROLOGUE

The first detector used for THz astronomy was a bolometer. Bolometers belong to the incoherent family of detectors. Such detectors are extremely sensitive to the power of an incident signal, but do not "phase-tag" the incoming photons as is done by coherent receivers. Incoherent detectors are well suited for conducting broadband observations of dust in the ISM (interstellar medium), and for performing low resolution spectroscopy of the Milky Way and external galaxies. In this chapter, three types of incoherent detectors will be introduced: the semiconductor bolometer, the superconducting transition edge sensor, and the microwave kinetic inductance detector.

7.1 INTRODUCTION

Unlike coherent detectors, incoherent detectors do not retain knowledge of the phase relationship of the photons that strike them. Such knowledge is only necessary to efficiently realize very high spectral resolution, R, systems (e.g., for observing Galactic spectral lines, where $R \geq 10^5$). For a host of astronomical observations, including dust continuum, dust polarimetry, external galaxies, and cosmic background studies, such high values of R are not necessary.

Not having to retain knowledge of the photon phase means incoherent, broadband continuum detectors are not subject to the quantum mechanical noise limit of coherent receivers (Shimoda et al., 1957; Zmuidzinas, 2002). A simplified block diagram of an incoherent detector system is shown in Figure 7.1. Light from the telescope enters a cryostat where it encounters a filter and/or dispersive optical system that selects the range of frequencies to be observed (Stacey, 2011). The filtered light is then conveyed to one or more detectors that

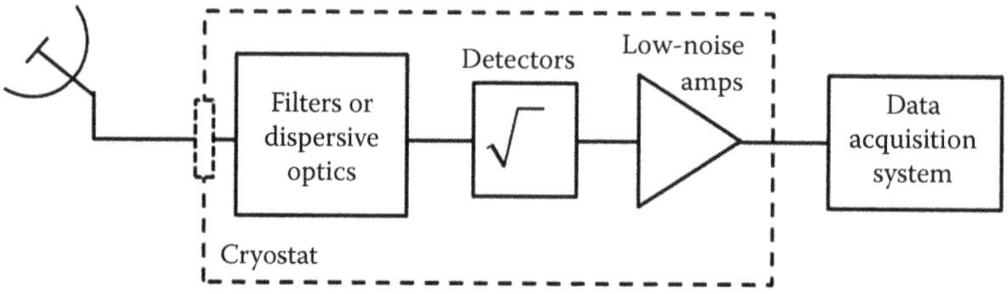

FIGURE 7.1 Simplified block diagram of an incoherent detection system. Light from the telescope enters a cryostat through a low-loss vacuum window and then passes through filters and/or dispersive optics to set the frequency range of observations. The light then illuminates one or more detectors, each of which produces an output voltage proportional to the incident power. The detector(s)'s output voltage is then amplified and recorded along with the corresponding telescope position and instrument parameters by a data acquisition system.

produce an output voltage proportional to the incident power. A data acquisition system then records and processes the data.

In this chapter we introduce the physics associated with 3 types of widely used incoherent detectors; semiconductor, superconducting transition edge, and superconducting microwave kinetic inductance (MKID). We then discuss how observational requirements drive the selection of one type of detector over another.

7.2 BOLOMETER BASICS

At THz frequencies, the formalisms (e.g., Gaussian beam approximation) and components (e.g., lenses, mirrors, substrate antennas, horns, and wire grids) used to design and build optical systems to couple photons to incoherent detectors are much the same as for coherent detectors (see Chapter 6). However, once they reach the detector, only the time-varying incident power level, $P_\nu(t)$, or time-dependent "heat," $Q(t)$, of the incident photons is of consequence, since the phase signature of the photon is not retained. This permits the performance of an incoherent detector to be described using thermodynamics. Following the approach of Rieke (2003), let us imagine a situation as shown in Figure 7.2, where incoming photons from a telescope strike the absorbing layer of an incoherent detector that is loosely coupled to a thermal bath of temperature, T_0. The detector achieves an equilibrium temperature, T_1, due to a combination of $P_\nu(t)$ and the power, P_0, it deposits into the thermal sink to which it is connected. The amount of incident power absorbed from the source by the detector is given by

$$\eta P_\nu(t) = \frac{dQ}{dt} = C\frac{dT_1}{dt} \tag{7.1}$$

where

$P_\nu(t)$ = time-dependent input power from telescope (watts)
η = fraction of incoming power absorbed by the detector (i.e., optical efficiency)

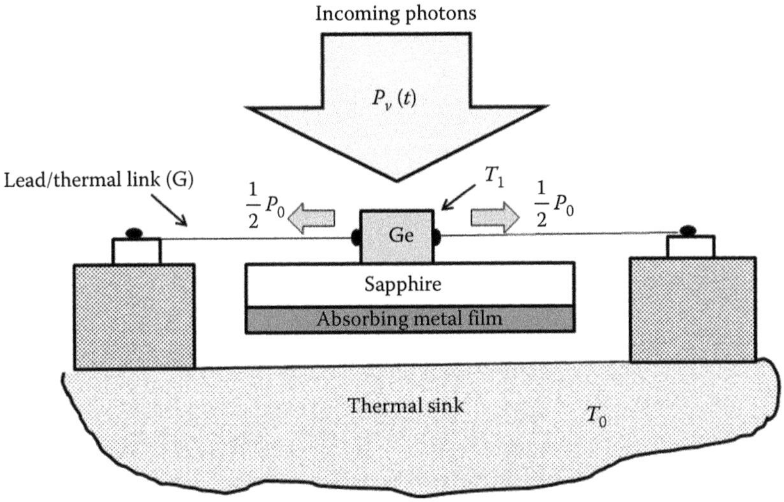

FIGURE 7.2 Semiconductor bolometer schematic. The bolometer depicted above is a germanium composite bolometer. Most of the incoming photons pass by the small, germanium detecting element and are absorbed by a blackened, thin metal film deposited on the back of a sapphire substrate. The film conducts the thermal energy of the collected photons into the sapphire substrate, which then conveys the energy to the germanium thermistor. Sapphire is chosen for its low thermal time constant. The voltage drop observed across the detector is proportional to the incident thermal power. The detector assembly is suspended on electrical/thermal leads. The leads provide a means of electrically biasing the detector. They also regulate the flow of thermal (cooling) power that reaches the assembly within a given amount of time, thereby setting the thermal time constant of the system.

C = heat capacity (joule/K)
Q = heat energy (joules)

The value of P_0 depends on the strength (or conductance, G) of the thermal connection.

$$P_0 = GT_1 \tag{7.2}$$

where
P_0 = power deposited by thermistor into thermal heat sink (watts)
G = thermal conductance (watts/K)
T_1 = detector temperature (K)

In the absence of bias current or optical loading, the bolometer is in thermal equilibrium with the bath via the thermal link G, with equal heat-flow in and out.

Combining Equations 7.1 and 7.2, we have

$$\begin{aligned} P_T(t) &= P_0 + \eta P_v(t) \\ &= GT_1 + C\frac{dT_1}{dt} \end{aligned} \tag{7.3}$$

The solution to Equation 7.3 is the time-dependent temperature of the detector, $T_1(t)$,

$$T_1(t) = \begin{cases} \frac{P_0}{G}, & t < 0 \\ \frac{P_0}{G} + \frac{\eta P_1}{G}(1 - e^{-t/(C/G)}), & t \geq 0 \end{cases} \tag{7.4}$$

The thermal time constant of the detector is, then,

$$\tau_{th} = \frac{C}{G} \tag{7.5}$$

For the case of $t = \tau_{th}$, Equation 7.4 becomes

$$\begin{aligned} T_1(t) &= \frac{P_0}{G} + \frac{\eta P_1}{G} \\ &= \frac{1}{G}(P_0 + \eta P_1) \end{aligned} \tag{7.6}$$

From an examination of Equations 7.5 and 7.6, we see that

1. More photons ($P_1 \Uparrow$) results in higher detector temperatures
2. A lower thermal conductance ($G \Downarrow$) will result in higher detector temperatures for a given number of input photons, at the expense of tolerating a longer-time constant

7.3 SEMICONDUCTOR BOLOMETERS

Unlike other types of detectors that utilize a semiconductor, bolometers do not rely directly on the detection of photons by the excitation of charge carriers by incident photons. Instead, they use an absorbing layer to convert the energy of incident photons into heat, which changes the resistivity of an underlying semiconductor thermometer. To obtain appropriate electrical properties at very low temperatures (<5 K), a sufficient amount of impurities (e.g., gallium, boron, arsenic, and phosphorous) must be added to the semiconductor's crystal lattice (in a process called doping), so there is a probability electrons can "hop" via quantum tunneling from one atom to another without first entering the conduction band. In comparison, conduction processes in a pure semiconductor will "freeze-out" at low temperatures. The resistance in a doped semiconductor bolometer is given by an expression of the form (Rieke, 2003; Hollister, 2006),

$$R_B(T) = R_0 e^{[\Delta/T_1]^{\xi}} \tag{7.7}$$

where

$R_B(T)$ = bolometer resistance as a function of temperature (Ω)
R_0 = characteristic resistance, depends on material and geometry (Ω)
Δ = energy gap temperature, 4–10 K, depending on material (K)
T_1 = bolometer temperature (K)
ξ = nonlinearity exponent (≈ 1/2)

Changes in R_B can be measured using the simple circuit of Figure 7.3. Since the bolometer resistance drops with increasing temperature, it has a negative temperature coefficient of resistance, $\alpha(T)$.

$$\alpha(T) = \frac{1}{R}\frac{dR}{dT} \tag{7.8}$$

As a consequence, since the bolometer bias current, I_B, can heat the device, it is possible for a thermal runaway condition to occur where the bolometer resistance is driven to lower and lower values, until it is fried by the bias circuit. To avoid this unhappy occurrence, a large load resistance, $R_L \gg R_B(T)$, is used in series with the bolometer, to limit the current that can flow through it under any circumstance. At a given bias voltage, V_{bias}, as the number of incoming photons goes up, R_B and V_B go down, while I_B is held constant.

The relationship between V_B, I_B, and the power, P_I, dissipated in the bolometer by the current, I_B, is

$$V_B = [P_I R_B(T)]^{1/2}$$
$$I_B = \left[\frac{P_I}{R_B(T)}\right]^{1/2} \tag{7.9}$$

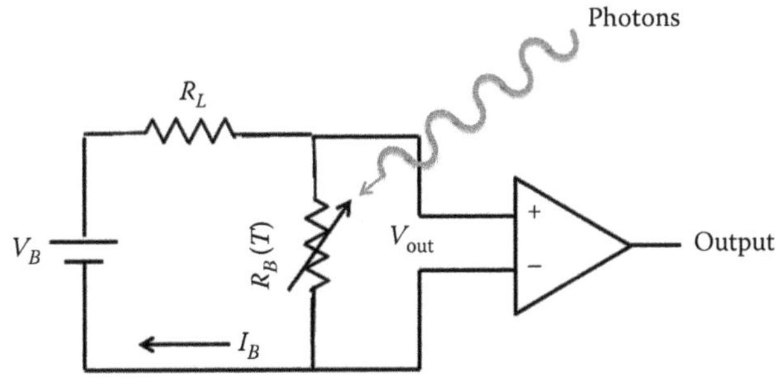

FIGURE 7.3 Bolometer bias and readout circuit. Incoming photons deposit their energy on the bolometer and change its resistance, $R_B(T)$. The change in resistance causes a change in voltage across the bolometer that is amplified and sent to the data acquisition system. A large load resistance, R_L, is placed in series with the bolometer to prevent thermal runaway from occurring.

Since the semiconductor bolometer resistivity is a function of incident power, the electrical time constant appears as a modified version of the thermal time constant of expression Equation 7.5 (Low, 1961).

$$\tau_e = \frac{C}{G - \alpha P_I} \tag{7.10}$$

The responsivity, S_E, of a bolometer is a measure of its voltage response to incident power and can be written as

$$S_E = \frac{\tau_e \alpha V_B}{C} = \frac{\alpha V_B}{G - \alpha P_I} \tag{7.11}$$

where

S_E = electrical responsivity (V/W)
τ_e = electrical time constant (s)
α = temperature coefficient of resistance (K^{-1})
C = heat capacity (joule/K)
V_B = bolometer voltage (V)
G = thermal conductance (watts/K)
P_I = power dissipated in bolometer (W)

The parameters α and G are not always available to the instrumentalist, but can be derived from the bolometer's I-V curve, such as that shown in Figure 7.4 (Rieke, 2003). As in the case of coherent detectors, the I-V curve is found to be nonlinear. In the parlance of incoherent detectors, the I-V curve is often referred to as the "load curve," and plotted as V-I. Refering to Figure 7.5, the bolometer is DC biased to the operating point V_o, I_o. The operating point resistance is $R_o(V_o/I_o)$. The slope of the line tangent to the operating point is the bolometer's impedance, given by

$$Z = \frac{dV}{dI} \tag{7.12}$$

Following Jones (1953), the responsivity can then be written as

$$S_E = \frac{Z - R_o}{2V_o} \tag{7.13}$$

The electrical responsivity, S_E, can be converted to a radiant responsivity, S_R, by multiplying Equation 7.13 by the bolometer's quantum efficiency, η,

$$S_R = \frac{\eta(Z - R_o)}{2V_o} \tag{7.14}$$

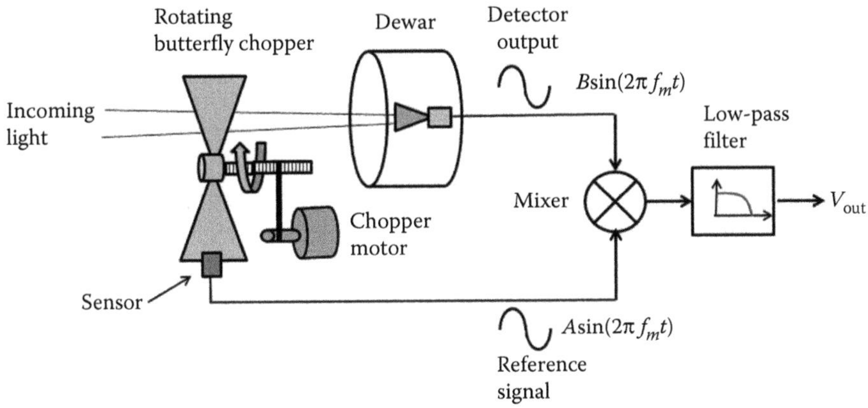

FIGURE 7.4 Schematic of synchronous detection system. The incoming light is modulated at a frequency f_m by the rotating load. The sinusoidally varying detector and reference signals are then multiplied together in a mixer, the product of which contains a copy of the modulated signal shifted to zero frequency (i.e., DC). All but the DC component is filtered out by a low pass filter.

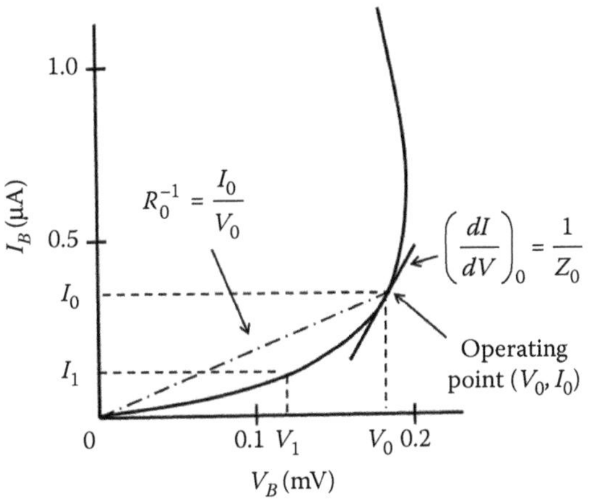

FIGURE 7.5 Bolometer load curve. A bolometer load curve is the current–voltage (I-V) characteristic of the detector. The operating point of the bolometer is often chosen to be at the knee of the load curve. The bolometer resistance at this point is R_0 (the inverse of the slope of the bias line) and its impedance is Z_0 (the inverse of the slope of a line tangent to the operating point). The curve will flatten out as the thermal load on the detector increases. To be consistent with the other I-V curves presented in this text, the load curve has been plotted with current on the y-axis and voltage on the x-axis. However, since bolometers are current biased, most load curves are plotted with current (independent variable) on the x-axis, and voltage (dependent variable) on the y-axis. (After Rieke, G., 2003, *Detection of Light: From Ultraviolet to the Submillimeter*, Cambridge University Press, London.)

The electrical and thermal time constants are then related through the expression,

$$\tau_e = \tau_{th}\left[\frac{Z + R_0}{2R_0}\right] \tag{7.15}$$

Examination of Equations 7.13 through 7.15 reveals that bolometer performance is not dependent on the frequency of the signal photons, except perhaps through the value of η.

The noise in incoherent detectors is typically described in terms of a noise equivalent power (NEP), with units of watts/Hz$^{1/2}$. The NEP is the signal power that results in a root-mean-square (rms) signal-to-noise ratio of one in a system with an electronic bandpass of 1 Hz (Rieke, 2003). The first large format THz bolometer array, submillimetre common-user bolometer array (SCUBA), utilized 131 composite germanium bolometers mounted in waveguide cavities with feedhorns. The typical bolometer NEP was 6×10^{-17} W/$\sqrt{\text{Hz}}$ at 450 μm (Holland et al., 1998).

The noise in a bolometer has contributions from three sources: (1) Johnson (or Nyquist) noise due to the thermal motions of charge carriers in a resistive device; (2) thermal noise due to fluctuations of entropy across the thermal link with conductance G between the detector and heat sink; and (3) the noise generated by the stream of incoming photons. Photon noise is due to the quantized nature of light, and the independence of photon detections. Each noise source is characterized by its own NEP,

$$\begin{aligned}
\text{NEP}_\text{J} &= (4kTR)^{1/2}\frac{1}{\eta}\left|\frac{Z + R_0}{Z - R_0}\right| = (4kT)^{1/2}\left(\frac{G}{\eta|\alpha|}\right)P_I^{-1/2} \\
\text{NEP}_\text{T} &= \frac{1}{\eta}(4kT^2G)^{1/2} \\
\text{NEP}_\text{ph} &= h\nu\left(\frac{2\varphi}{\eta}\right)^{1/2} = \left(\frac{2h\nu P_i}{\eta}\right)^{1/2}
\end{aligned} \tag{7.16}$$

where

NEP_J = noise equivalent power due to Johnson noise in detector (watts/Hz$^{1/2}$)
NEP_T = noise equivalent power due to thermal noise from thermal link (watts/Hz$^{1/2}$)
NEP_ph = noise equivalent power due to incoming photon noise (watts/Hz$^{1/2}$)
$\varphi = (P_i/h\nu)$ = number of photons per sec incident on the detector (Hz)
P_I = power dissipated in detector from bias circuit (watts)
P_i = power of incident photon stream (watts)
G = thermal conductance (watts/K)
α = temperature coefficient of resistance (K^{-1})
ν = frequency of operation (Hz)
η = fraction of incoming power that is absorbed (optical efficiency)

The total NEP of the bolometer is given by the quadratic sum of Equation 7.16.

$$\text{NEP} = (\text{NEP}_\text{J}^2 + \text{NEP}_\text{T}^2 + \text{NEP}_\text{ph}^2)^{1/2} \tag{7.17}$$

EXAMPLE 7.1

You are given the task of measuring the responsivity, S_E, of a detector/amplifier system at 3 THz using a calibrated black-body source. The detector has a diameter of 3.175 mm. The black-body source has a diameter of 2 mm and is heated to 1,263 K. The distance between the source and detector is 0.311 m. With a 50 GHz bandpass filter in front of the detector, the system has an output voltage of 1 mV. What is the system's voltage responsivity?

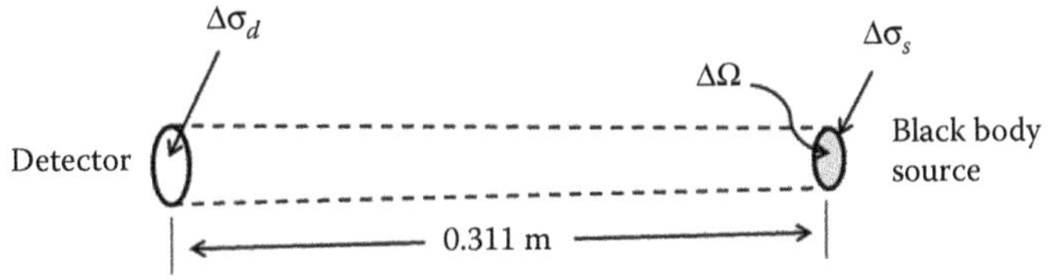

The experimental setup is shown in the above figure. The detector with surface area $\Delta\sigma_d$ is on the left, and the black-body source with area $\Delta\sigma_S$ is located 0.311 m to the right.

$$\Delta\sigma_d = \pi[(0.5)(2\times 10^{-3}\,\text{m})]^2 = 3.14\times 10^{-6}\,\text{m}^2$$
$$\Delta\sigma_S = \pi[(0.5)(3.175\times 10^{-3}\,\text{m})]^2 = 7.92\times 10^{-6}\,\text{m}^2$$

The solid angle, $\Delta\Omega_S$, subtended by the black-body source a distance $S = 0.311$ m from the detector is,

$$\Delta\Omega_d = \frac{\Delta\sigma_S}{S^2} = \frac{7.92\times 10^{-6}\,\text{m}^2}{(0.311\,\text{m})^2} = 8.18\times 10^{-5}\,\text{rad}^2.$$

The intensity of the black-body source is described by solving the Planck function (Equation 2.19), $B_\upsilon(T)$, for $\upsilon = 3\times 10^{12}$ Hz and $T = 1000$ K. Substitution into Equation 2.19 (using MKS units) yields

$$I_\upsilon = 2.58\times 10^{-12}\frac{\text{watts}}{\text{m}^2\text{rad}^2\text{Hz}}.$$

The total power incident over the surface area of the detector is, then,

$$P = I_\upsilon \Delta\upsilon \Delta\Omega_d \Delta\sigma_d = (2.58\times 10^{-12})(50\times 10^9)(8.18\times 10^{-5})(3.14\times 10^{-6})$$
$$= 3.3\times 10^{-11}\,\text{watts}$$

The detector responsivity, S_E, is then simply equal to the detector's output voltage, V_d, when exposed to P divided by the value of P.

$$S_E = \frac{V_o}{P} = \frac{1\times 10^{-3}\,\text{V}}{3.3\times 10^{-11}\,\text{watts}} = 3\times 10^7\frac{\text{V}}{\text{watts}}.$$

EXAMPLE 7.2

With no incident power, the detector–amplifier of Example 7.1 has a measured noise spectral density of

$$N_{d-a} = 1.83 \times 10^{-7} \frac{\text{V}}{\sqrt{\text{Hz}}}.$$

What is the detector–amplifier NEP?

The NEP is simply the ratio of the intrinsic detector–amplifier noise spectral density to the detector–amplifier responsivity,

$$\text{NEP} = \frac{N_{d-a}}{S_E} = \frac{1.83 \times 10^{-7}(\text{V}/\sqrt{\text{Hz}})}{3 \times 10^{7}(\text{V/watts})} = 6.1 \times 10^{-15} \frac{\text{watts}}{\sqrt{\text{Hz}}}.$$

EXAMPLE 7.3

A neutron doped germanium bolometer is mounted at the center of a free-standing silicon nitride mesh. A thin metallic film evaporated on the mesh absorbs radiation and provides an impedance match to free space. The mesh thermally isolates the bolometer from its surroundings. The bolometer is electrically biased using a circuit like that of Figure 7.5. The detector system has the following characteristics:

$$T = 1.5\,\text{K}$$
$$R_L = 1 \times 10^{7}\,\Omega$$
$$R_B = 1 \times 10^{6}\,\Omega$$
$$G = 1 \times 10^{-10}\,\text{W/K}$$
$$\eta = 0.7$$
$$\alpha = -2\text{K}^{-1}$$

If the bolometer is biased at 1.3 V, what will be its NEP? Assume the detector system is a cooled, space-based telescope.

The current through the bolometer, I_B, is determined by the bias voltage, V_B, and the load resistance, R_L.

$$I_B \approx \frac{V_B}{R_L} = \frac{1.3\text{V}}{1 \times 10^{7}\,\Omega} = 1.3 \times 10^{-7}\,\text{A}.$$

The electrical power dissipation, P_I, is, then,

$$P_I \approx I_B^2 R_B = (1.3 \times 10^{-7}\,\text{A})^2 (1 \times 10^{6}\,\Omega) = 1.7 \times 10^{-8}\,\text{W}$$

Substituting into Equation 7.16, we find the contribution to the device NEP in the Johnson noise limit is

$$\begin{aligned}\mathrm{NEP_J} &= [(4)(1.38\times10^{-23})(1.5)]^{1/2}\left(\frac{1\times10^{-10}}{(0.7)-2}\right)(1.7\times10^{-8})^{-1/2}\\ &= 5\times10^{-18}\,\mathrm{WHz^{-1/2}}\end{aligned}$$

The contribution to the NEP from thermal fluctuations in the thermal link is (Equation 7.16)

$$\begin{aligned}\mathrm{NEP_T} &= \frac{1}{0.7}[(4)(1.38\times10^{-23})(1.5)^2(1\times10^{-10})]^{1/2}\\ &= 1.6\times10^{-16}\,\mathrm{WHz^{-1/2}}\end{aligned}$$

Since the detector system is used on a cooled, space-based telescope, it is operating in the low background limit, where photon noise makes a negligible contribution to the overall system noise. In this case, the system NEP is equal to the quadratic sum of $\mathrm{NEP_J}$ and $\mathrm{NEP_T}$.

$$\begin{aligned}\mathrm{NEP} &= (\mathrm{NEP_J^2}+\mathrm{NEP_T^2})^{1/2}\\ &\approx 1.6\times10^{-16}\,\mathrm{WHz^{-1/2}}.\end{aligned}$$

Here, we can see that thermal noise from the thermal link is limiting performance. Operating the bolometer at lower temperatures (e.g., 300 mK) would significantly lower (i.e., improve) the system's NEP.

Another source of noise in bolometric systems is "$1/f$" or "pink noise" due to slow fluctuations in the thermal and electrical properties of materials within the system. As the term implies, the noise drops with frequency. For this reason, most bolometer systems are amplitude modulated, either by (1) periodically varying the device's bias current, (2) periodically placing an absorbing load into the incoming beam, or, if on a telescope, (3) using a nutating (i.e., chopping) secondary, to alternately look on and off a source. Modulation shifts the output of the bolometer to a frequency above which the contribution of $1/f$ noise is no longer significant. Usually, a modulation frequency of only a few Hz is required. A synchronous detector (often referred to as a lock-in amplifier) is then used to demodulate the bolometer output and produce a DC voltage proportional to the voltage drop observed across the detector due to the source being observed.

A block diagram of a synchronous detector utilizing a rotating chopper is shown in Figure 7.4. The incoming light is modulated at a frequency, f_m, by the rotating load. Here, f_m is measured by the passage of the chopper blade through a sensor. The detector output and reference signal are both sine waves of frequency f_m, but with different amplitudes. The reference signal has a fixed amplitude, A, while the detector signal amplitude, B, varies with the amount of incoming light. Similar to what occurs in a heterodyne receiver (see Equation 6.10), the two signals are multiplied together in a mixer, but here with the same frequency. The product yields one signal at zero frequency (i.e., DC), and another at twice the original frequency. A low pass filter connected to the output of the mixer removes all but the DC component, $AB/2$, in Equation 7.18, which is a copy of the modulated signal moved back to zero frequency (Orozco, 2014).

$$\underbrace{A\sin(2\pi f_m t)}_{\substack{\text{Reference}\\\text{signal}}} \times \underbrace{B\sin(2\pi f_m t)}_{\substack{\text{Detector}\\\text{output}}} = \underbrace{\frac{1}{2}AB}_{V_{out}} - \underbrace{\frac{1}{2}AB\cos(4\pi f_m t)}_{\text{Filtered-out}} \tag{7.18}$$

If the response of the detector is fast enough, square wave modulation can also be used.

Since synchronous detectors only recover the signal occurring at the chopping (i.e., reference) frequency, they suppress noise at other frequencies, often dramatically improving the signal-to-noise ratio of an observation. Synchronous detection plays a key role in many astronomical detection systems.

7.4 SUPERCONDUCTING INCOHERENT DETECTORS

Due to their sensitivity and ability to be readily fabricated into arrays, most incoherent astronomical instruments in use today utilize superconducting detectors. These fall into two categories, superconducting transition edge sensors (TES) and, more recently, microwave kinetic inductance detectors (MKID). The TES is a thermal device, with operating characteristics in some ways analogous to the semiconductor bolometer described above. The operation of an MKID is more analogous to an superconductor-insulator-superconductor (SIS) or hot electron bolometer (HEB) mixer (Section 7.5), with the output appearing as a power spectrum on a spectrometer.

7.4.1 TRANSITION EDGE SENSORS (TES)

An introduction to superconductivity is provided in Section 6.8.2. There, we learned that when some materials are cooled to sufficiently low temperatures, the transfer of acoustical energy between electrons passing through the material's crystal lattice will form a tenuous link between them. Two electrons bound in this way share the same energy state and are referred to as a Cooper pair. It is the existence of Cooper pairs that leads to the low resistivity observed in superconductors. When the link is severed, say, by the power absorbed from an incoming stream of photons, or by the stress of an applied DC voltage, the resistivity of the material reverts back to normal values. A TES sensor works by DC biasing a superconducting bridge in the transition region between being a superconductor and normal, such that only the slightest amount of power from an incoming photon is required to make the bridge swing more toward normal. As shown in Figure 7.6, the transition region between superconducting and normal resistances can be quite narrow.

The two greatest challenges to realizing practical TES detectors were providing a stable bias and reading them out. If current biasing is used, Joule heating can lead to thermal

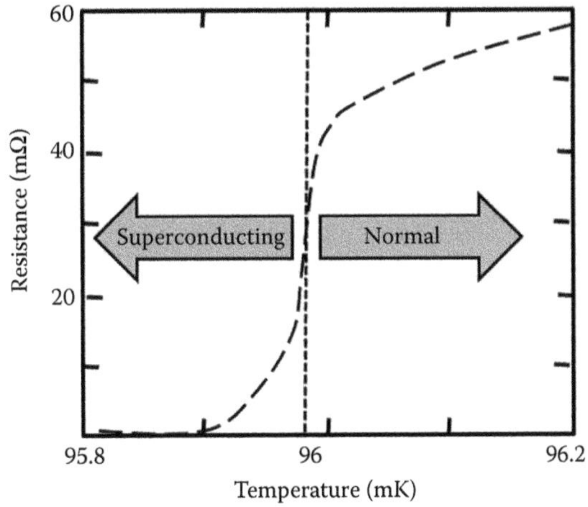

FIGURE 7.6 Nonlinear resistance at a superconducting transition. Above its critical temperature, T_c, (here, ~96 mK for a Mo/Cu bilayer) a superconductor goes "normal." The transition can be quite sharp, leading to the high sensitivity possible with a transition edge sensor (TES). (After Irwin, K. and Hilton, G., 2005, *Topics Appl. Phys.* 99, 63–149.)

runaway. Thermal runaway occurs when collisions between bias electrons and the crystal lattice make part of the device go normal, leading to greater resistivity and increased heating, until the whole device goes normal. To avoid this possibility, TESs are voltage biased instead of current biased. The read-out of TESs was initially challenging due to their inherently low resistance, less than a few Ohms. A number of output circuit approaches were attempted (see Irwin et al., 1995). By far the most successful approach has been to utilize magnetically coupled, superconducting quantum interference devices (SQUIDs). A schematic of a voltage biased TES with a SQUID readout is shown in Figure 7.7.

As in the case of the semiconductor bolometer, a voltage applied to a load resistor, R_L, is used to set the "dark" current, i_B, through the TES. The TES resides in a parallel branch of the circuit in series with an inductor, L. The TES acts as a resistor whose steady state value, R_o, is modulated by a dynamic component, R_{dyn}, whose value changes depending on the amount of power absorbed, such that

$$R_{dyn} = R_o(1 + \beta_I) \tag{7.19}$$

where

$\beta_I = \frac{I_o}{R_o} \frac{\partial R}{\partial I}\bigg|_{T_o}$ = current sensitivity (unitless)

I_o = bias current (A)

R_o = steady state resistance (Ω)

Variations in R_{dyn} cause proportional changes in i_B, leading to proportional variations in the amount of magnetic field, B_{dyn}, generated by the series inductor. The adjacent SQUID serves as a sensitive magnetometer that picks up these small changes in B_{dyn}, which

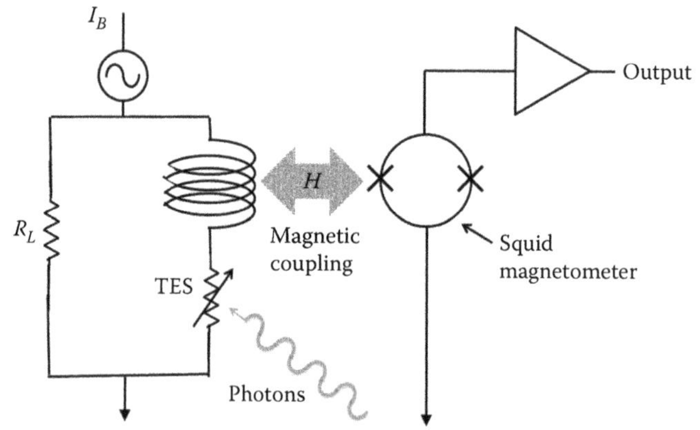

FIGURE 7.7 Schematic of a transition edge sensor (TES). Incoming photons change the resistivity of a TES biased on the superconducting transition. The impedance of a TES is very low, making it impractical to directly measure its output voltage with an amplifier. Instead, the TES is placed in series with a coil, which produces a magnetic field whose intensity, H, is slaved to I_B. Changes in the photon flux falling on the TES cause changes in H, which are detected by an adjacent squid magnetometer. The change in voltage on the squid is amplified and conveyed to the data acquisition system.

manifest themselves as voltage swings across the SQUID bridge, that are then amplified and read by a computer. The electrical time constant, τ_{el}, of the circuit is

$$\tau_{el} = \frac{L}{R_L + R_{dyn}} \tag{7.20}$$

The current biased thermal time constant, τ_l, is given by

$$\tau_l = \frac{\tau_{th}}{1 - g_{lp}} \tag{7.21}$$

where

$\tau_{th} = C/G$ = natural thermal time constant (sec), see Equation 7.5

$g_{lp} = \frac{P_{J_o}\alpha_I}{GT_o}$ = low frequency loop gain under constant current (unitless)

$P_{Jo} = I_o^2 R_o$ = steady state joule power (watts)

$\alpha_I = \frac{T_o}{R_o}\frac{\partial R}{\partial T}\Big|_{I_o}$ = temperature sensitivity (unitless)

Irwin and others have shown that the current responsivity (A/Watt), $s_I(\omega)$, of a TES to radiation at angular frequency, ω, can be expressed as (Irwin and Hilton, 2005)

$$s_I(\omega) = -\frac{1}{I_o R_o}\left(\frac{L}{\tau_{el} R_o g_{lp}} + \left(1 - \frac{R_L}{R_o}\right) + i\omega\frac{L\tau}{R_o g_{lp}}\left(\frac{1}{\tau_l} + \frac{1}{\tau_{el}}\right) - \frac{\omega^2\tau}{g_{lp}}\frac{L}{R_o}\right)^{-1} \tag{7.22}$$

The noise sources in a TES include the external Johnson noise voltage with power spectral density $S_{Vext}(\omega)$, internal Johnson noise voltage $S_{V_{\text{int}}}(\omega)$, noise power from thermal fluctuations $S_{P_{TFN}}$(É), and amplifier current noise $S_{I_{amp}}$(É). Assuming the noise sources are uncorrelated, they can be summed, such that

$$S_I(\omega) = S_{V_{ext}}(\omega)\,|\,Y_{ext}(\omega)\,|^2 + S_{V_{int}}(\omega)\,|\,Y_{\text{int}}(\omega)\,|^2 + S_{TFN}(\omega)\,|\,s_I(\omega)\,|^2 + S_{I_{amp}}(\omega) \qquad (7.23)$$

where

$S_I(\omega)$ = overall current noise ($A^2\ Hz^{-1}$)
$Y_{ext}(\omega)$ = admittance of external noise source (Ω^{-1})
$Y_{int}(\omega)$ = admittance of internal noise source (Ω^{-1})

Expressions for each noise term can be found in Irwin and Hilton (2005). The overall power spectral density of the power-referred noise in a TES is, then,

$$S_P(\omega) = \frac{S_I(\omega)}{\left|s_I(\omega)\right|^2} \qquad (7.24)$$

The NEP of a TES is, then, simply,

$$NEP(\omega) = \sqrt{S_P(\omega)} \qquad (7.25)$$

The impedance of a TES is generally much higher than that of the SQUID, such that noise terms associated with the SQUID can be neglected. Since the SQUID acts as a magnetometer, it should be shielded to prevent external magnetic fields from interfering with measurements. Also, as in the case of semiconductor bolometers, the sensitivity and thermal time constant of the TES is a function of the absorber and thermal conductance.

At THz frequencies, radiation is most often coupled to a TES by way of a broadband matching layer to free space, locating the device in a quarter wave cavity at the throat of a feedhorn, or by locating it at the apex of a planar antenna. To control thermal conductance, the TES is fabricated on a thin dielectric membrane (i.e., silicon nitride; Irwin and Hilton, 2005), in a pop-up structure (see Figure 7.8 and Dowell et al., 2003), or by locating it at the center of a spiderweb structure (Bock et al., 1995). NEP values as low as $\sim 2.5 \times 10^{-19}\ \text{W}/\sqrt{\text{Hz}}$ have been achieved with iridium (Ir) TES bolometer arrays (T_c = 10 mK, $\tau \sim 4.5$ ms; Beyer et al., 2011).

7.4.2 MICROWAVE KINETIC INDUCTANCE DETECTORS (MKID)

MKIDs, like all superconducting detectors, ultimately achieve their sensitivity through the breaking of Cooper pairs within a superconducting film. However, the mechanism by which pair-breaking is utilized is different than in other detectors. Here, it is the change in the impedance of a superconducting film due to pair-breaking that is used to deduce the incident photon flux. "Microwave" refers to the frequency range of the read-out electronics, not the observing frequency.

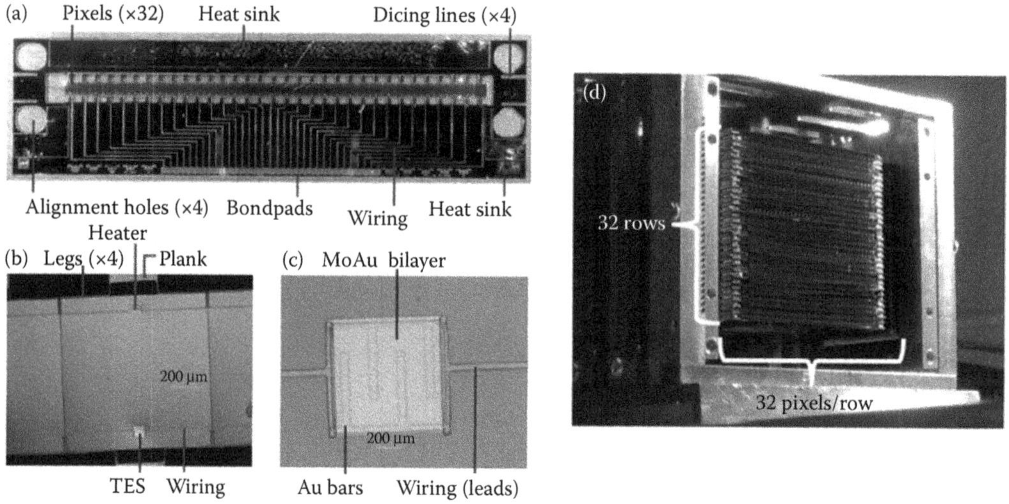

FIGURE 7.8 TES implementation: 32 × 32 TES pop-up array for THz astronomy. (a) Photograph of 1 × 32 row of pop-up bolometers with wiring layer. (b) Close-up of individual pixel with 1.055 mm² absorbing layer. (c) Close-up of TES device. (d) Stack of 32 rows, yielding a 1024 pixel array. The array is only ~32 mm on a side. (Courtesy of S.H. Mosley group, Goddard Spaceflight Center. With permission.)

When photons strike the surface of a superconductor cooled to below its critical temperature, $T < T_c$, the time-varying potential associated with the photons (see Equation 7.18) will accelerate the Cooper pair charge carriers in such a way as to reflect or absorb the incoming photons. As dictated by Newton's Second Law, this acceleration will be opposed by the inertia of the Cooper pairs, resulting in a phase lag between the incident electric field and that produced in opposition by the Cooper pairs on the superconductor's surface. The electrical phase lag is the same as would be produced by an inductor that opposes the finite rate of change of flux through it (i.e., Faraday's Law). Since the phase lag is rooted in the dynamics of Cooper pairs, it is referred to as a kinetic inductance, L_K. The value of L_K in a superconducting film can be computed by equating the total kinetic energy of the Cooper pairs, KE_C, with an equivalent inductive energy, IE_C (Annunziata et al., 2010).

$$\begin{aligned} KE_C &= IE_C \\ \frac{1}{2}(2m_e v^2)(n_{CP} l A) &= \frac{1}{2} L_K I^2 \end{aligned} \tag{7.26}$$

where

m_e = mass of an electron (kg)
v = average Cooper pair velocity (m s^{-1})
n_{CP} = number density of Cooper pairs (m^{-3})
l = length of film (m)
A = cross-sectional area of film (m^2)
L_K = effective kinetic inductance (Henrys)
I = supercurrent through film (A)

Since current can be defined as, $I = 2evn_{CP}A$, Equation 7.26 can be rewritten as

$$L_K = \left(\frac{m_e}{2n_{CP}e^2}\right)\left(\frac{l}{A}\right) \quad (7.27)$$

where
e = charge of an electron (C).

Using the Bardeen, Cooper, Schrieffer (BCS) theory of superconductivity in the low frequency limit ($h\nu = k_B T$), Equation 7.27 takes the form,

$$L_K = \left(\frac{l}{w}\right)\frac{R_{sq}h}{2\pi^2\Delta}\frac{1}{\tanh\left(\frac{\Delta}{2k_B T}\right)} \quad (7.28)$$

where
l = length of film (m)
w = film width (m)
R_{sq} = sheet resistance in nonsuperconducting state (Ω/square)
h = Planck's constant = 6.626×10^{-34} m² kg/s
$T \approx T_c$ = critical temperature of superconducting film (K)
$\Delta(T) \approx \Delta(0)1.74(1 - T/T_c)^{1/2}$ (Joules)
$\Delta(0) \approx 1.76\, k_B T_c$ (Joules)

Incident photons with energy, $E_p = h\nu_p$, greater than the Cooper pair binding energy, $2\Delta \approx 3.5k_B T_c$, will break Cooper pairs, causing the total number of Cooper pairs to decrease by Δn_{CP}. The breaking of Cooper pairs, in effect, decreases the device's superconductivity, thereby increasing both the film's surface resistance, R_{sq}, and the kinetic inductance, L_K. The inductive response to the incident radiation is generally several times greater than the resistivity response (Schlaerth et al., 2009). Therefore, by monitoring the change in L_K and R_S, the incident photon flux can be determined.

The value of L_K and the associated incident photon flux are determined by placing the film in parallel with a fixed capacitor, C, and measuring changes in the circuit's resonant frequency, f_R, (see Figure 7.9),

$$f_R = \frac{1}{2\pi\sqrt{(L_D + \delta L_K)C}} \quad (7.29)$$

where
f_R = resonant frequency of circuit (Hz)
L_D = "dark" circuit inductance (Henrys)
δL_K = change in kinetic inductance due to incident photons (Henrys)
C = fixed capacitance used in resonant circuit (F)

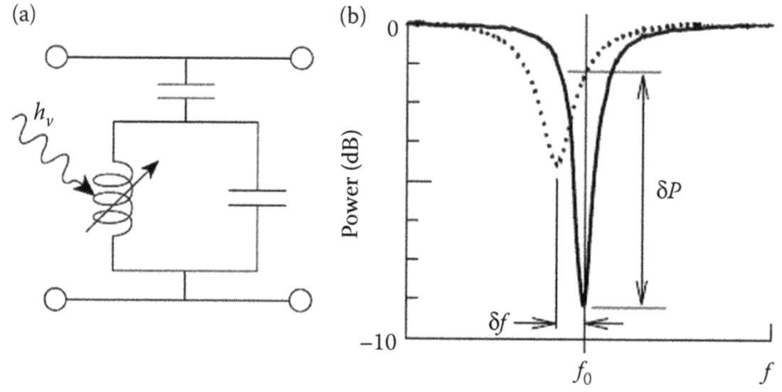

FIGURE 7.9 Schematic of a microwave kinetic inductance detector (MKID). (a) Incoming photons change the inductance and resistivity of a superconducting film within a resonant circuit. (b) The change in inductance produces a shift, δf in the circuit's resonant frequency, f_o. The accompanying change in film resistivity makes the new resonant dip shallower and broader. The value of δf and the change in the shape of the dip (or Q) can be used to determine the incident power. Unlike conventional bolometers, the output of an MKID appears on a spectrometer. (Courtesy of B. A. Mazin, University of California Santa Barbara; see also Day, P. et al., 2003, *Nature*, 425, 817. With permission.)

The value of C is typically chosen to put f_R at microwave frequencies. In astronomical instruments employing MKIDS, a frequency comb is generated, with one tone for each MKID resonator (plus additional off-resonance tones, which are used for noise measurement and subtraction). The MKID output response is amplified using a low-noise cryogenic amplifier (see Section 6.6). The complex transmission (denoted by the scattering parameter, S_{21}) at each MKID frequency is then measured (as in the lower right part of Figure 7.10). From these measurements, the value of δL_K for each MKID is computed.

Resonant circuits are often characterized in terms of a quality factor, or Q, defined as the ratio of energy stored in the circuit, to the energy dissipated in the circuit. For the case of the MKID circuit of Figure 7.9, Q can be defined as

$$Q = \frac{R_S}{\omega L_{tot}} \tag{7.30}$$

where

$L_{tot} = L_D + \delta L_K =$ total inductance (Henrys).

Since incident light increases the inductance of a superconductor, one would expect the observed resonance from an MKID to be shallower and wider when light is applied, than when no light is applied. This is the situation observed in Figure 7.9b.

Solving Equation 7.29 for δL_K, we have,

$$\delta L_K = \frac{1}{C}\left(\frac{1}{2\pi f_R}\right)^2 - L_D \tag{7.31}$$

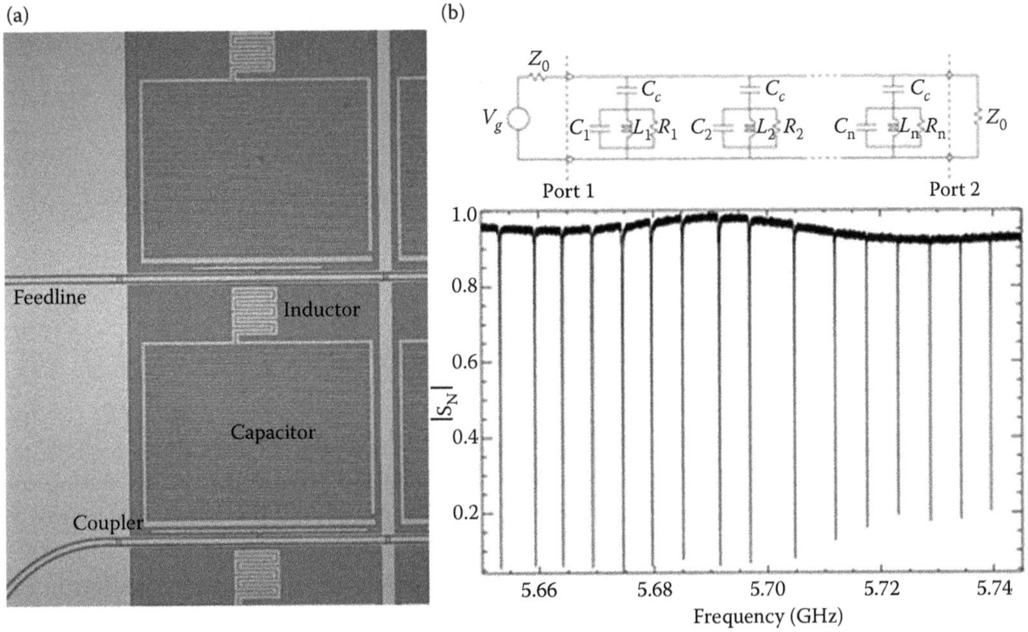

FIGURE 7.10 MKID array implementation. (a) Optical image of a lumped element MKID array made on 60 nm thick TiN film. (b) By changing the value of the capacitor, C_n, between pixels, the array can be read out using a single transmission line. Resonant dips from the array's pixels appear adjacent to each other on a spectrometer. (Courtesy of B. A. Mazin, University of California Santa Barbara.)

Substituting Equation 7.31 into the analog of Equation 7.27, for the case of a change in kinetic inductance due to a change in the number density of Cooper pairs, Δn_{CP}, we find,

$$\Delta n_{CP} = \frac{m_e}{2e^2}\left(\frac{l}{A}\right)\left[\frac{1}{C}\left(\frac{1}{2\pi f_R}\right)^2 - L_D\right]^{-1} \tag{7.32}$$

Let ΔP_{in} be the power incident on the MKID over the Cooper pair recombination time, τ_r. Since the value of Δn_{CP} is directly proportional to P_{in}, and inversely proportional to L_K, we have

$$\begin{aligned}\Delta P_{in} &= \frac{\alpha\eta}{\tau_r}\Delta n_{CP} \\ &= \frac{\alpha\eta}{\tau_r}\frac{m_e}{2e^2}\left(\frac{l}{A}\right)\left[\frac{1}{C}\left(\frac{1}{2\pi f_R}\right)^2 - L_D\right]^{-1}\end{aligned} \tag{7.33}$$

where

α = optical efficiency by which photons are absorbed by the radiator

$\eta = 0.57$ = efficiency of Cooper pair breaking by absorbed photons

The voltage associated with the binding energy of a Cooper pair is

$$\Delta V_{CP} = \frac{\Delta}{e} \approx 3.5\frac{k_B}{e}T_c \tag{7.34}$$

Dividing Equation 7.34 by Equation 7.33, we can derive a simple expression for the intrinsic voltage responsivity, s_V, of an MKID, as,

$$s_V = \frac{\Delta V_{CP}}{\Delta P_{in}} = \frac{7k_B e\tau_r T_c}{\alpha\eta m_e}\left(\frac{A}{l}\right)\left[\frac{1}{C}\left(\frac{1}{2\pi f_R}\right)^2 - L_D\right] \tag{7.35}$$

During astronomical observations (for any ground-based MKID instrument), there is sufficient background power (from the sky and telescope) to substantially shift the dark (no-photon) resonant frequencies, even when off-source. So observers take a frequency sweep across the resonators prior to an observation, then fit the resulting profiles to determine the resonance frequencies with background. This sweep is then used to generate a background reference frequency comb for use during observations. The observed on-source shift in frequency, δf, and the change in the shape of the resonant dip (or Q) are compared to the reference comb to determine the incident power, P_S, from the astronomical source.

The primary source of nonastronomical noise in MKIDs is phase noise arising in the resonators that manifests itself as frequency jitter, which, in turn, leads to uncertainties in measuring f_r and, therefore, ΔP_{in}. The excess phase noise has an $f_\nu = \nu^{-0.5}$ spectrum, and decreases as the microwave power in the read-out is increased by $f_\nu(P) = f_\nu P^{-0.5}$ (Mazin, 2014). The primary source of the noise is believed to be caused by two level systems (TLSs) on the surfaces of the metals and dielectrics of the resonator (Gao et al., 2008).

MKIDS are often made using coplanar waveguide (CPW) quarter wavelength resonators capacitively coupled to a feedline, as shown in Figure 7.10. MKID superconducting resonators have been successfully made using Nb, NbTiN, Ta, Re, Al, AlMn, Mo, PtSi, Ti, and Ir. With this architecture, many MKID pixels can be fabricated on a single substrate, with each pixel designed to have a different resonant frequency. This approach allows many pixels to be read out using a single transmission line and cryogenic low noise amplifier. The relative simplicity of the read-out electronics makes the realization of very large format MKID arrays possible. The measured NEP's of MKIDs (e.g., $\sim 3.8 \times 10^{-19}$ W/$\sqrt{\text{Hz}}$, de Visser et al., 2014) are competitive with TES detectors and is limited by quasiparticle fluctuations.

7.5 BACKGROUND NOISE LIMITED OPERATION

For many THz ground-based systems, the photon stream arriving at the detector contains a significant contribution from the atmosphere, such that NEP_{ph} dominates the system sensitivity. In this case, the observations made with the system are said to be *background noise limited.*

Due to the presence of water vapor, the Earth's atmosphere appears as a diluted back-body load to the detector system with a temperature,

$$T_{atm} = T_p(1 - e^{-\tau'}) \tag{7.36}$$

where

T_{atm} = apparent atmospheric temperature (K)
T_p = physical temperature of the atmosphere (K)
τ' = average optical depth of atmosphere over frequency interval of interest, $\Delta\nu$

Integrating the Planck function, $B_\nu(T_{atm})$, over an observational bandwidth, $\Delta\nu_{BW}$, centered on ν, gives the total atmospheric brightness, B',

$$B' = \int_{\nu - \Delta\nu_{BW}/2}^{\nu + \Delta\nu_{BW}/2} B_\nu(T_{atn})d\nu \tag{7.37}$$

The value of $\Delta\nu_{BW}$ is usually set by one or more filters in front of the bolometer. The total power, P_o, incident on the detector from the atmosphere is, then,

$$P_o = AB'\Delta\Omega \tag{7.38}$$

where

P_o = power incident on detector (W)
A = detector absorption cross-section (m^2)
$\Delta\Omega$ = solid angle from detector to atmosphere (rad^2)
B' = Planck function evaluated at temperature T_{atm} (K)

The photon flux, φ, is then given by

$$\varphi = \frac{P_o}{h\nu} \tag{7.39}$$

Substitution into Equation 7.16 yields

$$\text{NEP}_{\text{ph}} = (h\nu)^{\frac{1}{2}}\left(\frac{2AB'\Delta\Omega}{\eta}\right) \tag{7.40}$$

The above expression shows that the photon noise contribution to the system can be mitigated to some degree by increasing the detector quantum efficiency, and reducing the size and field of view of the detector.

The performance of an incoherent detector system on a particular telescope is sometimes characterized by a noise equivalent flux density (NEFD), corresponding to the flux density that produces a signal-to-noise-ratio (SNR) of one, in one second of integration.

$$\text{NEFD} = \frac{\text{NEP}}{\eta_c \eta_t A_e e^{-\tau_\nu A} \Delta\nu} \tag{7.41}$$

where

NEFD = noise equivalent flux density watts/m^2 Hz$^{3/2}$
NEP = noise equivalent power of system (watts/$\sqrt{\text{Hz}}$)
η_c = source/telescope coupling efficiency
η_t = telescope/detector coupling efficiency
A_e = effective aperture of telescope (m^2)
τ_ν = zenith optical depth of atmosphere at observing frequency
A = atmospheric air mass
$\Delta\nu$ = bandwidth of observation (Hz)

In the above expression, the atmospheric photon noise contribution to the NEP is excluded, since it is taken care of in the denominator. Adopting this definition for NEFD, the time, $\Delta\tau$, required to achieve a flux limit, S_o, on a telescope, is

$$\Delta\tau = \left[\frac{NEFD}{S_o}\right]^2 \tag{7.42}$$

where

$\Delta\tau$ = required integration time (s)
S_o = target flux density (watts/m^{2}Hz)

7.6 INSTRUMENT NOISE LIMITED OPERATION

For high-altitude balloon-borne or space-based missions, the absorptive effects of the atmosphere no longer dominate the system's NEP. In this case, the observations made with the system are said to be *instrument noise limited*. In some instances, particularly in the case of space-based telescopes, what sets the lower limit to the NEP may not be the detector itself, but the diluted black-body emission of optical elements leading up to it. These can include the telescope, dewar windows, infrared filters, and bandpass filters. As in the case of atmospheric noise, the greater the system bandwidth, $\Delta\nu_{BW}$, the more susceptible the detector will be to unwanted background noise, in this case, due to the optics themselves. In broadband systems, great effort is made to make the emission from optics as low as possible. This is accomplished by using optical components with low emissivity, Q_ν^{opt}. The emissivity is the ratio of radiated power from a material to that radiated by a true black body at the same temperature. For any given values of Q_ν^{opt}, an improvement in sensitivity can be achieved by cooling the optics. *In detector systems that utilize heterodyne receivers, the bandwidth is sufficiently narrow that the black-body emission from the optics does not limit system sensitivity and cooling of optical components is not essential.*

The contribution of the optical emissivity to the overall NEP is found by computing the value of NEP_{ph} due to the optics. Integrating the Planck function at T_{opt} over $\Delta\nu$, and multiplying by Q_ν^{opt}, gives the total optical component brightness, B', of

$$B' = Q_\nu^{opt} \int_{\nu-\Delta\nu_{BW}/2}^{\nu+\Delta\nu_{BW}/2} B_\nu(T_{opt})d\nu \approx Q_\nu^{opt} B_\nu(T_{opt})\Delta\nu_{BW} \tag{7.43}$$

where

ν = center frequency of observation (Hz)
T_{opt} = physical temperature of optics (K)
$\Delta\nu_{BW}$ = predetection bandwidth of system (Hz)
Q_ν^{opt} = emissivity of component

The value of Q_ν^{opt} varies between materials. Gold-coated optics can have a Q_ν^{opt} at THz/far-infrared (e.g., 164 μm) of 0.006. Likewise, aluminum coatings have been produced that achieve an emissivity of ~0.007. For optimum performance, the coating thickness should be ≥ 3 skin depths (see Equation 5.65) at the frequency of interest (Xu et al., 1996). The optics contribution to NEP_{ph} is then determined by substituting the appropriate value of B' into Equation 7.40.

If the optical system contains multiple n components, then the photon noise contribution from each, NEP_n^{ph}, can be computed and quadratically added to yield a total NEP_{ph}^2 to be used in Equation 7.17.

$$NEP_{ph}^2 = \sum_{n=1}^{n} (NEP_n^{ph})^2 \tag{7.44}$$

7.7 SENSITIVITY REQUIREMENTS

The push to higher incoherent detector sensitivities is driven principally by the desire to observe the spectral energy distributions of galaxies (see Section 1.5) at greater and greater distances. The distances to extragalactic objects is often specified in terms of their redshift, z, defined as

$$z = \frac{\nu_{rest} - \nu_{obs}}{\nu_{obs}} \tag{7.45}$$

where

ν_{rest} = frequency at which the photon was emitted (Hz)
ν_{obs} = frequency at which photon is observed (Hz)

The approximate distance to a galaxy is, then,

$$D \approx \frac{cz}{H_o} \tag{7.46}$$

where

D = distance to galaxy (Mpc)
z = redshift (unitless)
H_0 = Hubble's constant (67.8 km s^{-1} Mpc^{-1})
c = speed of light (3×10^5 km s^{-1})

In Figure 7.11, the SEDs (spectral energy distributions) of a galaxy with a bolometric luminosity of $L_{bol} = 10^{12} M_\odot$ is plotted as a function of wavelength assuming it lies at a $z = 0.4$, 1, 2.5, and 6. Along with the SEDs are plotted the sensitivities of several past, present, and proposed THz/far-infrared (FIR) telescopes and instruments, assuming a 1 h integration time. The lowest sensitivity curve (4 × 6 m, 4K2%) corresponds to what can be achieved with a 4 × 6 m off-axis telescope cooled to 4 K with only 2% emissivity, with an optimized grating using direct detectors, which add no appreciable noise (background limited). In this case, the sensitivity is set by the photon noise in the astrophysical backgrounds: zodiacal dust for $\lambda < 80$ μm, galactic cirrus between ~80 μm and 350 μm, and the cosmic microwave background (CMB) at longer wavelengths (Bradford, M., personal communication, 2014).

7.8 COMPARING HETERODYNE AND INCOHERENT DETECTOR SENSITIVITY

From Section 7.6, we know the rms noise level in a heterodyne receiver can be expressed in terms of temperature as

$$\Delta T_{\text{rms}} = \frac{T_{sys}}{\sqrt{B_H \Delta\tau}} \tag{7.47}$$

where

ΔT_{rms} = rms noise level (K)
T_{sys} = receiver noise temperature (K)
B_H = detection bandwidth (Hz)
$\Delta\tau$ = integration time (s)

For a source with temperature, T_S, the heterodyne signal-to-noise ratio (SNR_H) would be (Blundell and Tong, 1992),

$$SNR_H = \frac{T_S}{\Delta T_{\text{rms}}} = \frac{T_S (B_H \Delta\tau)^{1/2}}{T_{sys}} \tag{7.48}$$

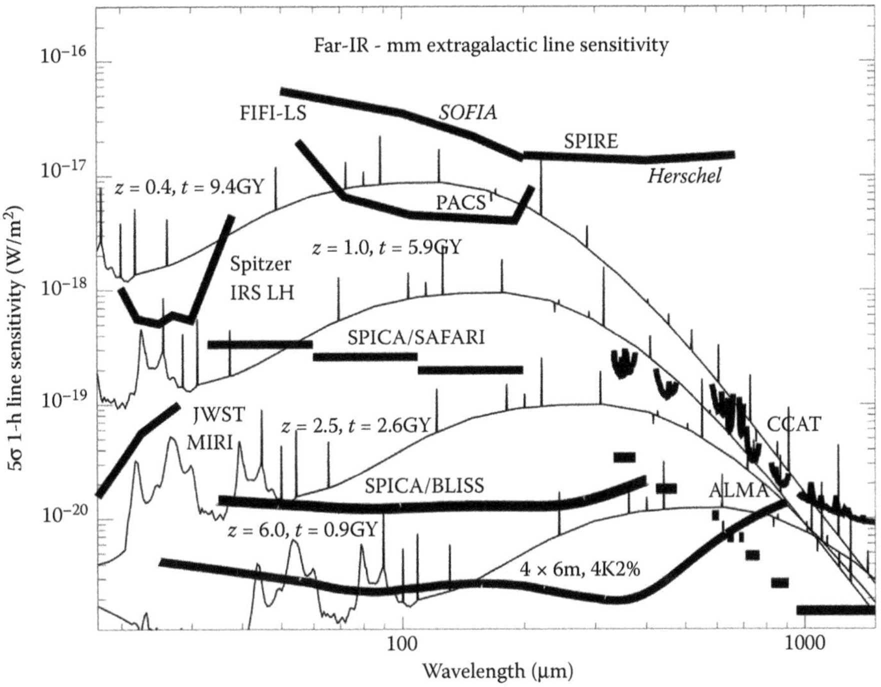

FIGURE 7.11 Instantaneous point source sensitivity (thick lines) of direct-detector platforms for moderate-resolution (R ~ 1000) spectroscopy in the far-IR through millimeter band. *Herschel* and *SOFIA* are limited by emission from their warm telescopes (and in *SOFIA's* case, the atmosphere). Far infrared field-imaging line spectrometer (FIFI-LS) and photoconductor array camera and spectrometer (PACS) are grating spectrometers, and spectral and photometric imaging receiver (SPIRE) is a Fourier-transform spectrometer (FTS). The FTS has a penalty relative to an optimized grating because the detector carries the full photon noise of the wide bandwidth, while the grating restricts the bandwidth on each detector. Space infrared telescope for cosmology and astrophysics (SPICA) is a 3-meter-class (here assumed 3.15 m) space telescope under development in Japan and Europe—it will be cooled to 6 K, which makes the telescope background small compared with the astrophysical background for wavelengths less than 250 microns. Spica far-infrared instrument (SAFARI) is a Fourier-transform spectrometer under development in Europe (led by Netherlands institute for space research (SRON) in the Netherlands). Background-limited infrared-submillimeter spectrograph (BLISS) is an optimized grating spectrometer under study at Jet propulsion laboratory (JPL) in the United States. The lowest (best performance) sensitivity curve (4 × 6 m, 4K2%) can be achieved with a 4 × 6 m off-axis telescope cooled to 4 K with only 2% emissivity, with an optimized grating using direct detectors which add no appreciable noise (background limited). This sensitivity is then set by the photon noise in the astrophysical backgrounds: zodiacal dust for $\lambda < 80$ μm, galactic cirrus between ~80 μm and 350 μm, and the cosmic microwave background (CMB) at longer wavelengths. Cerro Chajnantor Atacama telescope (CCAT) is a proposed 25 m submillimeter telescope at a high Andean peak. CCAT and ALMA (Atacama large millimeter array) benefit from a larger collecting area, but note the discrete nature of the spectral bands available at even mountain-top sites. Overplotted are redshifted model galaxy spectra assuming an underlying luminosity of $10^{12} L_{\odot}$. Grating spectrometers on cold telescopes can be used to study galaxies from the Universe's first billion years. (Figure and caption courtesy of M. Bradford. With permission.)

The corresponding expression for bolometer signal-to-noise ratio (SNR_B) would be,

$$SNR_B = \frac{mk_B T_S B_B (\Delta\tau)^{\frac{1}{2}}}{NEP} \tag{7.49}$$

where

m = 1 or 2, depending on whether single or dual polarization detection
B_B = predetection bolometer bandwidth (Hz)
NEP = bolometer noise equivalent power (W $Hz^{-1/2}$)

Taking the ratio of Equation 7.48 and Equation 7.49, yields

$$\frac{SNR_B}{SNR_H} = \frac{mk_B B_B T_{sys}}{NEP(B_H)^{1/2}} \tag{7.50}$$

Typical values for a broadband, ground-based, background limited systems at 345 GHz are

$NEP = 7 \times 10^{-17}$ W $Hz^{-1/2}$ (SCUBA-2)
B_B = 35 GHz (width of atmospheric window)
T_{sys} = 500 K (SIS)
B_H = 4 GHz (full IF bandwidth)
m = 1 (single polarization)

Substitution into Equation 7.50 shows that incoherent systems have a ~54× advantage for conducting broadband (e.g., dust continuum) observations for a given integration time over a heterodyne receiver. However, for observations where high spectral resolution (e.g., $R = \nu_{obs}/\Delta\nu$ ~10^5–10^6 for galactic clouds) is required for spectral line work, heterodyne systems have superior sensitivity. Setting $B_B \sim B_H \sim 1$ MHz, and keeping all other parameters the same, we find heterodyne receivers have a significant advantage, with $SNR_H \sim 10 \times SNR_B$.

Due to the limited spatial resolution of telescopes, together with large rotational velocities, the emission lines from external galaxies can appear 100–200 km/s wide. In these instances, resolving powers of ~10^3 to ~10^4 are sufficient for low-resolution spectroscopy. At an observing frequency of 345 GHz, a resolving power 10^4 corresponds to 30 km/s and a $B_B = B_H = 30$ MHz. To perform low-resolution spectroscopy with an incoherent system, a dispersive device is placed in front of the detector (see Figure 7.12). Fabry–Perot interferometers, Fourier transform spectrometers (FTS), or gratings have been used for this purpose (see Stacey, 2011).

The turn-over point bandwidth, B_{TO}, between choosing an incoherent or heterodyne detection system for a particular observation can be found from Equation 7.49, by setting $B_{TO} = B_B = B_H$ and $SNR_B/SNR_H = 1$, in which case,

$$B_{TO} = \left[\frac{NEP}{mkT_{sys}}\right]^2 \tag{7.51}$$

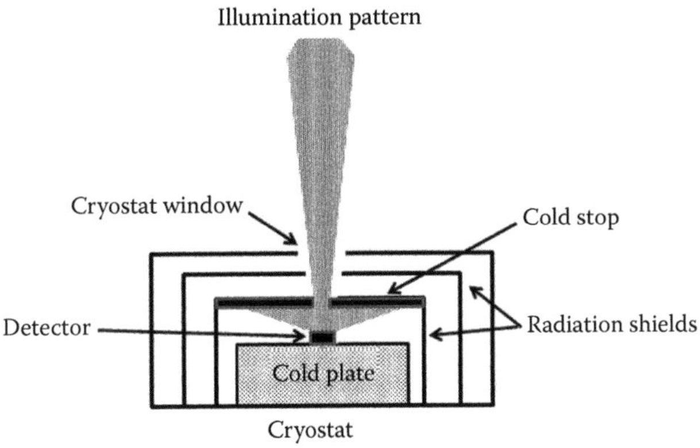

FIGURE 7.12 Filled detector array with cold stop. Here the detector is one or more "naked" pixels, that is, pixels without a feedhorn or lens between them and the outside of the cryostat. The characteristics of the incoming beam (i.e., telescope illumination pattern) are determined by the size of the entrance window in the Cold Stop. Since the Cold Stop fills much of the detector's field of view, maximum sensitivity to the telescope's signal is achieved when the Cold Stop is operated at as low a temperature as possible.

When the desired frequency resolution, $\Delta\nu$, is $<B_{TO}$, a heterodyne receiver should be used. If $\Delta\nu$ is $>B_{TO}$, then an incoherent detector system is a better choice.

7.9 INCOHERENT ARRAY CONSIDERATIONS

The first bolometers arrays (e.g., SCUBA-1, Holland et al., 1998) were made by packing handcrafted, feedhorn coupled pixels into a focal plane. When the cost or difficulty of fabricating pixels is high, the most efficient means for coupling photons to the detector should be used. At THz frequencies, feedhorns are well-suited for this purpose. As discussed in Chapter 5, to keep the cross-talk between adjacent feedhorns to the ≈1% level (see Figure 5.38), the horn apertures should have a diameter of ≈3ω, where ω is the Gaussian beamwaist at the horn aperture. Such a packing arrangement yields an "every-other" diffraction limited beam footprint on the sky (see Figure 5.38). Assuming a 10 dB Gaussian beam edge taper on the telescope's primary, substitution into Equation 5.93 shows that

$$3\omega \approx 2\left(\frac{f_e}{D_p}\right)\lambda \tag{7.52}$$

where

$\frac{f_e}{D_p}$ = the telescope's effective $f\#$ in the focal plane

Therefore, a 3ω separation of feedhorns in a focal plane is equivalent to the $2f\#\lambda$ feedhorn center-to-center spacing prescription often used in geometric optics. In order to

Nyquist sample all the spatial frequencies within a patch of sky using an array with 3ω spaced horns requires either 16 repointings at half-beam increments, or the use of drift scans (see Chapter 8).

Unlike the case of a heterodyne receiver, an incoherent detector does not require the injection of a local oscillator tone at its input, or a wideband intermediate frequency (IF) amplifier at its output. Also, incoherent detectors, and their associated read-out electronics, can now be fabricated in large numbers at relatively low cost, using semiconductor processing techniques. This leads to the possibility of realizing incoherent arrays with 1000's of pixels (see Figure 7.8). In such arrays, the feedhorns can be dispensed with, and the bolometer elements closed-packed with a center-to-center spacing of $0.5f\#\lambda$ or less, referred to as a "filled array" (see Figure 7.8 and Dowell et al., 2003). With no feedhorn, the pixel size determines the acceptance angle of the incoming light, which can be quite broad, $\approx \pi$ rad^2. Therefore, one or more cold stops are needed within the cryostat (see Figure 7.12), to limit the amount of stray radiation reaching the detectors. Unlike the case of a detector with a feedhorn, the resulting telescope illumination profile for each detector will contain multiple Gaussian beam modes and, as a consequence, yield a sharper main beam with a higher sidelobe level. Therefore, filled arrays with many detectors provide higher point source sensitivity, but at the expense of beam purity. When operated close to the background limit, the filled array is 3.5 times faster than a 3ω spaced feedhorn array for the identification of point sources. For a given number of detectors, or if the location of a point source is known, a feedhorn coupled array provides greater mapping efficiency than a filled array (Griffin et al., 2002). For the same number of pixels, a feedhorn coupled array will have a larger field of view. A larger field of view is beneficial when mapping extended objects, but a disadvantage when mapping sources smaller than the array's footprint on the sky.

CONCLUSION

In this chapter, we have discussed the physics and properties of three types of incoherent detectors. Of the three types, superconducting transition edge sensors (TESs) are the most sensitive. Microwave kinetic inductance detectors (MKIDs) are currently approaching the sensitivity of transition edge sensors (TESs) and have the advantage that they are easier to fabricate and implement into large format arrays. For low spectral resolution observations of galactic and extragalactic objects, incoherent detectors have greater sensitivity than coherent (e.g., heterodyne) detectors, while the reverse is true when high spectral resolution is required. In the next chapter, observational and calibration techniques for THz astronomy will be discussed.

PROBLEMS

1. If the electrical power dissipation in a bolometer operating at 4 K is 1×10^{-9} W, and it has a thermal conductivity of 1×10^{-10} W/K, what is its NEP in the thermal noise limit? Assume an optical coupling efficiency of 0.8.

2. A detector has a responsivity of 8×10^7 V/W and spectral noise density of 3×10^{-8} V/$\sqrt{\text{Hz}}$. What is its NEP?

3. What is the kinetic inductance of an MKID made from NbN "wire" with a length l of 500 μm, width w of 0.1 μm, sheet resistivity of 20 Ω/square, and $T_c = 9.3$ K? Assume an operating temperature of 6 K.

4. What will be the observed frequency shift on the output of an MKID device if its capacitance is 100 pF, kinetic inductance is 22 nH, and all other inductances are negligible?

5. Given the operating parameters provided in the chapter for a SCUBA-2 pixel ($NEP = 7 \times 10^{-17}$ WHz$^{-1/2}$, $B_B = 35$ GHz), and a ground-based heterodyne receiver operating at 345 GHz with a $T_{sys} = 500$ K and $B_H = 4$ GHz, what is the turn-over bandwidth for deciding which type of instrument (incoherent or heterodyne) to use for an observation?

6. Using a bolometer system with an NEP = 3×10^{-17} W Hz$^{-1/2}$, how long would you need to integrate to detect a 10 mJy source to an SNR of 10 on a 10 m diameter telescope? Make the following assumptions: the telescope has an aperture efficiency of 0.8, the source and detector coupling efficiencies are each 0.7, the zenith optical depth is 0.2, and you are observing at 2 air masses through a 35 GHz bandpass filter.

REFERENCES

Annunziata, A., Santavicca, D., Frunzio, L., Catelani, G., Rooks, M., Frydman, A., and prober, D., 2010, Tunable superconducting nanoinductors, *Nanotechnology*, 21, 445202.

Beyer, A.D., Kenyon, M.E., Echternach, P. M. et al., 2011, Ultra-sensitive transition-edge sensors for the Background Limited Infrared/Sub-mm Spectrograph (BLISS), *J. Low Temp. Phys.*, 2012 167, 182–187.

Blundell, R. and Tong, C. Y. E., 1992, Submillimeter receivers for radio astronomy, *Proc. IEEE*, 80(11), 1702–1720.

Bock, J., Chen, D., Mauskopf, P., and Lange, A., 1995, A novel bolometer for infrared and millimeter-wave astrophysics, *Space Sci. Rev.*, 74, 229–235.

Day, P., LeDuc, H., Mazin, B., Vayonakis, A., and Zimuidzinas, J., 2003, A broadband superconducting detector suitable for use in large arrays, *Nature*, 425, 817.

de Visser, P., Baselmans, J., Bueno, J., and Klapwijk, T., 2014. Demonstration of an NEP of 3.8×10^{-19} W/Hz 1/2 at 1.54 THz in multiplexible superconducting microresonator detectors. http://www.irmmw-thz2014.org/sites/default/files/R2_A-28.4_de%20Visser.pdf.

Dowell, C. D., Allen, C. A., Babu, R. S. et al., 2003, SHARC II: A Caltech submillimeter observatory facility camera with 384 pixels, *Proc.* SPIE, 4855, 73.

Gao, J., Zmuidzinas, J., Vayonakis, A., Day, P., Mazin, B., and LeDuc, H., 2008, Equivalence of the effects on the complex conductivity of superconductor due to temperature change and external pair breaking, *J. Low Temp. Phys.*, 151, 557–563.

Griffin, M. J., Bock, J. J., and Gear, W. K., 2002, Relative performance of filled and feedhorn-coupled focal plane architectures, *Appl. Opt.*, 41(31), 6543.

Holland, W., Robson, E., Gear, W. et al., 1998, SCUBA: A common-user submillimetre camera operating on the James Clerk Maxwell Telescope, http://arxiv.org/abs/astro-ph/9809122.
Hollister, M., 2006, An introduction to the thermal modelling and characterisation of semiconductor bolometers. http://www.roe.ac.uk/ifa/postgrad/pedagogy/2006_hollister.pdf
Irwin, K. and Hilton, G., 2005, Transition-edge sensors, in *Cryogenic Particle Detection* C. Enss (ed.), *Topics Appl. Phys.* Springer-Verlag, Berlin, Heidelberg. 99, 63–149.
Irwin, K. D., Nam, S. W., Cabrera, B., Chugg, B., Park, G. S. Welty, R. P., and Martinis, J. M., 1995, A self-biasing cryogenic particle detector utilizing electrothermal feedback and a SQUID readout. *IEEE Trans. Appl. Supercond.*, 5(2 pt.3), 2690–2693.
Jones, R., 1953, *Advances in Electronics,* Academic Press Inc., New York, V, 1.
Low, F., 1961, Low-temperature germanium bolometer, *J. Opt. Soc. Am.*, 51(11), 1300.
Mazin, B., 2014, http://web.physics.ucsb.edu/~bmazin/Mazin_Lab/MKIDs.html.
Orozco, L., 2014, Synchronous detectors facilitate precision, low-level measurements, *Analog Dialogue* 48-11 (analog.com/analogdialogue).
Rieke, G., 2003, *Detection of Light: From Ultraviolet to the Submillimeter*, Cambridge University Press, London.
Schlaerth, J., Golwala, S., Zmuidzinas, J. et al., 2009, Sensitivity optimization of millimeter/submillimeter MKID camera pixel device design, *AIP Conference Proceedings*, 1185, 180–183.
Shimoda, K., Takahasi, H., and Townes, C. H., 1957, Fluctutations in amplification of quanta with application to maser amplifiers, *J. Phys. Soc. Japan,* 12, 686–700.
Stacey, G. J., 2011, THz low resolution spectroscopy for astronomy, *IEEE Trans. Terahertz Sci. Technol.*, 1(1), 241.
Xu, J., Lange, A., and Boch, J., 1996, Far-infrared emissivity measurements of reflective surfaces, *Proc. 30th ESLAB Symp.*, Noordwijk, the Netherlands, ESA SP-388, p. 69.
Zmuidzinas, J., 2002, "Coherent Detection and SIS Mixers", http://www.submm.caltech.edu/~jonas/tex/papers/pdf/2002-Monterey-Zmuidzinas_a.pdf.

8

TERAHERTZ OBSERVING TECHNIQUES

PROLOGUE

Whether it be from the ground, stratosphere, or space, observing time at THz frequencies is a precious commodity. An efficient observing strategy is needed to make the most of the available time. In the case of ground-based observations, large, time-varying atmospheric opacities can make data calibration particularly challenging. In this chapter, observational strategies and calibration techniques commonly used in millimeter-wave and THz astronomy will be discussed.

8.1 INTRODUCTION

Observing at THz frequencies is challenging. This is particularly true from the ground, where any losses or instabilities in the optics and receiver are compounded by fluctuations in atmospheric transparency. Over the past few decades, observing and calibration techniques have been developed to maximize the efficiency and efficacy of THz observations. The choice of which techniques to use depends on the type of observation to be performed (e.g., spectral line or continuum), atmospheric conditions, instrument stability, number of detector pixels, and size of the region to be mapped.

The output power of a detection system, P_R, is converted to a time-varying voltage, V_{out}, by running it through a square law detector, such that

$$V_{out} \propto P_R \propto (V_R)^2 \tag{8.1}$$

where
  V_R = voltage associated with detected signal (V).

Assuming all contributions to P_R can be described by emission from one or more black bodies, we can define an equivalent, linear, black body temperature scale for the output power, such that,

$$V_{out} \propto P_R = kT_{sys}\beta \tag{8.2}$$

where
  P_R = power associated with the downconverted signal (W)
  T_{sys} = equivalent black-body system noise temperature (K)
  β = predetection bandwidth (Hz)
  k = Boltzmann's constant = $1.3806488 \times 10^{-23}$ m^{2}kg s^{-2}K^{-1}

Rearranging Equation 8.2, we find

$$T_{sys} = \frac{V_{out}}{gk\beta} \tag{8.3}$$

where
  g = voltage to temperature conversion factor of receiver (V/K).

For a particular receiver system, the value of g can be determined by injecting a broadband source of known temperature, ΔT_{cal}, in front of the receiver, and monitoring the resulting change in output voltage, ΔV_{out}.

$$g = \frac{\Delta V_{out}}{\Delta T_{cal}} \tag{8.4}$$

At frequencies ≤50 GHz, a noise tube is commonly used as a calibration source. At higher frequencies, the fundamental temperature calibration is performed by putting hot and/or cold loads in front of the receiver. Typically, the hot load is a piece of absorbing material at ambient temperature ($T_H \approx 290$ K) large enough to fully block the receiver's beam and the cold load is the same material cooled by a refrigerator or dipped in liquid nitrogen ($T_C \approx 77$ K). In this case,

$$g = \frac{[V_{Rx} + V_H] - [V_{Rx} + V_C]}{T_H - T_C} \tag{8.5}$$

where
  V_{Rx} = receiver output voltage with no load in front (V)
  V_H = receiver output voltage with a hot (e.g., ambient) temperature load in front (V)
  V_C = receiver output voltage with a cold (e.g., liquid nitrogen) temperature load in front (V)

How often the value of g should be measured/calibrated is set by the stability of the observing system, which includes the receiver itself, the optical system (e.g., the telescope and intermediate optics), and the atmosphere (if there is one). The Allan time, Δt_A, of the system (see Section 6.11) can be used as a guide to determine the time interval between calibrations, Δt_C, and, therefore, how long the integration time, Δt_{LOS}, can be along a given line of sight before switching to a reference position or calibration source.

$$\Delta t_{LOS} \leq \Delta t_C \leq \Delta t_A \tag{8.6}$$

For ground-based systems, the value Δt_A is almost always set by the atmosphere. For orbital and balloon-borne telescopes, Δt_A is set by the stability of the receiver itself. The total time required to perform an observation can be determined from the Dicke radiometer equation,

$$\Delta T_{\text{rms}} = K \frac{T_{sys}}{\sqrt{B_d \Delta \tau_{ON}}} \tag{8.7}$$

where

T_{sys} = equivalent black-body system noise temperature (K)
B_d = post detection bandwidth (Hz)
$\Delta\tau_{ON}$ = integration time along an LOS (line of sight)
K = observing constant
ΔT_{rms} = rms noise level of observation ≈(1/3)$T_{peak\text{-}to\text{-}peak}$ noise level (K)

8.2 OBSERVING STRATEGIES

The two most common observing strategies used at THz frequencies are absolute position switching (APS) and On-the-Fly (OTF) mapping. OTF observing has been the standard technique used for continuum mapping at millimeter wavelengths for over 40 years (e.g., Haslam, 1974). A third strategy used at lower frequencies for spectral line observations, which shows promise in the THz regime, is frequency switching (FS). All three techniques involve observing the line of sight (LOS) to a source at a frequency of interest and then measuring a nearby LOS and/or frequency free of source emission.

8.2.1 ABSOLUTE POSITION SWITCHING

As described in Section 2.2, a standard observing mode in THz astronomy is absolute position switching (APS) (Gordon and Meeks, 1968). It is particularly useful for observations of sources of limited extent, that is, a few main beams across. Here the intensity along a LOS to the object of interest (e.g., an interstellar cloud), I_ν^{ON}, is measured for a time, $\Delta\tau_{ON}$, as well as the intensity along a LOS just off the object, I_ν^{OFF}. Since these intensities are being observed through a telescope beam of solid angle, Ω_b, they are converted to fluxes, F_ν,

$$F_\nu^{ON} = I_\nu^{ON}\Omega_b$$
$$F_\nu^{OFF} = I_\nu^{OFF}\Omega_b \tag{8.8}$$

The job of the detection system is to convert these observed fluxes into measurable powers,

$$P_\nu^{ON} \propto F_\nu^{ON}$$
$$P_\nu^{OFF} \propto F_\nu^{OFF} \tag{8.9}$$

Both P_ν^{ON} and P_ν^{OFF} contain the noise power associated with the detection system, optical losses, and, in the case of all but space-based telescopes, the atmosphere. In the ideal case, by subtracting the two powers, we can recover what we are after, the power associated with the astrophysical source, P_ν^S,

$$P_\nu^S = P_\nu^{ON} - P_\nu^{OFF} \tag{8.10}$$

For maximum suppression of common mode noise associated with atmospheric and receiver instabilities, the ON and OFF LOSs should be as close as possible.

As an illustration, let us imagine we are using a heterodyne receiver to observe a strong spectral line toward a molecular cloud (e.g., from Orion OMC-1). The output of the receiver as viewed by the back-end spectrometer will show the receiver bandpass (i.e., power spectrum) with the spectral line riding on top of it (see Figure 8.1). When looking off the interstellar cloud, the receiver bandpass will look approximately the same, but with no spectral line (see Figure 8.2). The difference between the two power spectra (ON and OFF) renders the spectral line. If our goal is to measure the broadband, incoherent emission from a dust cloud or planet, then there may be no spectral line. In this case, P_ν^S reflects the difference in the continuum level between the ON and OFF source.

For the common case, where equal time is spent ON and OFF the source (e.g., $\Delta t_{ON} = \Delta t_{OFF}$), the value for the observing constant in Equation 8.10 is $K = \sqrt{2} \cdot \sqrt{2} = 2$. One $\sqrt{2}$ originates because only (1/2) the time is spent ON source, and the other from subtracting uncorrelated signals. Taking into account the overhead in time associated with moving the telescope from the ON to the OFF position, K is in reality >2. If conditions are stable, then it may be possible to share the same OFF scan with multiple ON scans, M. In

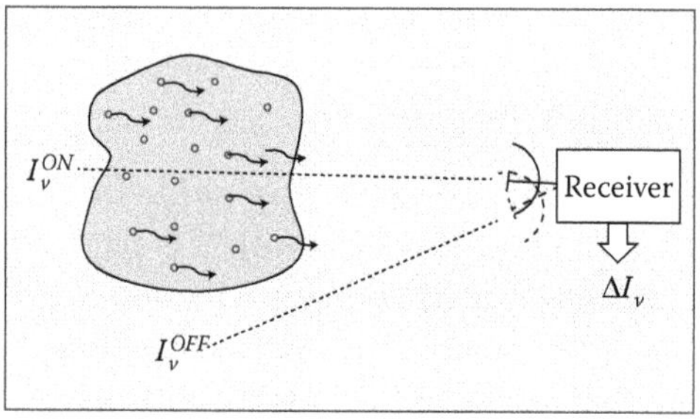

FIGURE 8.1 Absolute position switching (APS) is the most common observing mode. The telescope first look ON and then just OFF the source.

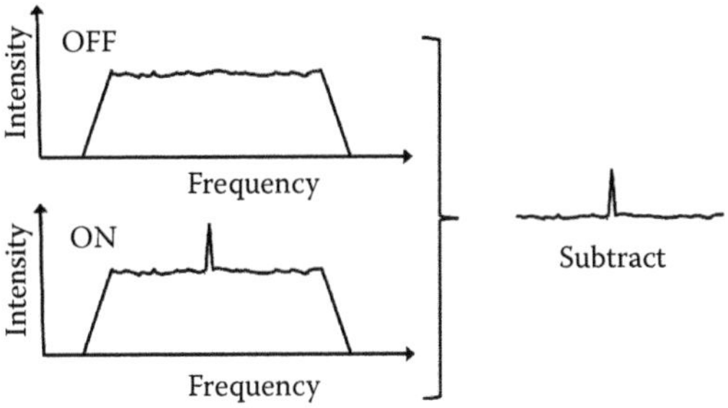

FIGURE 8.2 Absolute position switched (APS) observation. Looking ON source the receiver bandpass includes the spectral line from the source, as well as noise from the Earth's atmosphere (if there is any) and the receiver. The OFF position contains noise from the atmosphere and receiver, but no spectral line. Subtracting the OFF from the ON yields the spectrum of the source.

this case, the observing constant becomes $K = 1 + \sqrt{M}$. In the limit of large M, $K \rightarrow 1$, the ideal case (Emerson, 1997).

8.2.2 ON-THE-FLY MAPPING

For mapping extended sources, that is, > a few beams across, and conducting large-scale surveys, On-the-Fly (OTF) mapping is the preferred observing technique. Here, the telescope beam is scanned rapidly back and forth across the target source recording data and the telescope's position almost continually (see Figure 8.3). There are numerous advantages to this approach (Emerson, 1997):

1. The overhead of moving the telescope is nearly eliminated, with useful data being obtained even as the telescope is accelerating and decelerating.

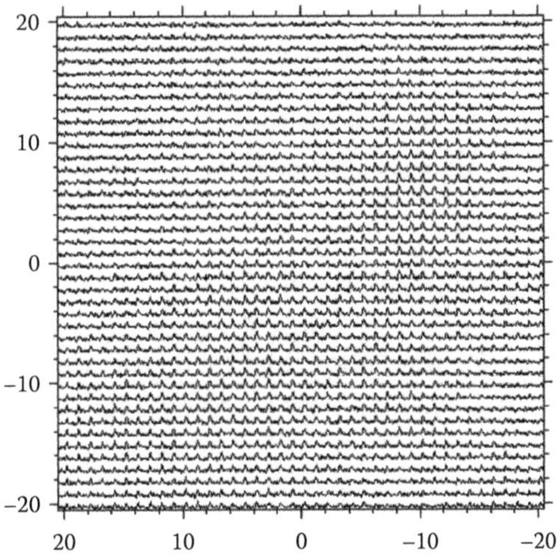

FIGURE 8.3 On-the-Fly (OTF) map of ^{12}CO J = 3 → 2 emission in Orion made at the Arizona Radio Observatory. Spectra are taken as the telescope drifts across the object. Each row of spectra share a common OFF position. The OTF technique can increase the mapping speed of a telescope by as much as an order of magnitude compared to absolute position switching (APS).

2. With overhead reduced, coverage of a field can be completed relatively rapidly, reducing the effect of time variability in weather, pointing, or instrument stability on system performance.
3. With OTF mapping, there can be 100's of ON scans for each OFF scan, so that the observation constant $K \rightarrow 1$.

For mapping extended sources, the OTF approach yields an effective gain in observing speed approaching an order of magnitude over the APS approach, with improved data quality.

How often the data must be sampled during a scan is a function of the telescope's scan speed and beam size. Depending on the edge-taper on the primary (set by the receiver optics) the sampling should be between 2.4 and 2.5 points per full-width-half-maximum (FWHM) beam. Undersampling will result in a loss of higher spatial frequencies sampled by the telescope and, therefore, result in reduced angular resolution within the map.

8.2.3 FREQUENCY SWITCHING

Frequency switching (FS) is, in principle, the most efficient observing mode for spectral line work, and the one used by Ewen and Purcell to make the first detection of the neutral hydrogen (HI) line at 1420 MHz (Ewen and Purcell, 1951). In this mode of operation, the OFF spectrum is obtained, not by moving the telescope, but by offsetting the center frequency of the receiver by $\Delta \upsilon_{OFF}$ to an emission-free region of the receiver's bandpass. In practice, this can be achieved by switching the frequency of the first local oscillator (LO) or, if present, the second LO. The minimum frequency offset is ~2 to 3 times the expected spectral line-width. If the frequency switching is performed in-band, such that $\Delta \upsilon_{OFF}$ lesser than the receiver's intermediate frequency (IF) bandwidth, $\Delta \upsilon_{IF}$, then the line will appear twice, once in emission and once in absorption. The part of the receiver's bandpass that occurs outside the shifted absorption line is used as the spectral "OFF" for the ON frequency spectrum. Likewise, the part of the receiver baseline that occurs outside the ON emission line when the ON frequency is used, serves as the spectral "OFF" for the absorption line. This means that at the end of a frequency switched scan, there are two independent spectral line measurements, (ON−OFF)/OFFs, that can be coadded (once the absorption spectrum is inverted) to reduce noise. Therefore, frequency switched observations have an observing constant of $K = \sqrt{2}$, leading to an improvement in observing speed over absolute position switching of at least a factor of 2, with no dead time due to waiting for the telescope to complete a move. Also, there is no need to locate a clean, emission-free region for the OFF, which is important when observing extended emission regions. There are two disadvantages to the FS technique (Figure 8.4):

1. The receiver's intermediate frequency bandwidth, $\Delta \upsilon_{IF}$, must be sufficient to contain both the switched and unswitched spectra.
2. There could be incomplete subtraction of standing waves within the spectrometer due to the underlying gain profiles of the spectral "OFFs" not being truly representative of conditions when the corresponding spectral "ONs" were being taken (likely due to the frequency shift). Incomplete subtraction can lead to standing waves appearing in the calibrated baselines.

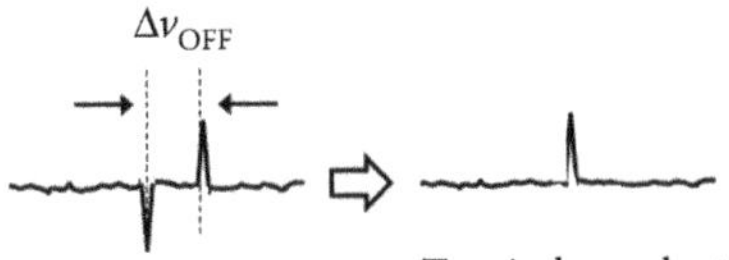

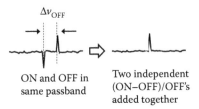

FIGURE 8.4 Frequency switched (FS) observation. Looking ON source the receiver bandpass includes the spectral line from the source, as well as noise from the Earth's atmosphere and the receiver. In FS mode the telescope does not move to an OFF position as it does with APS observing. Instead, the telescope stays pointed ON source and switches the receiver frequency by an amount, $\Delta\upsilon_{OFF}$. The switched part of spectrum at the nominal line frequency can then be used as the OFF. If $\Delta\upsilon_{OFF}$ is less than the IF bandpass, then two lines will appear as shown above, one in emission and the other in absorption. The two measurements can be coadded to reduce noise and increase observing efficiency.

The first disadvantage can be addressed by designing a receiver system with a $\Delta\upsilon_{IF}$ six to ten times the expected line-width. The $\Delta\upsilon_{IF}$ of most modern Schottky, superconductor-insulator-superconductor (SIS), and hot electron bolometer (HEB) receivers is ≥4 GHz, which is wide enough to support THz FS observations along most LOSs in the galaxy, except, perhaps, toward the galactic center (where rotational broadening is the greatest). The standing waves referred to in the second disadvantage can be largely eliminated by: (a) modulating the telescope focus by $\pm\lambda/8$ during alternate integrations (to modulate the standing wave frequency) and (b) using a scattering cone at the center of the secondary (to redirect reflected LO power from the receiver out of the telescope beam) (Mangum, 2006). Residual standing waves can be overcome by employing a dual-beam switched, image restoration algorithm similar to that for mapping extended sources, but applied in the spectral domain (Emerson, 1997; White et al., 2003).

8.3 RECEIVER CALIBRATION

Up to this point, the observed source power, P_ν^S, remains uncalibrated. To interpret P_ν^S in an astrophysical context, we must develop a calibration procedure. At THz frequencies and millimeter-wavelengths, it is the variability of the column density of precipitable water vapor that dominates both the system sensitivity and stability for ground-based observatories (see Figure 1.1). Since the early days of millimeter-wave astronomy, the "chopper-wheel" calibration technique has been used as a convenient way to compensate for both fluctuations in the gain and atmospheric absorption within an observing system (e.g., Penzias and Burrus, 1973). Chopper-wheel calibration is performed by inserting an ambient temperature load of temperature, T_{amb}, into the signal beam for a time, Δt_{load}, every Δt_C, and monitoring the change in the receiver's output voltage, V_C. The value of Δt_{load} depends on the receiver sensitivity, but is typically on the order of a few seconds. The assumption is made that

$$T_{amb} \approx T_{atm} \approx T_{ground} \tag{8.11}$$

where

T_{amb} = temperature of chopper-wheel load (K)
T_{atm} = mean temperature of atmosphere (K)
T_{ground} = temperature of surroundings behind the telescope (K)

Since most of the atmospheric absorption occurs due to water vapor in the lower atmosphere, the above assumption is valid to first order. Once the load is removed, the observer then integrates on blank sky (i.e., toward the OFF position) for an equal number of seconds, Δt_{load}, as toward the calibration load.

When the telescope is looking toward blank sky (see Figure 8.5), it will have an output voltage, $V_{OFF} = V_{sky}$, where, following the formalism of Phillips (1989), we have

$$V_{sky} = g\left[\underbrace{\alpha\beta(1-e^{-\tau})T_{amb}}_{\text{atmosphere}} + \underbrace{(1-\alpha)T_{amb}}_{\textit{hot spillover}} + \underbrace{\alpha(1-\beta)(1-e^{-\tau})T_{amb}}_{\textit{cold spillover}} + \underbrace{T_{Rx}}_{\textit{receiver}}\right] \quad (8.12)$$

where

α = hot spillover efficiency = 1—fraction of power falling off the primary on ground, dome, etc.
β = cold spillover efficiency = 1—fraction of power falling on sky which does not form part of the beam
T_{Rx} = receiver double sideband noise temperature (K)
T_{amb} = ambient load temperature (K)
τ = atmospheric opacity at zenith angle of observation = τ_{zenith}AM
τ_{zenith} = atmospheric opacity at zenith

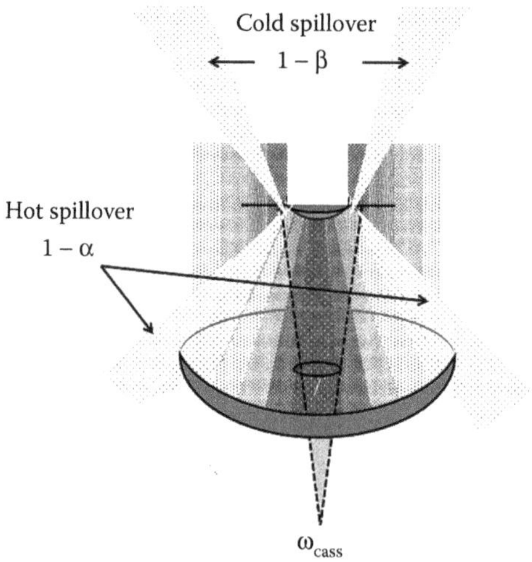

FIGURE 8.5 When the detection system views the subreflector of a telescope with a Gaussian beam, some fraction, 1 − β, of the beam will spill over the edge of the subreflector on to cold sky. Another fraction, 1 − α, of the beam will be reflected by the outer edge of the subreflector, but spill over the edge of the primary and terminate on the relatively hot area behind the telescope. Both the cold and hot spillover contribute to the overall system noise temperature, T_{sys}.

AM = s(z) = air mass
z = zenith distance (= 90° – elevation angle)
V_{sky} = receiver output voltage when looking at blank sky
g = voltage to temperature conversion factor of receiver (V/K)

Simplifying, we find,

$$V_{OFF} = V_{sky} = g[T_{Rx} + (1 - \alpha e^{-\tau})T_{amb}] \tag{8.13}$$

When we look ON source, then the receiver output voltage, V_{ON}, is

$$V_{ON} = g[T_{Rx} + (1 - \alpha e^{-\tau})T_{amb} + \alpha\beta\gamma T_S e^{-\tau}] \tag{8.14}$$

where
T_S = source temperature (K)
γ = source/beam coupling efficiency

After a calibration cycle, the instrument computer calculates the ratio,

$$\text{Cal} = \frac{V_{amb} - V_{OFF}}{V_{OFF}} = \frac{\alpha g T_{amb} e^{-\tau}}{V_{OFF}} = \frac{\alpha T_{amb} e^{-\tau}}{T_{RX} + (1 - \alpha e^{-\tau})T_{amb}} \tag{8.15}$$

where
$V_{amb} = gT_{amb}$ = receiver's output voltage when looking at ambient (i.e., chopper-wheel) load (V)

After observing an equal amount of time ON and OFF source, the instrument computer calculates the ratio,

$$\text{Scan} = \frac{V_{ON} - V_{OFF}}{V_{OFF}} = \frac{\alpha g\beta\gamma T_s e^{-\tau}}{V_{OFF}} \tag{8.16}$$

Taking the ratio of Scan to Cal, we find the effect of the atmosphere conveniently drops out of the expression

$$\frac{\text{Scan}}{\text{Cal}} = \frac{\beta\gamma T_S}{T_{amb}} \tag{8.17}$$

The source temperature, T_S, can now be found by rearranging the above expression.

$$T_S^{DSB} = \frac{T_{amb}}{\beta\gamma} \times \frac{\text{Scan}}{\text{Cal}}; \quad \text{for double sideband (DSB) data} \tag{8.18a}$$

$$T_S^{SSB} = \frac{2T_{amb}}{\beta\gamma} \times \frac{\text{Scan}}{\text{Cal}}; \quad \text{for single sideband (SSB) data} \tag{8.18b}$$

What the computer will plot (as a function of velocity, channel number, or frequency) at most observatories is the quantity,

$$T_A^* = 2T_{amb} \times \frac{\text{Scan}}{\text{Cal}} \tag{8.19}$$

T_A^* is known as the antenna temperature and is usually assumed to be SSB, since a spectral line occurs in only one sideband. T_A^* is the source temperature corrected for the atmosphere and warm spillover. The atmospheric correction occurs automatically, since the Cal signal is inversely proportional to opacity. So, according to the above expression, as the Cal signal gets less (meaning there is more atmospheric absorption present) the chopper-wheel calibration will compensate by increasing the value of T_A^*. To also correct for cold spillover and source coupling efficiencies, divide by $\beta\gamma$,

$$T_S = \frac{T_A^*}{\beta\gamma} \tag{8.20}$$

The product $\beta\gamma$ is often referred to as the main beam efficiency, η_{mb}.

$$T_S \approx T_{mb} = \frac{T_A^*}{\eta_{mb}} \tag{8.21}$$

where

T_{mb} = main beam temperature (K)

The value of η_{mb} depends on how big your object is relative to the telescope main beam. For sources a few times larger than the main beam, η_{mb} can be estimated using the Moon.

$$\eta_{mb} \approx \eta_{moon} = \beta\gamma = \frac{T_A^*(\text{Moon})}{T_{\text{Moon}}} \tag{8.22}$$

where

T_{Moon} = physical temperature of the Moon at observation frequency (K)

For sources comparable to the telescope main beam size, η_{mb} can be measured using a planet,

$$\eta_{mb} = \beta\gamma = \frac{T_A^*(\text{Planet})}{T_{\text{Planet}}} \times \left[1 - \exp\left(\frac{-D^2}{\theta_{mb}^2}\ln 2\right)\right]^{-1} \tag{8.23}$$

where

D = planet diameter on day of observation, found in an ephemeris (")

θ_{mb} = full-width-half-maximum (FWHM) of main beam (")

T_{Planet} = physical temperature of planet at observation frequency (K)

For sources smaller than the size of the smallest available planet (e.g., Mars), the value of η_{mb}^{planet} can be corrected for beam dilution according to (Mangum, 1993).

$$\eta_{mb}^{small} = \eta_{mb}^{planet}\left(\frac{\theta_S}{\theta_{planet}}\right)^2 \quad (8.24)$$

where

η_{mb}^{small} = main beam efficiency for small source
η_{mb}^{planet} = main beam efficiency measured on planet
θ_S = FWHM size of small source (")
θ_{planet} = FWHM size of small planet (")

Typically, the size of a telescope's main beam is determined by scanning a planet in azimuth. The telescope's true beamsize is then computed by deconvolving the observed FWHM, and the planetary disk using the expression (Urquhart et al., 2010),

$$\theta_{mb} = \sqrt{\left(\theta_{obs}^2 - \frac{\ln 2}{2}\theta_{planet}^2\right)} \quad (8.25)$$

where

θ_{mb} = full-width-half-maximum (FWHM) of main beam (")
θ_{obs} = full-width-half-maximum (FWHM) of observed beam (")
θ_{planet} = full-width-half-maximum (FWHM) of planet (")

Since at THz frequencies η_{moon} typically includes the main beam, sidelobes, and all or part of the telescope's error beam, one finds that ($\eta_{mb}/\eta_{moon} < 1$). For example, a value of ($\eta_{mb}/\eta_{moon} < 0.8$) would indicate 20% or more of the photons collected by the telescope are originating from outside the main beam.

Using this nomenclature, the system noise temperature at a given zenith distance can be written as

$$T_{sys} = \frac{T_{Rx} + (1 - \alpha e^{-\tau})T_{amb}}{\alpha e^{-\tau}} \quad (8.26)$$

Recalling the above expression for Cal (Equation. 8.15), we have,

$$T_{sys} = \frac{T_{amb}}{\text{Cal}} \quad (8.27)$$

For orbital and balloon-borne telescopes, the atmospheric optical depth, $\tau \to 0$, with the result,

$$T_{sys} \Rightarrow \frac{1}{\alpha}[T_{Rx} + (1 - \alpha)T_{amb}] \quad (8.28)$$

If the telescope is well-designed and sufficiently underilluminated, then $\alpha \rightarrow 1$ and, as might be expected,

$$T_{sys} \approx T_{Rx} \tag{8.29}$$

The chopper-wheel calibration technique is a simple, elegant solution to a complex problem, and can provide calibration accuracy in the range of 5%–10% for ground-based observatories. Simplifying assumptions that lead to inaccuracy include: (1) a receiver sideband ratio of one; (2) equal atmospheric absorption at the two sidebands frequencies, and (3) that the atmosphere is at the temperature of the ambient load. On the last point, the atmosphere will, on average, be colder than a ground-based ambient load, leading to a tendency to overestimate the brightness of a spectral line (see Equation 8.19). The calibration accuracy of the technique can be improved significantly by using a two-temperature-load system instead of a single ambient load. This allows the receiver and sky noise contributions to be separated out, which can lead to an improvement in calibration accuracy to as good as ~1% (Mangum, 2000; Jewell, 2002).

EXAMPLE 8.1

You have just installed a 230 GHz DSB (double sideband) receiver on a 10 m telescope and need to measure the system's main beam efficiency, η_{mb}.

a. What is the full-width at half-maximum (FWHM) beam size of the telescope?

$$\theta_{FWHM} \approx 1.2\frac{\lambda}{D} \approx 1.2\left(\frac{1.3\times10^{-3}\,\text{m}}{10\,\text{m}}\right) \approx 1.56\times10^{-4}\ \text{rad} \Rightarrow 32''$$

b. After checking a planetarium program and an ephemeris, you find Saturn is currently well above the horizon and has a diameter, D, of 18″; a good match to the telescope's beam size.
c. You connect a voltmeter to the total power output of the receiver's IF chain and move the telescope to the vicinity of Saturn. Using the voltmeter, you measure:

$$V_{Sky} = 5.6\ \text{V}$$
$$V_{Sat} = 6.2\ \text{V}$$
$$V_{amb} = 16.9\ \text{V}$$
$$V_C = 6.7\ \text{V}$$

where

V_{Sky} = receiver's output voltage when telescope is looking at blank sky adjacent to Saturn

V_{Sat} = receiver's output voltage when telescope is looking directly at Saturn

V_{amb} = receiver's output voltage when looking at an ambient (hot) load at ~280 K

V_C = receiver's output voltage when looking at a cold load at ~80 K

Using the above measurements and Equations 8.15 and 8.16, values for the system CAL and SCAN are found to be

$$\mathrm{CAL} = \frac{16.9 - 5.6\,\mathrm{V}}{5.6\,\mathrm{V}} = 2.018$$

$$\mathrm{SCAN} = \frac{6.2 - 5.6\,\mathrm{V}}{5.6\,\mathrm{V}} = 0.107$$

d. We can use Equation 8.18 to calculate a value of T_A^* for Saturn. Since Saturn emits photons as a thermal black body, its emission shows up in both sidebands of the receiver, so we use Equation 8.18a. The temperature of the ambient (chopper) load is measured to be 280 K.

$$T_A^* = T_{amb} \times \frac{\mathrm{SCAN}}{\mathrm{CAL}} = 280\,\mathrm{K} \times \frac{0.107}{2.018} = 14.85\,\mathrm{K}$$

e. From Table 8.4, the true temperature of Saturn, T_{Sat}, is taken to be

$$T_{Sat} = 126\ \mathrm{K}$$

f. Substitution into Equation 8.23 yields,

$$\begin{aligned}\eta_{mb} &\approx \frac{14.85\,\mathrm{K}}{126\,\mathrm{K}}\left[1 - \exp\left(-\left(\frac{18''}{32''}\right)^2 \ln(2)\right)\right]^{-1} \\ &\approx (0.118)(5.08) \\ &\approx 0.6\end{aligned}$$

8.4 ESTIMATING ATMOSPHERIC OPTICAL DEPTH

The sensitivity of THz ground-based observations is often background-limited due to absorption of incoming photons by the Earth's atmospheric water vapor. The amount of precipitable water vapor (PWV) above a given location is simply the height of water (in millimeters) that could be condensed out of a square centimeter column projected upward through the atmosphere. Measuring the atmospheric optical depth, τ, at or near the frequency of observation, is important for both calibration and determining if the observation is even possible. Rewriting Equation 8.26 for small values of τ, we find,

$$T_{sys} = \frac{1}{\alpha}(T_{RX} + \tau\alpha T_{amb}), \quad \text{for } \tau \ll 1 \tag{8.30}$$

where

$$\tau = \tau_{zenith} \times \text{ air mass } = \tau_{zenith}\sec(z)$$

$$z = 90° - \text{elevation} = \text{zenith distance (°)}$$

An air mass (AM) of one occurs at $z = 0°$. At a $z = 60°$, the path length through the atmosphere (and the associated absorption) is twice that as at zenith. Substitution yields

$$\begin{aligned} T_{sys} &= \frac{1}{\alpha}(T_{RX} + \alpha T_{amb}\tau_{zenith}\sec(z)) \\ &= \frac{T_{RX}}{\alpha} + T_{amb}\tau_{zenith}\sec(z) \end{aligned} \tag{8.31}$$

Rearranging, we find,

$$\begin{aligned} \frac{T_{sys}}{T_{amb}} &= \frac{T_{RX}}{\alpha T_{amb}} + \tau_{zenith}\sec(z) \\ &= \frac{T_{RX}}{\alpha T_{amb}} + \tau_{zenith}\text{AM} \\ &= \frac{T_{RX}}{\alpha T_{amb}} + \tau \end{aligned} \tag{8.32}$$

The form of Equation 8.32 is that of the equation of a line, with y-intercept, $b = (T_{RX}/\alpha T_{amb})$, and slope, $m = \tau_{zenith}$. Therefore, all we need to do to determine the zenith atmospheric optical depth, τ_{zenith}, is to measure the T_{sys} as a function of AM on blank sky and calculate its slope. Since, in the above expression, both T_{RX} and T_{amb} are independently measured, the hot spillover efficiency, α, of the system, can also be determined.

Many ground-based THz observatories (e.g., the Caltech Submillimeter Observatory, CSO, and the Heinrich Hertz Submillimeter Observatory, HHSMT) have dedicated radiometers with an elevation scanning mirror to provide routine monitoring of the optical depth, that is, a "sky-tipper." A typical frequency of operation for a sky-tipper is 225 GHz, which corresponds to the center of a broad atmospheric window and is close in frequency to the ^{12}CO J = 2 → 1 line at 230 GHz. Figure 8.6 is a plot of one such measurement made at the HHSMT (elevation ~3200 m) under moderate sky conditions, yielding a $\tau_{zenith}^{225} \approx 0.12$. The atmospheric transmission, Γ_{atm}^{ν}, can then be readily found from

$$\Gamma_{atm}^{\nu} = e^{-\tau_{zenith}^{\nu}\sec(z)} \tag{8.33}$$

For a given value of τ_{zenith}^{ν}, atmospheric models (e.g., atmospheric transmission (ATRAN)) can be used to predict the transmission at other frequencies (see Table 8.1 and Figure 8.7). By combining the results of theoretical models with empirical data, Masson (1994) derived

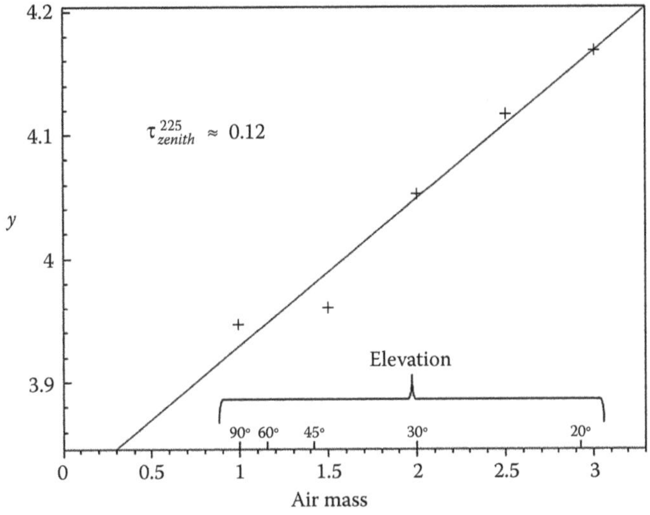

FIGURE 8.6 Sky-tipper fit to elevation scan at 225 GHz.

the following relationships between PWV and 225 GHz optical depth for the Caltech Submillimeter Observatory (CSO) site on Mauna Kea (elevation ~4070 m):

$$\tau^{225}_{zenith} \approx 0.01 + 0.04 \times \text{PWV} \tag{8.34}$$

where

PWV = precipitable water vapor column (mm)

The offset of 0.01 is due to the presence of an atmospheric oxygen absorption line. The opacity at 345 GHz is, then,

$$\tau^{345}_{zenith} \approx 0.05 + 2.5 \times \tau^{225}_{zenith} \tag{8.35}$$

and, in the higher frequency atmospheric windows,

$$\tau^{820}_{zenith} \approx \tau^{690}_{zenith} \approx \tau^{490}_{zenith} \approx 20 \times (\tau^{225}_{zenith} - 0.01) \tag{8.36}$$

TABLE 8.1 Optical Depth Relationship between Atmospheric Windows

Spectral Line	Frequency (GHz)	τ_{zenith}
^{12}CO $J = 2 \rightarrow 1$	230	0.17
^{12}CO $J = 3 \rightarrow 1$	345	0.41
^{12}CO $J = 4 \rightarrow 1$	460	1.33
[CI]	809	2.20

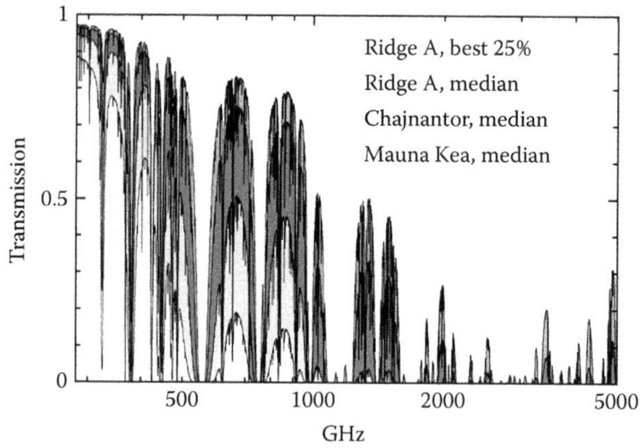

FIGURE 8.7 Plot of atmospheric transmission for ground-based observatories. (Courtesy of C. Kulesa.) Top curve is for the best transmission,"Ridge A, best 25%," followed by "Ridge A, median," "Chajnantor, median," and lowest is "Mauna Kea, median." A very low column density of precipitable water vapor, for example, PWV ≤ 0.5 mm, is required to observe at THz frequencies. The best ground-based sites (e.g., Dome A/Ridge A) are capable of achieving a PWV of ~60 μm for ~30 days a year. At stratospheric altitudes (above ~40,000 ft) PWV values of <10 μm are typical. At such low PWV's the THz atmospheric transmission, Γ_{atm}^{THz} is ≥90%, making airborne and balloon-borne observatories second only to space-based platforms for conducting sensitive THz studies of the cosmos (see Figure 1.1). (Figure Credit: http://soral.as.arizona.edu/HEAT/site/. With permission.)

EXAMPLE 8.2

The 225 GHz tipping radiometer at an observatory measures a zenith atmospheric opacity of $\tau^{225}{}_{zenith} \approx 0.08$. What will be the atmospheric transmission at a zenith angle of 45° at 345 GHz?

Substituting into Equation 8.35, we find,

$$\tau_{zenith}^{345} \approx 0.05 + 2.5(0.08) \approx 0.25$$

For a zenith angle (= 90° – elevation) of 30°, the atmospheric transmission will be (see Equation 8.33),

$$\Gamma_{atm}^{345} = \exp[-(0.25)s(45^\circ)] = 0.70 \Rightarrow 70\%$$

EXAMPLE 8.3

If we wanted to observe at 810 GHz on the same night as Example 8.2, what would the zenith atmospheric transmission be?

Using Equation 8.36, we find the zenith opacity to be,

$$\tau_{zenith}^{810} \approx 20 \times (0.08 - 0.01) \approx 1.4$$

The corresponding zenith atmospheric transmission is, then,

$$\Gamma_{zenith}^{810} \approx \exp[-1.4] \approx 0.25 \Rightarrow 25\%$$

8.5 THz BRIGHTNESS TEMPERATURE OF PLANETARY BODIES

The thermal emission from planetary bodies within the solar system has been used for decades to calibrate millimeter-wave and far-infrared observations. These bodies include Venus, the Moon, Mars, Jupiter, Saturn, Uranus, and Neptune. The thermal emission associated with bodies having little or no atmosphere (e.g., the Moon and Mars) is relatively simple to predict, while the Jovian planets require more detailed models. Thermal emission can be quoted in terms of brightness temperature, T_b(K), or flux, S_ν(Jy), which are related through the expression,

$$\frac{S_\nu(\text{Jy})}{T_b(K)} \approx 1 \times 10^{26} \frac{2k}{A_p} \tag{8.37}$$

where
k = Boltzmann constant = 1.38×10^{-23} m^{2}kg s^{-2}K^{-1}
A_p = physical area of telescope (m^2)

8.5.1 THE MOON

The physical temperature of the Moon's surface, as seen from the Earth, varies during the lunar cycle, and is a function of the electric and thermal properties of its surface materials. These properties have been studied in depth by Krotikov and Troitskii (1964) and Linsky (1966, 1973). Using these results, Mangum (1993) derives the following expression for the wavelength (in mm) and time-dependent temperature of the Moon:

$$T_{Moon}(\lambda) = T_0(0) + \frac{0.77T_0(0)}{\left(1 + 0.48\lambda + 0.11\lambda^2\right)^{\frac{1}{2}}} \times \cos\left[\chi - \xi_1(\lambda,0,0)\right] \tag{8.38}$$

where

$$\chi \equiv \Phi - \phi_1 = \left(\frac{\text{days since new Moon}}{\text{period of Moon}}\right) \times 360° - 180°$$

$$\xi_1(\lambda,0,0) \equiv \tan^{-1}\left(\frac{0.24\lambda}{1 + 0.24\lambda}\right)$$

period of Moon = 29.53 days

Values of $T_0(0)$ appropriate for infrared through radio wavelengths are provided in Table 8.2 (from Linsky, 1973; Mangum, 1993).

TABLE 8.2 Values of $T_0(0)$ for Moon

λ (μm)	12	17.5	30	100	300	1000	3000
ν(THz)	25	17.1	10	3	1	0.3	0.1
$T_0(0)$	213.9 ± 6.0	212.7 ± 6.2	213.7 ± 8.7	216.1 ± 6.9	217.6 ± 8.1	219.1 ± 6.3	227.7 ± 8.9

8.5.2 MARS

Mars is relatively bright, compact, solid, does not undergo phases like the Moon and Venus, and has ≤ 1% the atmosphere of the Earth, making it a nearly ideal flux calibrator at THz frequencies. A thermal model of Mars, derived from spacecraft observations, was first made by Neugebauer et al. (1971) and then used to derive THz/far-IR brightness temperatures by Wright (1976, 2007). Mars is found to closely resemble a black-body radiator. Table 8.3 is an excerpt of Mars brightness temperatures as a function of wavelength for the years 2015 through 2020 from Wright (2007). The modified Julian day (MJD) in the table is the Julian Day (JD)−2440000.

At longer wavelengths (i.e., lower frequencies) the Martian brightness temperature, T_{Mars}, is well represented by (Ulich, 1981)

$$T_{Mars}(R) = T(R_0)\left(\frac{R_0}{R}\right)^{\frac{1}{2}} \tag{8.39}$$

where

R_0 = 1.524 AU = mean Martian heliocentric distance
R = heliocentric distance to Mars at time of observation (AU)
$T(R_0)$ = 206.8 ± 5.8 K

8.5.3 VENUS, JUPITER, URANUS, AND NEPTUNE

Table 8.4 lists observed THz brightness temperatures of Venus, Jupiter, Uranus, and Neptune. The measurements are based on broadband continuum observations made with bolometric detectors, most with postdetection bandwidths of order 100 GHz or greater. The observations of Serabyn and Weistein (1996) are the exception, since they used a Fourier transform spectrometer (FTS) before the bolometer. Also, all observations except Goldin et al. (1996), which were performed from a balloon-borne telescope, were made from the ground. The broadband measurements are sensitive to the presence of molecular

TABLE 8.3 Mars Brightness Temperatures versus Wavelength for 2015–2020

MJD	10 μm	20 μm	34 μm	65 μm	100 μm	150 μm	250 μm	350 μm
17240	245	238	231	225	221	219	217	216
17400	243	235	228	221	218	215	213	212
17600	243	231	223	216	212	210	207	206
17800	250	239	232	225	221	219	217	216
18000	243	235	229	223	219	217	215	214
18200	248	238	231	223	220	217	215	214
18400	250	237	229	221	218	215	213	211
18600	238	228	221	214	210	208	205	204
18800	246	238	232	225	222	219	217	216
19000	260	250	244	237	233	230	228	227
19200	243	232	225	218	214	212	210	209

TABLE 8.4 Planetary Brightness Temperatures

Planet	Wavelength (μm)	Frequency <THz>	Brightness Temperature (K)	Reference[a]
Venus	750	0.400	323 ± 41	1
	450	0.668	277 ± 24	1
	30–300	1.818	240 ± 8	2
	114–196	1.935	302 ± 27	3
	71–94	3.636	328 ± 14	3
	47–67	5.263	255 ± 7	3
	31–38	8.696	244 ± 7	3
Jupiter	1744	0.172	169 ± 2	6
	1332	0.227	170.9 ± 3.9	5
	1049	0.286	166 ± 2	6
	890	0.337	174.3 ± 5.1	5
	750	0.400	162 ± 18	1
	609	0.492	137 ± 2	6
	445	0.675	148.5 ± 4.9	5
	450	0.667	146 ± 8	1
	444	0.675	134 ± 2	6
	410	0.732	139 ± 4	4
	345	0.870	139.5 ± 3.8	5
	30–300	1.818	133 ± 3	2
Saturn	1744	0.172	141 ± 3	6
	1049	0.286	126 ± 2	6
	609	0.492	112 ± 2	6
	445	0.675	112 ± 2	6
	410	0.732	95 ± 2	4
	114–196	1.935	82 ± 4	3
	71–94	3.636	81.5 ± 3	3
	47–67	5.263	79 ± 1	3
	31–38	8.696	81 ± 1	3
	30–300	1.818	85 ± 2	2
Uranus	1200	0.2–0.3	92	7
	750	0.400	94 ± 16	1
	410	0.732	71 ± 9	4
Neptune	1200	0.2–0.3	86.2	7
	750	0.400	117 ± 24	1

[a] 1, Whitcomb et al. (1979); 2, Wright (1976); 3, Courtin et al. (1979); 4, Lowenstein et al. (1977); 5, Griffin et al. (1986); 6, Goldin et al. (1996); 7, Serabyn and Weistein (1996).

absorption features from planetary atmospheres. In the case of Jupiter and Saturn, these features can decrease the estimated brightness temperature by as much as 15% for Jupiter and 40% for Saturn (Lellouch et al., 1984; Mangum, 1993). The orientation of Saturn's rings at high inclination can add to the apparent size of the planet and increase estimates of the planet's brightness temperature by as much as ~10% compared to measurements at low ring inclination angles (Epstein et al., 1980; Mangum, 1993). Venus is a bright, nearby source, but its proximity to the Sun often puts it in the solar avoidance zone of most telescopes (both ground- and space-based).

The outer planets, Uranus and Neptune, are favorite absolute flux calibrators for space-based missions such as *Herschel*, where they were used to calibrate the spectral and photometric imaging receiver (SPIRE) and heterodyne instrument for the far-infrared (HIFI) instruments. Due to their distance, the planets are nearly point-like in THz telescope beams (3.3″–4.1″ for Uranus and 2.2″–2.4″ for Neptune), and have reasonably well understood THz spectra. Brightness temperature calibration curves of Uranus and Neptune compiled from models and observations for the *Herschel* mission (Marston, A., personal communication, 2014; Bendo et al., 2013) are shown in Figure 8.8.

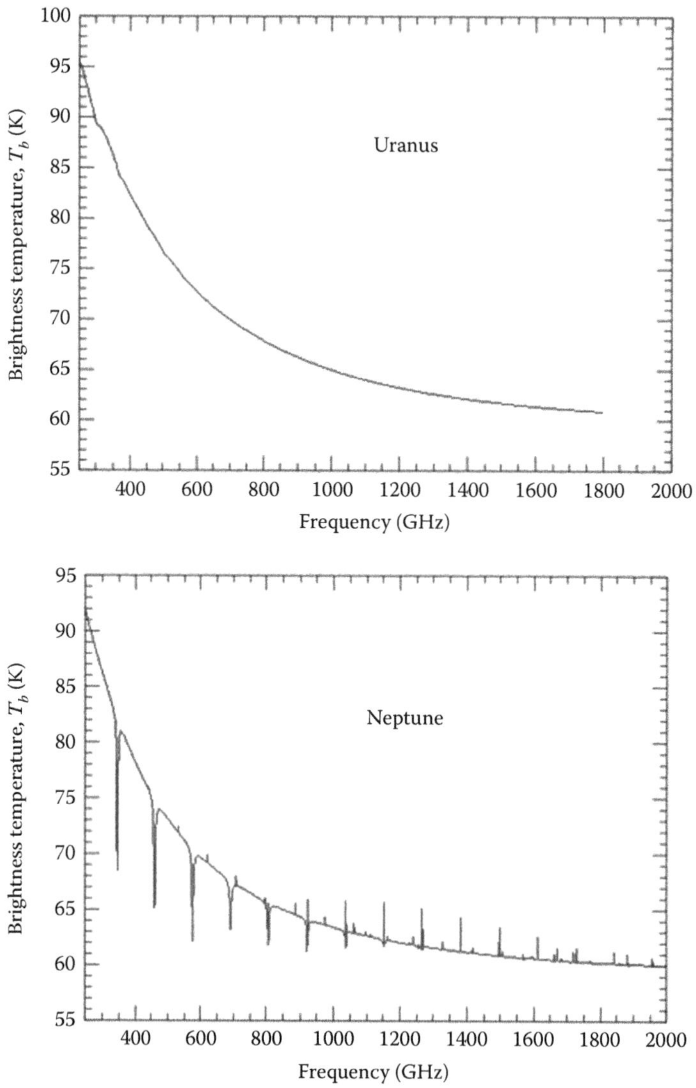

FIGURE 8.8 Uranus and Neptune brightness temperatures at THz frequencies. Compiled for the *Herschel* mission. (Courtesy of Marston, A., personal communication, 2014. With permission.)

CONCLUSION

Over the past ~40 years, a variety of observing and calibration strategies have been developed to deal with the challenges of obtaining and reducing millimeter-wave and, more recently, THz astronomical data. These include absolute position switching, On-the-Fly mapping, frequency switching, and the chopper-wheel calibration technique. The planets are particularly useful in providing absolute calibration, and deriving telescope performance parameters. In the next chapter, we will investigate how what we have learned about telescopes, optics, receiver systems, and calibration, can be synthesized into realizing high angular resolution, THz interferometers.

PROBLEMS

1. A common observational technique employed at THz frequencies is absolute position switching (APS).

    a. Under what circumstances might you consider using the On-the-Fly (OTF) mapping technique instead? Why?

    b. In what scenario might frequency switching be an appropriate observational technique to employ?

2. You have a 230 GHz receiver on a 10 m telescope, and are in the process of determining the η_{moon} of the system. Using a square law detector (e.g., voltmeter) monitoring the receiver's total power output, you measure,
   V_{amb} = 1.060 V = voltage measured when looking at an ambient temperature load.
   V_{moon} = 1.095 V = voltage measured when looking at the Moon.
   V_{Sky} = 0.387 V = voltage measured when looking at blank sky.
   The temperature of the ambient load is 273 K. The Moon is full.
   What is η_{moon}?

3. Using the same 230 GHz receiver/telescope as in Problem 1, you decide to determine the main beam efficiency on an object closer to the size of your main beam, in this case Saturn. Using a square law detector monitoring the receiver's total power output, you measure,
   V_{amb} = 1.81 V = voltage measured when looking at an ambient temperature load.
   V_{Saturn} = 0.66 V = voltage measured when looking at Saturn.
   V_{sky} = 0.60 V = voltage measured when looking at blank sky.
   The temperature of the ambient load is 287 K. On this day, the diameter of Saturn is 16″ (found using an ephemeris).
   What is η_{mb}?

4. Taking nothing for granted, you perform a manual sky-dip to determine the atmospheric optical depth at your observational frequency, 230 GHz. Using a voltmeter

monitoring the receiver's total power output as a function of air mass (AM), you measure,

AM	V_{Sky} (V)
1.1	0.394
1.24	0.422
1.68	0.448
1.92	0.475
2.16	0.496
2.31	0.514

When looking at an ambient temperature load, the output power of the receiver is $V_{amb} = 0.467$ V. Similarly, the output power looking at a cold (77 K) load is $V_C = 1.051$ V.

From this information, calculate the atmospheric optical depth at zenith.

5. How much worse would you estimate the zenith atmospheric transmission to be at 690 GHz compared to 230 GHz on the same night as problem 4?

REFERENCES

Bendo, G. J., Griffin, M. J., Bock, J. J. et al., 2013, Flux calibration of the *Herschel_*-SPIRE photometer, *MNRAS* 433, 3062–3078.

Courtin, R., Lena, P., de Muizon, M., Rouan, D., Nicollier, C., and Wijnbergen, J., 1979, Far-infrared photometry of planets—Saturn and Venus, *Icarus*, 38, 411.

Emerson, D. T., 1997, Increasing the yield of our telescopes, *IAUS*, 170, 207.

Epstein, E. E., Janssen, M. A., Cuzzi, J. N., Fogarty, W. G., and Mottmann, J., 1980, Saturn's rings—3-mm observations and derived properties, *Icarus*, 41, 103.

Ewen, H. I. and Purcell, E. M., 1951, Observation of a line in the galactic radio spectrum: Radiation from galactic hydrogen at 1,420 Mc/sec, *Nature*, 168, 356.

Goldin, A., Kowitt, M., Cheng, E. et al., 1996, Whole-disk observations of jupiter, saturn, and mars in millimeter/submillimeter bands, *Ap. J.*, 488L, 161.

Gordon, M. and Meeks M., 1968, Observations of 8-GHz continuum and hydrogen recombination lines in the orion nebula, *Ap. J.*, 152, 417.

Griffin, M. J., Ade, P. A. R., Orton, G. S., Robson, E. I., Gear, W. K., Nolt, I. G., and Radostitz, J. V., 1986, Submillimeter and millimeter observations of Jupiter , *Icarus*, 65, 244.

Haslam, C. G. T., 1974, NOD2 A general system of analysis for radioastronomy, *A&AAS*, 15, 333.

Jewell, P., 2002, Millimeter wave calibration techniques, *ASP Conference Series*, 278, 313.

Krotikov, V. D. and Troitskii, V. S., 1964, Radio emission and nature of the moon, *Soviet Phys.-Usp.* (Eng. Transl.), 6, 841.

Lellouch, E., Encrenaz, and Combes, M., 1984, The detectability of minor atmospheric species in the far infrared spectra of Jupiter and Saturn, *Astron. Astrophys.*, 140, 405–413.

Linsky, J. L., 1966, Models of the lunar surface including temperature-dependent thermal properties, *Icarus*, 5, 606.

Linsky, J. L., 1973, The moon as a proposed radiometric standard for microwave and infrared observations of extended sources, *Ap. Js.*, 25, 163.

Lowenstein, R., Harper, D. A., Moseley, S. H. et al., 1977, Far-infrared and submillimeter observations of the planets, *Icarus*, 31, 315.

Mangum, J., 1993, Main-beam efficiency measurements of the caltech, *Publications of the Astronomical Society of the Pacific*, 105, 117–122.
Mangum, J., 2000, Amplitude calibration at millimeter and submillimeter wavelengths, *ALMA Memo* 318.
Mangum, J., 2006, Observing Modes Used in Radio Astronomy, https://safe.nrao.edu/wiki/pub/Main/RadioTutorial/radio-obs-modes.pdf.
Masson, C. R., 1994, Atmospheric effects and calibrations, in *Astronomy with Millimeter and Submillimeter Wave Interferometry*, APS Conference Series, M. Ishiguro and Wm. J. Welch (eds.), 59, 87.
Neugebauer, G., Münch, G., Kieffer, H., Chase, S. C., Jr., and Miner, E., 1971, Mariner 1969 infrared radiometer results: Temperatures and thermal properties of the martian surface, *A. J.*, 76, 719.
Penzias, A. and Burrus, C., 1973, Millimeter-wavelength radio astronomy techniques, *Annu. Rev. Astro. Astrophys.*, 11, 51–72.
Phillips, T. G., 1989, Calibration at the CSO, CSO Cookbook.
Serabyn, E. and Weistein, E., 1996, Calibration of planetary brightness temperature spectra at near-millimeter and submillimeter wavelengths with a Fourier-transform spectrometer, *Appl. Opt.*, 35(16), 2752.
Ulich, B. L., 1981, Millimeter-wavelength continuum calibration sources, *A. J.*, 86, 1619.
Urquhart, J., Hoare, M., Purcell, C. et al., 2010, Characterisation of the Mopra radio telescope at 16–50 GHz, *Publications of the Astronomical Society of Australia*, 27, 321–330.
Whitcomb, S., Hildebrand, R., Keene, J., Steining, R., and Harper, D., 1979, Submillimeter brightness temperatures of Venus, Jupiter, Uranus, and Neptune, *Icarus*, 38, 75.
White, G., Araki, M., Greaves, J., Ohishi, M., and Higginbottom, 2003, A spectral survey of the Orion Nebula from 455–507 GHz, *A&A*, 407, 589–607.
Wright, E., 1976, Recalibration of the far-infrared brightness temperatures of the planets, *Ap. J.*, 210, 250–253.
Wright, E., 2007, Infrared Brightness Temperature of Mars, 1983–2103, arXiv:astro-ph/0703640.

9

THz INTERFEROMETRY

PROLOGUE

Radio interferometry was born out of the rapid evolution in microwave science and technology that occurred during World War II. An interferometer combines the light collected by two or more telescopes to create a power pattern sensitive to angular size scales proportional to the projected separation (i.e., baselines) between them. Over the past ~60 years interferometric techniques initially developed for radio frequencies have steadily progressed in usage up into the THz domain. Over the next decade, THz interferometers will play a key role in disentangling the physical processes that lead to star, planet, and galaxy formation. This chapter introduces the history, technology, and data reduction techniques of interferometry.

9.1 INTRODUCTION

The size of the smallest angular structure that can be resolved by a telescope operating in the diffraction limit is given by the Rayleigh Criterion,

$$\begin{aligned}\Omega_R &= 1.22\frac{\lambda}{D} \\ &\approx \Omega_{FWHM} \end{aligned} \tag{9.1}$$

where,

Ω_R = minimum detectable angular feature (rad)

Ω_{FWHM} = full-width-at-half-maximum beam size (rad), for telescope with ~14 dB edge taper (see Equation 5.90)

λ = wavelength of operation (m)
D = telescope diameter (m)

Figure 9.1 plots Ω_{FWHM} for telescope apertures as a function of wavelength and frequency for sizes ranging from 1 m to 30 km. At THz frequencies most galactic, and many nearby extragalactic, sources can be resolved with single-aperture telescopes with aperture diameters in the 1–30 m range. Beyond 30 m, the difficulty and cost of realizing a THz telescope is currently intractable. Radio astronomers, working at lower frequencies, where the wavelengths are much longer, encountered this dilemma early on, and came up with a solution: interferometry (Ryle and Vonberg, 1946).

From Equation 9.1, we see that angular resolution is only a function of wavelength and aperture. The purpose of the underlying optics, either reflective or refractive, is to convey the photons crossing the aperture to the detector, while preserving knowledge of their amplitude and phase. Once at the detector, the photons interfere with each other to produce the telescope's beam. Before this interference occurs, the telescope beam does not exist. Most telescopes utilize a parabolic mirror to bring photons to a focus. The path lengths traveled by all photons reflecting off the parabolic surface to the focus, are equal, meaning, they arrive at the same time. The same beam can be realized by replacing the large parabola with smaller, more easily fabricated parabolas that, when placed side-by-side, have the same effective diameter, D, as the large parabola. Each smaller parabola has its own coherent detection system that preserves knowledge of the amplitude and phase of the arriving photons (see Figure 9.2). This knowledge can then be used either in real time

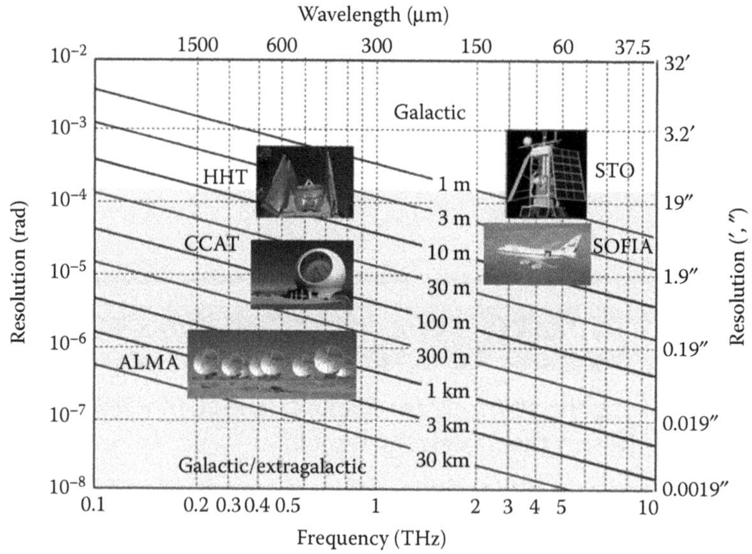

FIGURE 9.1 Telescope angular resolution as a function of aperture size and wavelength. At THz frequencies most galactic molecular clouds and some nearby galaxies can be resolved at the 1 arcmin level using telescope apertures of ~1 m. For observations requiring angular resolutions better than ~5 arcsec, interferometers are needed. Acronyms: HHT—Heinrich Hertz telescope, STO—stratospheric teraHertz observatory, CCAT—Cerro Chajnantor Atacama telescope, SOFIA—stratospheric observatory for infrared astronomy, and ALMA—Atacama large millimeter array.

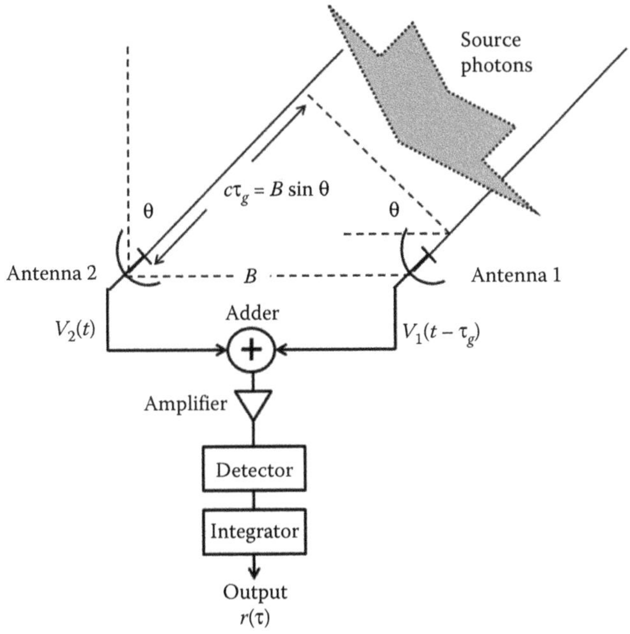

FIGURE 9.2 Adding interferometer. The first radio interferometers were formed by simply adding the signals from two antennas together, either before or after an amplifier. The two antennas can be thought of as fragments of a larger antenna with a diameter equal to the baseline *B*. The projected baseline changes as the source first rises and then sets, helping to fill in the missing pieces of the larger antenna.

or later (even years later!) to allow the photons to interfere yielding the telescope's beam at the time of the observation. The ability to synthesize a large telescope from a number of smaller ones by preserving knowledge of the amplitude and phase of arriving photons is how interferometry works.

If coherent receivers are used at each telescope, knowledge of both the amplitude and phase-difference between photons arriving at the antennas is retained through the down-conversion process. The receiver's local oscillator "phase tags" the photons in the mixing process (see Section 6.2). In an incoherent detection system, knowledge of the phase relationship between photons is lost after detection, requiring that any desired interference of photons take place at the signal frequency. Examples of interferometers that combine signals at the signal frequency, ν, are the Michelson interferometer and simple adding interferometer.

9.2 SIMPLE ADDING INTERFEROMETER

In order to investigate this process further, let us consider the case of the two-element interferometer of Figure 9.2. The two telescopes are located on an east–west baseline, with a separation, *B*, and are observing a distant rising point source at zenith angle, θ. Due to a longer path length, the arrival time of photons at the western telescope is delayed by a time interval, τ_g, compared to those photons arriving at the eastern telescope.

$$\tau_g = \frac{B}{c}\sin\theta \tag{9.2}$$

where

τ_g = geometric signal delay time (s)
B = baseline distance between telescope 1 and 2 (m)
θ = zenith angle (= 90° – elevation)

Let us call the time-dependent voltage waveform associated with photons arriving at telescope 1 (the eastern telescope) V_1 and those at telescope 2 (the western telescope) V_2, where

$$\begin{aligned} V_1 &\propto E_0\cos[\omega(t-\tau_g)] \\ V_2 &\propto E_0\cos[\omega t] \end{aligned} \tag{9.3}$$

where E_0 is the peak voltage. Now, in order to synthesize our interferometer beam, we must combine V_1 and V_2. This can be done either by adding or multiplying them together. The first interferometers were adding interferometers, essentially a radio version of the Michelson interferometer (Ryle and Vonberg, 1946). In an adding interferometer, the signals are combined in a coaxial "T" or hybrid, before reaching the receiver. The output response, $r(t)$, of the detector is

$$\begin{aligned} r(t) &\propto (V_1+V_2)^2 \\ &\propto V_1^2 + 2V_1V_2 + V_2^2 \end{aligned} \tag{9.4}$$

The V_1^2 and V_2^2 terms produce voltage offsets in the detector output, proportional to the received power on each individual telescope. The central term is proportional to the power associated with the product of V_1 and V_2

$$V_1V_2 = E^2\cos[\omega t] + E^2\cos\left[2\omega\left(t-\frac{1}{2}\tau_g\right)\right] \tag{9.5}$$

If the point source in Figure 9.2 were not to change in elevation, then the projection of the baseline on the incoming phase front ($B\cos\theta$) would stay constant, as would the time delay τ_g resulting from it. An east–west baseline of fixed length, B, will sample the sky angular frequency associated with each single aperture,

$$\Omega_{el} \approx 1.2\frac{\lambda}{D_{el}}, \tag{9.6}$$

and an interferometer angular frequency,

$$\Omega_{\tau_g} = \frac{1}{s_\lambda} = \frac{\lambda}{B_p} = \frac{\lambda}{B\cos\theta} = \frac{c}{\nu B\cos\theta}. \tag{9.7}$$

where

Ω_{el} = FWHM beam size of interferometer element with aperture diameter
Ω_{τ_g} = beamwidth between first nulls, BWFN, in array pattern (rad)
D_{el} = diameter of element telescope aperture (m)
s_λ = baseline length in units of wavelength, λ, (m)
B_p = projected baseline on incoming phase front (m)
B = baseline distance between telescopes 1 and 2 (m)
θ = zenith angle of source ($<\pi/2$) (rad)

The beam size associated with the individual antennas, Ω_{el}, defines the field of view (FOV) of the interferometer, while the value of Ω_{τ_g} denotes the small-scale angular sensitivity provided by the interferometer within the FOV.

Astronomical objects are often extended with complicated spatial structures, composed of many spatial frequencies. Many baselines of different lengths are needed to provide the range of spatial frequencies required to adequately sample a source. For a specific antenna separation, B, the Earth's rotation will cause the projected baseline, B_p, to vary as an object rises and sets in the sky, as will Ω_{τ_g}. Figure 9.3 shows the strip chart trace of output power versus time, from the simple adding interferometer of Ryle (1952). The interferometer consists of two Yagi antennas pointed at a fixed position in the sky. Two sources appear in the trace; the one on the left is Cygnus A (external galaxy) and the one on the right is Cassiopeia A (supernova remnant). The observed sinusoidal ripple is due to the source emission picked up by one antenna going in and out of phase with the emission picked up by the second antenna, as the source first rises and then sets. These are the interference fringes, mathematically the result of the V_1V_2 term of Equations 9.4 and 9.5. The fringe taper (or envelope function) is due to a convolution of the instrument

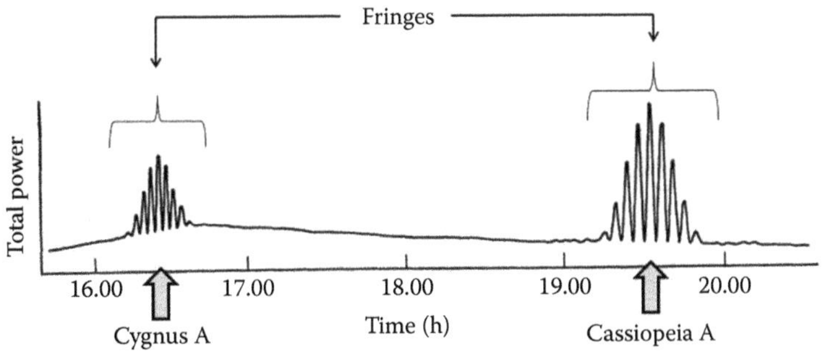

FIGURE 9.3 Strip chart output of Ryle's adding interferometer (Ryle, 1952). As time progresses from left to right, two sources pass through the beam of the interferometer; first the extragalactic source Cygnus A and then the supernova remnant Cassiopeia A. The antenna response is a convolution of the power pattern of an individual interferometer element with fringes produced by the constructive and destructive interference of the combined antenna voltages. The narrow fringes are what give an interferometer its resolving power. The width of a fringe is proportional to the projected separation (or baseline) between the antenna pair at the time of the measurement. (After Ryle, M., 1952, A new radio interferometer and its application to the observation of weak radio stars, *RSPSA*, 211, 351, by permission of the Royal Society.)

passband, the beam pattern of the individual antennas, and the source distribution. In modern interferometers, the antenna elements track objects in azimuth and elevation, so that the peak in each antenna's power pattern (and, therefore, the antenna's FOV) remains fixed on the source.

The null depths in the observed fringes of Figure 9.3 go all the way to the noise floor, indicating that the angular size of the astronomical source that passed through them was smaller in extent than the fringes themselves. In this case, the source is unresolved by the interferometer. Often, sources are a composite of point-like objects, corresponding to small angular frequencies, and more extended objects, corresponding to larger angular frequencies. In such an instance, the observed fringes will look more like those of Figure 9.4, where the null depths do not reach the noise floor. In this instance, the interferometer is said to partially resolve the source. If no fringes are observed, it means the characteristic angular frequencies associated with the source are much larger than the fringe width, Ω_{τ_g}, and the source is said to be "resolved-out" by the interferometer. The behavior of interference fringes produced by astronomical objects was first discussed by Michelson (1890, 1920) while measuring stellar diameters at optical wavelengths. As a way of quantifying the relative amplitude of fringes, he defined a fringe visibility, V_M, as (Thompson et al., 1991).

$$V_M = \frac{\text{brightness of maxima} - \text{brightness of minima}}{\text{brightness of maxima} + \text{brightness of minima}} = \frac{S_{\text{max}} - S_{\text{min}}}{S_{\text{max}} + S_{\text{min}}} \tag{9.8}$$

where

S_{max} = maximum observed source flux (Jy)
S_{min} = minimum observed source flux (jy)

With this definition,

$$\begin{aligned} V_M &= 1 \Rightarrow \text{source unresolved} \\ 1 > V_M &> 0 \Rightarrow \text{source partially resolved} \\ V_M &= 0 \Rightarrow \text{source resolved-out} \end{aligned}$$

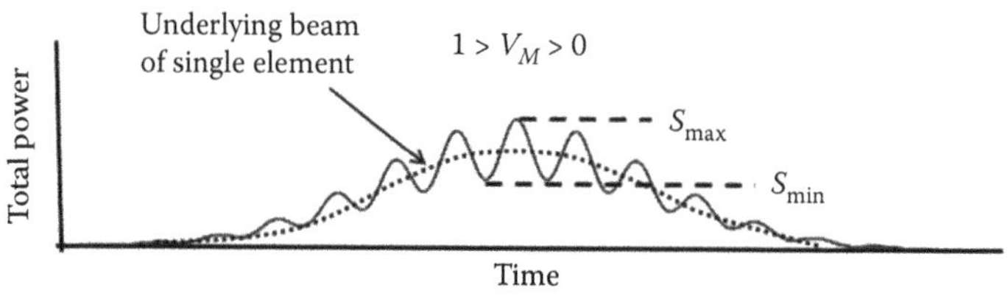

FIGURE 9.4 Visibility of an object. If the angular extent of a source is much smaller than the fringe width of the interferometer, the source is unresolved and has a visibility, V_M, of 1 (see Figure 9.3). If the source angular extent is much larger than the fringe width, the source is resolved-out and $V_M = 0$. Often the distribution of size scales within an object is such that some parts are resolved and others are not. This is the situation in the strip chart recording above, where V_M is between 0 and 1.

EXAMPLE 9.1

What is the field of view (FOV) of an interferometer composed of 12 m diameter dish antennas at 0.810 THz?

The FOV of an interferometer is the same as the beam size of one of its antennas at the frequency of observation.

$$\lambda = \frac{c}{\upsilon} = \frac{3 \times 10^8 \text{ m/s}}{0.81 \times 10^{12} \text{ HZ}} = 3.7 \times 10^{-4} \text{ m}$$

$$FOV \approx 1.2 \frac{\lambda}{D_{element}} \approx 1.2 \left(\frac{3.7 \times 10^{-4} \text{ m}}{12 \text{ m}} \right)$$
$$\approx 3.7 \times 10^{-5} \text{ rad}$$
$$\Rightarrow 7.64''$$

EXAMPLE 9.2

You have a two-element interferometer operating at 0.81 THz. What baseline, B, is needed to resolve a 100 AU protostellar disk 160 pc away?

We can use the small angle formula to determine the angular extent, θ_D, of the disk.

$$\theta_D \approx \frac{s}{d} \approx \frac{100 \text{ AU}(1.5 \times 10^{13} \text{cm/AU})}{160 \text{ pc}(3.1 \times 10^{18} \text{ cm/pc})} \approx 3.02 \times 10^{-6} \text{ rad} \Rightarrow 0.62''$$

In order to resolve the object, the interferometer should achieve an angular resolution on the sky $\Omega_{\tau_g} \leq (\theta_D/2)$. Assuming the object passes through zenith, from Equation 9.7 the required separation between the two antennas is,

$$B \geq \frac{2\lambda}{\Omega_{\tau_g} \cos\theta} \approx \frac{2(3.7 \times 10^{-4} \text{ m})}{(1.51 \times 10^{-6} \text{ rad})\cos(0^\circ)} \approx 490 \text{ m}$$

EXAMPLE 9.3

What would be the required baseline, in the above example, if the source only rose to an elevation of 30°?

In this case, $\theta = 90^\circ - \text{elevation} = 90^\circ - 30^\circ = 60^\circ$, yielding a required baseline of,

$$B \geq \frac{490 \text{ m}}{\cos(60^\circ)} \approx 980 \text{ m}.$$

9.3 PHASE SWITCHED INTERFEROMETER

The sensitivity and resolution of early adding interferometers was limited by the need to have the antennas close together, in order to limit signal loss in the transmission lines connecting them. The addition of an amplifier in each line would help solve this problem, but due to the V_1^2 and V_2^2 terms in Equation 9.4, the component of the signal, due to the correlated signal from the source, V_1V_2, would likely be washed out by the stronger, uncorrelated, amplifier noise. This problem was addressed by Ryle (1952), who introduced the concept of a phase switching interferometer. In this system, the phase of one arm of the interferometer is periodically reversed. The output of the detector then varies between $(V_1 + V_2)^2$ and $(V_1 - V_2)^2$. A synchronous ("lock-in") detector is then used to subtract the two signals, leaving only the cross correlation term, $4V_1V_2$, eliminating the troublesome, uncorrelated, V_1^2 and V_2^2 terms. With the elimination of these terms, amplifiers could be used to boost signal sensitivity and allow antennas to be separated by far greater distances, thereby improving angular resolution. A block diagram of a phase switching interferometer and its output is shown in Figures 9.5 and 9.6. The phase switching approach to interferometry is analogous to the absolute position switching approach used in single dish observations (see Section 8.2.1).

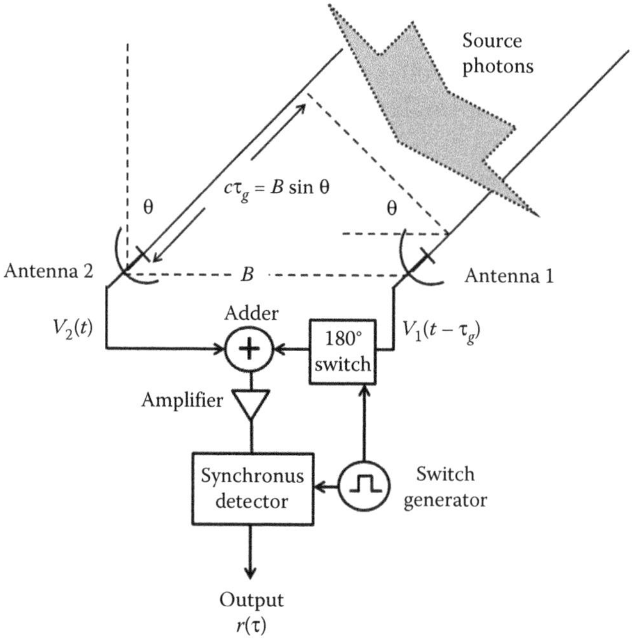

FIGURE 9.5 Phase-switched interferometer. In the case of the adding interferometer of Figure 9.1 the part of the received power that is associated with the interference fringes (i.e., correlated noise) is often small compared to the total output power of the receivers. Indeed, the uncorrelated noise due to gain variations within the receivers is often greater than the correlated noise signal. The correlated noise can be recovered from the uncorrelated noise by subtracting the output of one interferometer element from that of the other on a time scale shorter than that of the gain variations (i.e., the Allan time, see Section 6.11) and synchronously detecting the difference.

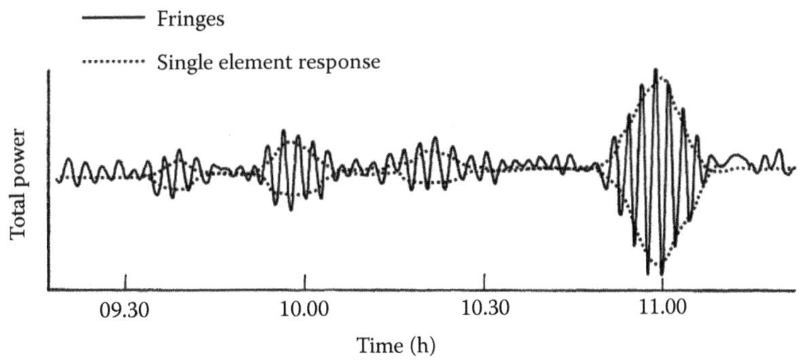

FIGURE 9.6 Output of a phase-switched interferometer. In a phase-switched interferometer the uncorrelated noise within the receiver output has been subtracted-out, leaving only the correlated, interference fringes (i.e., correlated noise). The fringe amplitude envelope reflects the shape of the power pattern of a single antenna element. (After Ryle, M., 1952, A new radio interferometer and its application to the observation of weak radio stars, *RSPSA*, 211, 351, by permission of the Royal Society.)

9.4 CORRELATION INTERFEROMETER

Another approach available to modern interferometry is to multiply the voltage outputs V_1 and V_2 directly, and integrate their product. These operations are performed by a correlator. Correlators can be either analog or digital. Almost all modern correlators are based on digital logic. For a *point source*, V_2 is just a time delayed version of V_1. In this case, the output of a correlator is mathematically equivalent to

$$r(t) = \frac{1}{2T}\int_{-T}^{T} V(t)V(t-\tau)dt \tag{9.9}$$

where

$V(t) = V_1$ = time dependent voltage waveform from Antenna 1 (V)
$V(t-\tau) = V_2$ = time dependent voltage waveform from Antenna 2 (V)
$\tau = \tau_g - \tau_i$ = total signal delay time between (s)
τ_g = geometric delay time (s)
τ_i = instrumental delay time (s)
$2T$ = integration time (s)

The correlator output is a measure of how alike the signals received by the two antennas are when one takes into account the time delay between them. This is equivalent to saying the correlator output is a measure of the mutual coherence of the signals measured by the antennas. By retaining the amplitude and phase information of the correlator output of an extended source over time, the source's spatial coherence function can be determined and used to derive the source structure.

During an observation, the correlator output is summed over time $2T$ and the result recorded in a scan file, along with the corresponding time and telescope position.

The correlator is then reset and another scan commences. If we assume that our integration time, $2T$, is long compared to the time scale of signal variations (~the reciprocal of the correlator bandwidth, $\Delta\nu_c$), then the above equation can be rewritten as,

$$r(\tau) = \frac{1}{2T}\int_{-T}^{T} V(t)V(t-\tau)dt \tag{9.10}$$

which is the familiar autocorrelation function (see Equation 6.75). Here, we have assumed the time averaged amplitude of the cosmic signal is constant over the passband of the receivers. For extended sources, the distinction between V_1 and V_2 is retained. In this case, Equation 9.10 is a complex cross-correlation function.

As discussed in Section 6.10.2, the signal from a receiver can be modeled as a superposition of sinusoidal functions of varying amplitude and frequency. The power spectrum of the signal is an x–y plot showing the amount of power contained in each frequency component of the signal. The Wiener–Kinchin relation states that the power spectrum of a signal can be obtained from performing a Fourier transform of the autocorrelation function of that signal (see Section 6.10.2.3 in this work and Section 2.2 of Thompson et al., 1991). Therefore, by taking the Fourier transform of Equation 9.10, we can obtain the power spectrum of the spatial frequencies that make up the astronomical source

$$\left|H(\nu)\right|^2 = \int_{-\infty}^{\infty} r(\tau)e^{-j2\pi\nu\tau}d\tau \tag{9.11}$$

where

$H(\nu)$ = voltage response of the input signal to the correlator's multiplier (V)

$|H(\nu)|^2$ = power spectrum of the input signal to the correlator's multiplier (W)

The range of spatial frequencies covered by the power spectrum of an interferometer is limited by the number and length of interferometer baselines, the receiver's passband, and the sampling speed of the correlator itself.

Substituting Equation 9.3 into Equation 9.10 we have,

$$r(\tau) \propto \frac{E_0^2}{T}\int_{0}^{\pi} \cos[\omega t]\cos[\omega(t-\tau)]dt \tag{9.12}$$

For $T \gg 2\pi/\omega$, the above equation can be rewritten as

$$r(\tau) \propto \frac{\omega}{2\pi}E_0^2\int_{0}^{2\pi} \cos[\omega t]\cos[\omega(t-\tau)]dt. \tag{9.13}$$

where $\omega = 2\pi\nu$ = angular frequency (rad/s).

Performing the integration yields the result

$$r(\tau) \propto \frac{1}{2} E_0^2 \cos(\omega t). \tag{9.14}$$

Since the average received monochromatic flux $\langle S_o \rangle$ is defined as

$$\langle S_o \rangle = \frac{1}{2} c\varepsilon_0 E_0^2 \tag{9.15}$$

where

$\langle \, \rangle$ = denotes the time average over at least one cycle
S_o = observed flux (watts m^{-2} Hz^{-1})
c = speed of light (3×10^8 m/s)
ε_0 = permittivity of free space (= 8.85×10^{-12} m^{-3} kg^{-1} s^4 A^2)
E_o = the peak voltage (V)

Substitution then yields

$$r(\tau) \propto \langle S_o \rangle \cos(\omega t). \tag{9.16}$$

9.5 PHASOR EQUATION FOR INTERFEROMETRY

The autocorrelation equation of interferometer (Equation 9.16) can be rewritten in a convenient phasor form. To see this, let us first put the solution to Maxwell's wave equation in phasor form. For time-varying electric fields, the wave equation can be written as

$$\frac{\partial^2 E_y}{\partial t^2} = v^2 \frac{\partial^2 E_y}{\partial x^2} \tag{9.17}$$

where

E_y = time varying electric field y component (V/m)
$v = 1/\mu\varepsilon = c$ = speed of light (m/s)
x = direction of propagation (m)

For the case of a wave traveling in the positive x direction, there are numerous possible solutions to the wave equation. Four such solutions are (Kraus and Carver, 1973):

$$\begin{aligned} E_y &= \sin(\beta x - \omega t) \\ E_y &= \sin(\omega t - \beta x) \\ E_y &= \cos(\beta x - \omega t) \\ E_y &= \cos(\omega t - \beta x) \end{aligned} \tag{9.18}$$

where

$\omega = 2\pi f$ = angular frequency (rad/s)
β = propagation constant (rad/m)

Equation 9.18 shows trigonometric solutions to the wave equation. They can also be expressed in exponential form as

$$E_y = E_o e^{j(\omega t \pm \beta x)} \tag{9.19}$$

where it is understood that the instantaneous value of the electric field can be given by either the real or imaginary part of the exponential function. The real part (Re) is

$$E_y = E_o \operatorname{Re} e^{j(\omega t - \beta x)} = E_o \cos(\omega t - \beta x) \tag{9.20}$$

Similarly, the imaginary (Im) part is written as

$$E_y = E_o \operatorname{Im} e^{j(\omega t - \beta x)} = E_o \sin(\omega t - \beta x) \tag{9.21}$$

Adopting the form of Equation 9.20, we can write Equation 9.16 in phasor form,

$$r(\tau) \propto \langle S_o \rangle e^{j\omega\tau} \tag{9.22}$$

where τ is the difference between the geometrical and instrumental delays τ_g and τ_i. Following Rohlfs (1986), if $\vec{B}$ is the baseline vector and $\hat{s}$ the source unit vector toward a direction on the celestial sphere θ, ϕ (see Figure 9.7),

$$\tau = \tau_g - \tau_i = \frac{1}{c}\vec{B} \cdot \hat{s} - \tau_i \tag{9.23}$$

The flux $\langle S_o \rangle$ is what we get at the terminals of our antennas. What we really want is the source brightness distribution, $I_\nu(\theta,\phi)$, which, for a single-aperture telescope, is related to $\langle S_o \rangle$ through (see Section 5.1)

$$S_o = \iint_{source} I_\nu(\theta,\phi) P_\nu(\theta,\phi) d\Omega, \tag{9.24}$$

where

$I_\nu(\theta,\phi)$ = source brightness distribution (watts m^{-2} Hz^{-1} rad^{-2})
$P_\nu(\theta,\phi)$ = normalized power pattern of individual antenna
$d\Omega$ = element of solid angle (= $\sin\theta \, d\theta \, d\phi$ rad^2)

Rewriting Equation 9.24 in terms of the source unit vector, $\hat{s}$, we get

$$S_o = \int_{source} I_\nu(\hat{s}) P_\nu(\hat{s}) d\hat{s} \tag{9.25}$$

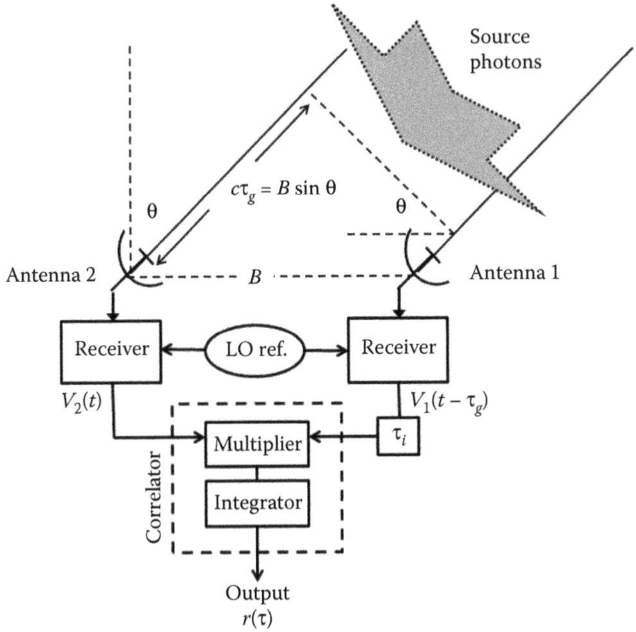

FIGURE 9.7 Correlator interferometer. In an adding interferometer the voltage signals from the two antennas are summed before they encounter a square law detector. What comes out of the detector is proportional to the quadratic function, $(V_1 + V_2)^2$. For interferometry, only the product term of the function, $\propto V_1V_2$ is important. In a modern interferometer, instead of adding V_1 and V_2 directly, they are instead detected/digitized using an A/D converter, and then multiplied together and integrated over some specified time. These processing functions take place within the correlator, the output of which is the correlation function, $r(\tau)$. In order to maintain knowledge of the phase relationship (i.e., photon arrival time) between V_1 and V_2, the two receiver front-ends utilize the same LO (local oscillator) phase reference. The LO, in essence, provides a "time stamp" for the incoming photons.

Substituting Equations 9.23 and 9.24 into Equation 9.22 yields,

$$R_\nu(\vec{B}) = A_e \iint_{source} I_\nu(\hat{s})\, P_\nu(\hat{s}) \exp\left[i\omega\left(\frac{1}{c}\vec{B}\cdot\hat{s} - \tau_i\right)\right] d\hat{s} \tag{9.26}$$

where A_e = effective collecting area of each antenna (m^2) (see Section 5.1).

The above expression is the phasor form of the interferometer equation.

9.6 APERTURE SYNTHESIS OF EXTENDED SOURCES

To obtain the intensity distribution of the source, I_ν, the interferometer must observe the source with enough baselines, $\vec{B}(s)$, to adequately sample the spatial frequencies that characterize its structure. Let $\hat{s}$ represent the unit vector pointing to the desired location in an extended source, and $\hat{s}_0$ the unit vector pointing toward the center of the object. The unit vector $\hat{\sigma}$ represents the offset between $\hat{s}$ and $\hat{s}_0$ (see Figure 9.8):

$$\hat{s} = \hat{s}_0 + \hat{\sigma} \tag{9.27}$$

Substituting Equation 9.27 into Equation 9.26 yields (Rohlfs, 1986)

$$R_\nu(\vec{B}) = A_e \exp\left[i\omega\left(\frac{1}{c}\vec{B}\cdot\hat{s}_0 - \tau_i\right)\right] \iint_{source} P_\nu(\hat{\sigma})\, I_\nu(\hat{\sigma}) \exp\left(i\frac{\omega}{c}\vec{B}\cdot\hat{\sigma}\right) d\hat{\sigma} \tag{9.28}$$

The leading factor describes the plane wave that defines the phase center of the map. The remaining integral describes the interaction of the source distribution with the beam pattern of the interferometer and is called the monochromatic visibility function, V_ν:

$$V_\nu(\vec{B}) = \iint_{source} P_\nu(\hat{\sigma}) I_\nu(\hat{\sigma}) \exp\left(i\frac{\omega}{c}\vec{B}\cdot\hat{\sigma}\right) d\hat{\sigma} \tag{9.29}$$

where

V_ν = monochromatic visibility function toward $\vec{B}$ (normalized)
P_ν = power pattern of individual antenna toward $\vec{B}$ (normalized)
I_ν = intensity distribution of the source toward $\vec{B}$ (normalized)
$\vec{B}$ = baseline vector toward direction on the celestial sphere θ,ϕ (m)
$\hat{\sigma}$ = angular offset between source phase center and direction θ,ϕ (rad)
$\omega = 2\pi\nu$ = angular frequency of observation (rad/s)

The above expression is a more rigorous version of Equation 9.8.

Our goal is to construct a map of the source distribution, I_ν, from knowledge of the spatial frequencies that make up the two-dimensional structure of the object. Let us define a "spatial frequency" coordinate system such that,

$$\frac{\omega}{2\pi c}\vec{B} \Leftrightarrow (u,v,w) \tag{9.30}$$

where u, v, and w are measured in units of the wavelength of light being observed, $\lambda = 2\pi c/\omega$. In this coordinate system (see Figure 9.8), the plane described by (0,0,1) is the plane which is tangent to the celestial sphere at s_0, with u being the east–west, x-type coordinate, and v being the north–south, y-type coordinate. The xy-plane is a flat projection of the celestial sphere onto the map phase center. Here, w is the z-type coordinate, projecting up from the center of the plane. The w component is nulled out by setting the geometric delay to zero, using a variable delay line, or a delay offset in the cross-correlator. The delay tracking has to be precise to a fraction of the reciprocal

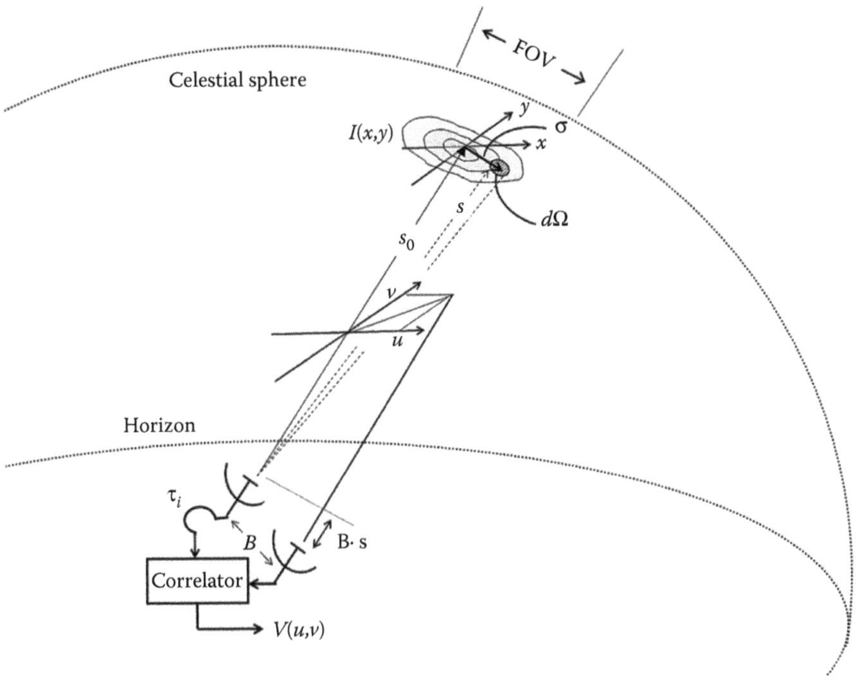

FIGURE 9.8 Aperture synthesis of extended sources. An extended source can be imagined as being spread in two dimensions, x and y, on the surface of the celestial sphere with an intensity distribution $I(x,y)$. The source structure in x and y can be represented by a superposition of spatial components with a given frequency and amplitude. When observed by an interferometer, the spatial frequency being sampled is determined by the length of the projected baseline ($\hat{B} \cdot \hat{\sigma}$) in x (referred to as u) and y (referred to as v). The field-of-view (FOV) of the interferometer is set by the full-width-half-maximum (FWHM) of an individual interferometer element. Within that FOV, the angular resolution that is achieved ($d\Omega$) is set by $\hat{B} \cdot \hat{\sigma}$. The quantity τ_i is the instrumental delay. The output of the correlator, $V(u,v)$, is the complex visibility function of the object, which describes the interaction of the source distribution with the interferometer's beam pattern. The observed source intensity distribution, $I^o_\nu(x,y)$, is obtained by taking the inverse Fourier transform of $V(u,v)$.

of the intermediate frequency (IF) bandwidth to preserve coherence of the two signals. The visibility can then be written such that,

$$V_\nu(u,v) = \iint P_\nu(x,y) I_\nu(x,y) e^{-i2\pi(ux+vy)}\, dx\, dy \tag{9.31}$$

Taking the inverse Fourier transform of $V_\nu(u,v)$, we obtain,

$$I^o_\nu(x,y) = P_\nu(x,y) I_\nu(x,y) = \iint V(u,v) e^{-i2\pi(ux+vy)}\, du\, dv \tag{9.32}$$

where $I^o_\nu(x,y)$ is the long sought after source intensity distribution within the FOV of the interferometer.

9.7 FILLING IN *uv* SPACE

The purpose of interferometry is to synthesize the beam of a big telescope from the beams of two or more smaller ones. This process is referred to as aperture synthesis. A single, large, telescope instantaneously fills in its entire *uv*-plane. When using an interferometer, each pair of telescopes will provide one baseline and one point in the *uv*-plane at a given point in time. In the usual case of an Earth-based interferometer, projection effects will cause the perceived distance between a pair of telescopes (and the associated time delay) to vary over a 24 h period. The continuously changing baseline results in the pair of telescopes producing a track in the *uv*-plane (see Figure 9.9a). The more telescopes there are making up the interferometer, the greater the number of baseline pairs, and the faster the *uv*-plane can be filled in. Once the telescopes have traced their tracks in the *uv*-plane, they can be physically moved and reconfigured to fill in the remaining space in the *uv*-plane to an acceptable level. For an east–west interferometer, the values of u and v, as a function of time, can be found from

$$
\begin{aligned}
u &= \frac{D}{\lambda}\cos t \\
v &= \frac{D}{\lambda}\sin t \sin \delta_o
\end{aligned}
\tag{9.33}
$$

where

D = distance between telescopes (m)

λ = observing wavelength (m)

t = local hour angle of field, that is, angle between local meridian and source (°; –East, +West)

δ_o = declination of field center (°)

9.8 TRANSFORMING THE VISIBILITY FUNCTION

Even if the *uv*-plane is undersampled, an observed source intensity distribution, $I_\nu^o(x,y)$, can be computed for any values of x,y. The better the sampling of the *uv*-plane, the more faithful $I_\nu^o(x,y)$ will be to the true source distribution, $I_\nu(x,y)$. Following the approach outlined by Rohlfs (1986), let us call the measured source intensity at these points $I_\nu^D(x,y)$, where the sum of these points over the source is

$$
I_\nu^D(x,y) = \sum_k g(u_k, v_k) V(u_k, v_k) e^{-i2\pi(u_k x + v_k y)}
\tag{9.34}
$$

where

$g(u,v)$ = normalized weighting function, sometimes called the gating or apodization function.

The form of the function can be selected to shape the effective beam (called the "dirty beam") and sidelobe level within the map. An image created from $I_\nu^D(x,y)$ with the dirty

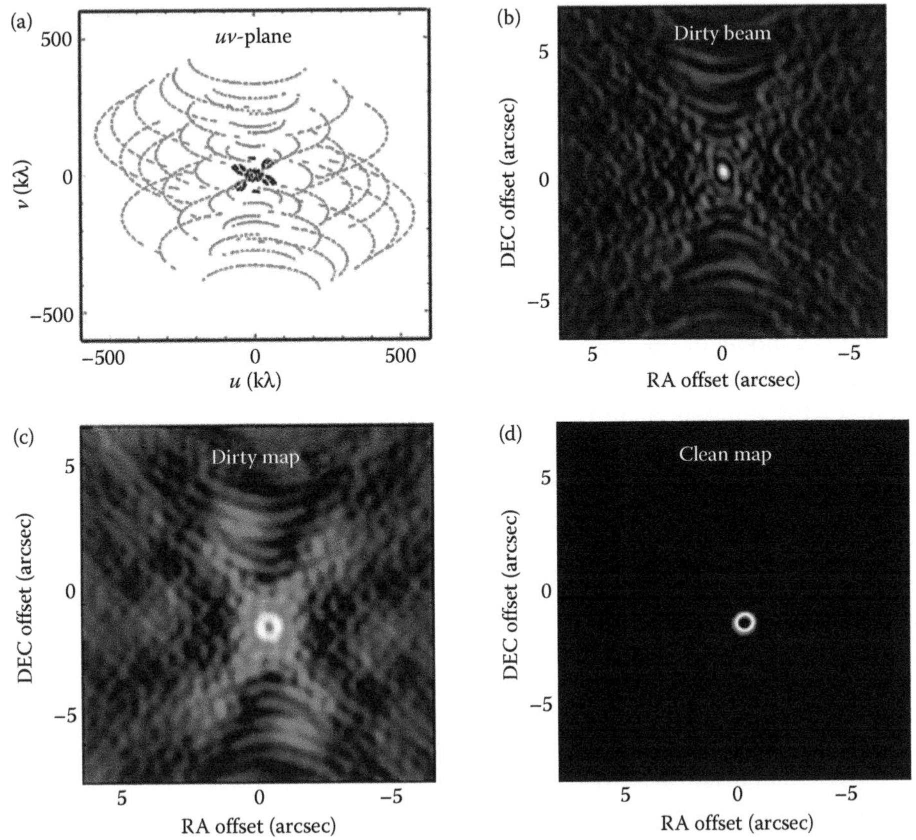

FIGURE 9.9 Imaging with an interferometer. (a) The *uv*-plane indicates which spatial frequencies of the source have been measured within the FOV of the interferometer. For a given physical spacing of interferometer elements, the projected baselines between pairs of interferometer elements will change as the object rises and sets. This change in projected baselines appears as "tracks" in *uv*-space. The more the *uv*-plane is filled in, the closer the resulting map will be to the true source distribution. (b) The undersampled portions of the *uv*-plane correspond to the "missing pieces" of the large telescope we are trying to synthesize with the interferometer. Due to the presence of these missing pieces, the synthesized beam is incomplete and is referred to as "dirty." (c) A "dirty map" is a source image made using a dirty beam. (d) A "clean-map" is produced by filling in the missing parts of the *uv*-plane with best guess estimates. (Adapted from Lundgren, A., 2013, *ALMA Cycle 2 Technical Handbook Version 1.1*, ALMA ISBN 978-3-923524-66-2; Wilner, D. J., 2010, Imaging and Deconvolution, presentation at the 12th NRAO Aperture Synthesis Summer School.)

beam is called the "dirty map" of the source. The dirty map can be viewed as a convolution of a fully sampled source distribution, $I_\nu^o(x,y)$, with the dirty beam of the interferometer, that is, its point spread function, $P_D(x,y)$. Mathematically, this process is described by the relation,

$$I_\nu^D = P_D(x,y) \otimes I_\nu^D(x,y) \tag{9.35}$$

where

$$P_D(x,y) = \sum_k g(u_k,v_k)e^{i2\pi(u_k x+v_k y)}.$$

An example of a dirty beam is shown in Figure 9.9b. The dirty map of a source made from the uv-plane and corresponding dirty beam is shown in Figure 9.9c. Clearly, we would like to do better than this. One method, called CLEAN, was developed by Hogbom to fill in the missing parts of the uv-plane and produce a better looking map. The CLEAN process creates a model of the true source intensity distribution, $I_\nu(x,y)$, as a set of point sources with amplitudes, A_i, extending over the full FOV of the interferometer. CLEAN then iterates on the values of $A_i(x,y)$ until the difference, $\Delta I_R(x,y)$, between the model $I_\nu(x,y)$ and $I_\nu^D(x,y)$ is less than some threshold, typically the expected root-mean-squared (rms) uncertainty in the measured intensities.

$$\Delta I_\nu^{\text{rms}} \approx \Delta I_R(x,y), \quad \text{for } \Delta I_R(x,y) = I_\nu^D(x,y) - \sum_i A_i P_D(x - x_i, y - y_i) \tag{9.36}$$

where

$I_\nu^D(x,y)$ = dirty source intensity distribution
$\sum_i A_i P_D(x - x_i, y - y_i)$ = model (i.e., fake) source distribution
$\Delta I_R(x,y)$ = residual brightness distribution
$\Delta I_\nu^{\text{rms}}$ = error in measured intensities

Once this target difference is achieved, a "clean" map is obtained by convolving the grid of point source amplitudes with a smoothing function, typically chosen to be a Gaussian with a full-width at half-maximum (FWFM) equal to that of the dirty beam. A cleaned version of the dirty map of Figure 9.9c is shown in Figure 9.9d. An example of the power of interferometry and the CLEAN process is the ALMA (Atacama large millimeter array) image of the protoplanetary disk around HL Tau, located 130 pc from Earth (Figure 9.10).

9.9 MAP NOISE LEVEL

The rms noise level of the flux density within an aperture synthesis map (such as Figure 9.9) is given by Thompson et al. (1991) to be

$$\Delta S_\nu^{\text{rms}} = \frac{2kT_{sys}}{A_s \eta_Q \sqrt{n_a(n_a - 1)\Delta\nu_{IF}\tau_0}} \frac{w_{\text{rms}}}{w_{\text{mean}}} \tag{9.37}$$

where

$\Delta S_\nu^{\text{rms}}$ = rms noise level in map flux density (watts/m^2 Hz)
T_{sys} = system noise temperature of an interferometer element (K)

FIGURE 9.10 ALMA image of the protoplanetary disk around HL Tau, located 130 pc from Earth. The image was taken using 25–30 antennas at a maximum baseline of 15.24 km at 233 GHz, with an integration time of 4.5 h. The resulting angular resolution of 0.035″ (or about 5 AU) is sufficient to resolve the 235 AU disk into concentric dust/gas rings. The rings may indicate the presence of recently formed protoplanets in the gaps in between. (Credit by ALMA (ESO/NAOJ/NRAO.)

A_s = effective collecting area of single antenna (m^2)
η_Q = quantization efficiency (=0.64 $\rightarrow$ 1 depending on number of bits used)
n_a = number of antennas in array
τ_o = total integration time (s)
$\Delta\nu_{IF}$ = IF bandwidth (Hz)
$w_{\text{mean}} = (1/n_d)\Sigma w_i$ = mean weighting factor
$w_{\text{rms}} = [(1/n_d)\Sigma w_i^2]^{1/2}$ = rms weighting factor
w_i = normalized weight of each point in uv-plane
$n_d = n_p(\tau_o/\tau_a)$ = total number of independent data points in uv-plane
$n_p = (1/2)n_a(n_a - 1)$ = number of antenna baseline pairs used
τ_a = data averaging time in correlator (s)
$k = 1.38 \times 10^{-23}$ joule/K = Boltzmann's constant

If all the weighting factors, w_i, for points in the uv-plane are equal, then the ratio $w_{\text{rms}}/w_{\text{mean}} = 1$. This is referred to as the natural weighting function. In this case, Equation 9.37 is the same as that for a single dish antenna with aperture = $A_s\sqrt{n_a(n_a - 1)}$ (see Equation 6.3).

9.10 PHASE CLOSURE OR SELF-CALIBRATION

Interferometry is all about retaining knowledge of the phase of the photons arriving from the source. Anything, whether atmospheric or instrumental, that leads to a loss of this knowledge, causes blurring in the reconstructed source image. Jennison (1958) described a technique by which phase errors can be largely eliminated (Kraus, 1986). Three or more antennas are needed to employ the technique. The larger the number of antennas, the better the technique works (Ekers, 1983). Let A, B, and C represent three antennas of a switched or correlation style interferometer, with baseline pairs AB, BC, and AC. Antenna A is designated as the phase reference.

The measured phase at the correlator has contributions from a number of sources. At a given instant of time, these are

1. θ_{AB}, θ_{BC}, and θ_{AC}: The phases of the complex Fourier transforms of the three baseline pairs due to source structure.
2. $\alpha_{\omega B}$ and $\alpha_{\omega C}$: The time-variable phase rotation of the interferometer relative to A at antennas B and C due to changes in the source hour angle.
3. ϕ_B and ϕ_C: The geometric phase angle produced by the source position relative to B and C.
4. δ_B and δ_C: The atmospheric and/or instrumental phase instability errors at B and C.

For each baseline pair the phase output is then:

$$\text{AB phase: } \Psi_{AB} = \theta_{AB} + \alpha_B + \phi_B + \delta_B \tag{9.38}$$

$$\text{BC phase: } \Psi_{BC} = \theta_{BC} + \alpha_C - \alpha_B + \phi_C - \phi_B + \delta_C - \delta_B \tag{9.39}$$

$$\text{AC phase: } \Psi_{AC} = \theta_{AC} + \alpha_C + \phi_C + \delta_C \tag{9.40}$$

By properly combining the above expressions, it is possible to cancel out rotation, position, and phase error, to recover phase information of the source alone, Ψ_S,

$$\begin{aligned} \Psi_S &= (\Psi_{AB} + \Psi_{BC}) - \Psi_{AC} \\ &= \theta_{AB} + \theta_{BC} - \theta_{AC} \end{aligned} \tag{9.41}$$

This process of "self-calibration" can be used to remove both phase errors and gain variations between antennas. Self-calibration is possible because there are more complex visibility functions available between pairs of antennas than there are internal degrees of freedom needed to solve for the amplitude (i.e., gain) and phase variations. The minimum number of antennas needed to self-calibrate phase (i.e., achieve phase closure) is three. The number of antennas required for amplitude closure is four. Having more antennas reduces the signal-to-noise of the amplitude and phase error measurements (see discussion of self-calibration in Thompson et al., 1991; Ekers, 1983).

9.11 PHASE ERROR AT THz FREQUENCIES

At frequencies above ~100 GHz, one of the primary factors limiting the angular resolution achievable with an interferometer is the fluctuation in measured phases caused by the atmosphere, that is, "THz seeing." At these short wavelengths, atmospheric phase fluctuations may be severe enough to make phase closure between interferometer elements and image reconstruction difficult or impossible (see Figure 9.9). As in the case of optical seeing, these fluctuations are due to refractive effects within the Earth's atmosphere. Water vapor is the principal cause of these fluctuations at THz frequencies for the following two reasons (Nikolic et al., 2008):

1. Water vapor in the troposphere is poorly mixed along different lines of sight to the telescopes.
2. Water vapor has a high refractive index, with each millimeter of precipitable water vapor (PWV) capable of retarding the phase of incoming photons by 7 mm of path-length (e.g., 20 wavelengths or 125 radians at 850 GHz) or more.

Phase fluctuations may also be caused to a lesser degree by fluctuations of temperature of the dry air within each telescope's line-of-sight (LOS) (Stirling et al., 2005). All ground-based THz telescopes suffer to a greater or lesser extent from the absorptive effects of atmospheric water vapor. For THz interferometers, these effects are compounded by the added dimension of phase fluctuations. Therefore, it is imperative for THz interferometers to be located at the highest, driest, and coldest sites possible. Cold is good, because, when atmospheric water is frozen, it will only scatter photons out of the telescope beams, while liquid water both absorbs and alters the phase of the photons that pass through it. One such interferometer is the Atacama large millimeter array located on the Chajnantor Plateau at an altitude of 5000 m in northern Chile. ALMA consists of 66 antennas and is designed to achieve an angular resolution as fine as 0.005 arcsec using a 10 km baseline. To achieve this goal requires very precise measurement and correction of atmospheric and instrumental phase errors. ALMA uses the following two phase correction strategies:

1. Fast switching: A common strategy in use on aperture synthesis telescopes for phase calibration is to observe a point-like phase calibrator (e.g., a quasar) several times during a track (i.e., an observing session, where a source is observed over an extended period, e.g., from rise to set). In fast-switching mode, phase calibration is performed, for example, every 10–20 s instead of on-time scales of minutes or hours.
2. Radiometric phase correction using 183 GHz water vapor radiometers. Here, independent 183 GHz water vapor radiometers are used to measure the precipital water vapor (PWV) along each telescope's LOS to the source. The measurements are then used to compute and correct for phase shifts due to the water vapor column above each antenna.

With these two strategies in place, ALMA is capable of achieving milli-arcsecond resolution at THz frequencies.

EXAMPLE 9.4

An interferometer consists of 54 antennas, each with a diameter of 12 meters and an aperture efficiency of 80%. The system noise temperature achieved at each antenna is 300 K. Assuming natural weighting and a 4 GHz IF bandwidth, what will be the rms noise level achieved in a map in 10 min of integration? The correlator efficiency is 88%.

We can use Equation 9.37 to estimate the noise level in the map using the following quantities:

$$T_{sys} = 300\,\text{K}$$

$$A_S = \eta_A \pi \left(\frac{D}{2}\right)^2 = (0.8)\pi \left(\frac{12\text{ m}}{2}\right)^2 = 90.5\text{ m}^2$$

$$\eta_Q = 0.88$$

$$n_a = 54$$

$$\Delta\nu_{IF} = 4 \times 10^9\,\text{Hz}$$

$$\tau_o = 10 \times 60\,\text{s}$$

$k = 1.38 \times 10^{-23}$ joule/K = Boltzmann's constant

$w_{\text{rms}}/w_{\text{mean}} \Rightarrow$ natural weighting

Substitution into Equation 9.37 then yields,

$$\begin{aligned} \Delta S_\nu^{\text{rms}} &= 1.25 \times 10^{-30}\ (\text{watts/m}^2\,\text{Hz}) \\ &= 1.25 \times 10^{-4}\ \text{Jy} \\ &\approx 0.13\ \text{mJy}. \end{aligned}$$

PROBLEMS

1. What is the field of view (FOV) of an interferometer composed of 7 m diameter dish antennas operating at 0.65 THz?

2. You have a two-element interferometer oriented East–West operating at 345 GHz. What is the highest angular resolution the interferometer can achieve with a 15 km baseline if the source you are observing rises due East?

3. What would be the highest angular resolution you could achieve with the interferometer of Problem 2 if the source you are interested in transited at an elevation of 45°?

4. An interferometer consists of eight antennas, each with a diameter of 6 m and an aperture efficiency of 70% at 345 GHz. The system noise temperature achieved at each antenna is 400 K double sideband (DSB). Assuming natural weighting and a 2 GHz IF bandwidth, what will be the rms noise level achieved over the FOV in 10 min of integration? The correlator efficiency is 88%.

5. An interferometer operating on Ridge A in Antarctica is composed of six antennas, each with a diameter of 0.5 m and an aperture efficiency of 80%. The longest baseline is 100 m. Each telescope has a receiver operating at 2 THz with a DSB noise temperature of 1000 K. The atmospheric transmission is 20% (see Figure 8.7). The correlator efficiency is 88%. The load ambient temperature is 200 K (it's cold!) and the telescope hot spillover efficiency, α, is 0.85.

   a. What would be the highest angular resolution achievable with the interferometer?

   b. What is the FOV of the interferometer?

   c. What rms noise level could you achieve over the FOV of the interferometer in 10 min of integration?

   d. If this interferometer were in space, what would be the rms noise level achievable in the same period of time?

REFERENCES

Ekers, R. D., 1983, The almost serendipitous discovery of self-calibration, in *Proceedings of the Workshop on Serendipitous Discoveries in Radio Astronomy*, K. Kellermann and B. Sheets (eds), NRAO, Green Bank, WV.

Jennison, R., 1958, A phase sensitive interferometer technique for the measurement of the Fourier transforms of spatial brightness distributions of small angular extent, *MNRAS*, 118, 276.

Kraus, J. and Carver, K., 1973, *Electromagnetics*, McGraw-Hill, New York.

Kraus, J. D., 1986, *Radio Astronomy*, Cygnus-Quasar Books, Powell, Ohio, pp. 6–49.

Lundgren, A., 2013, *ALMA Cycle 2 Technical Handbook Version 1.1*, ALMA ISBN 978-3-923524-66-2.

Michelson, A., 1890, Application of interference methods to astronomical measurements, *Philosophical Magazine*, 30, 1.

Michelson, A., 1920, On the application of interference methods to astronomical measurements, *Ap. J.*, 51, 257.

Nikolic, B., Richer, J. S., and Hills, R. E., 2008, Simulating atmospheric phase errors, phase correction and the impact on ALMA Science, *ALMA Memo* 582.

Rohlfs, K., 1986, *Tools of Radio Astronomy*, Springer-Verlag, Berlin.

Ryle, M., 1952, A new radio interferometer and its application to the observation of weak radio stars, *RSPSA*, 211, 351.

Ryle, M. and Vonberg, D., 1946, Solar radiation on 175 Mc/s, *Nature*, 158, 339–340.

Stirling A., Richer J. S., Hills R. E., and Lock A., 2005, Turbulence simulations of dry and wet phase fluctuations at chajnantor. Part I: The daytime convective boundary layer, *ALMA Memo Series*, 517, The ALMA Project.

Thompson, A., Moran, J., and Swenson, G., 1991, *Interferometry and Synthesis in Radio Astronomy*, Wiley, New York.

Wilner, D. J., 2010, Imaging and deconvolution, *Presentation at the 12th NRAO Aperture Synthesis Summer School.*

ANSWERS TO PROBLEMS

CHAPTER 1

1. The atmospheric transmission is 0% at 4.2 km, ~85% at 14 km, and 94% at 32 km.
2. The angular extent of Orion Nebula on the sky ~1°.
3. The Jeans mass of the cloud is $M_J \approx 22.5M_\odot$.
4. From Equation 1.11, we find for an $A_V = 20$, that $N_H \approx 3.6 \times 10^{22}$ (H atoms/cm^{-2}).
5. From Figure 1.8, we have in order of occurrence: C^+, C, CO, and S.
6. From Table 1.2, columns E.P. and n_{H_2} we find $^{12}\mathrm{CO}\, J = 6 \rightarrow 5, J = 7 \rightarrow 6, J = 9 \rightarrow 8, J = 11 \rightarrow 10, J = 13 \rightarrow 12$, $[\mathrm{OI}]\,^3P_1 \rightarrow {}^3P_2$, and $[\mathrm{OI}]\,^3P_0 \rightarrow {}^3P_1$.
7. $\tau_{\mathrm{Formation}} \leq 0.1\%$ of total lifetime.
8. $I_{\mathrm{CII}} = 3.6$ K km/s.
9. $L_{\mathrm{NII}} = 0.026L_\odot$.
10. The computed frequency of ^{12}CO $J = 12 \rightarrow 11$ line is 1.382×10^{12} Hz $\Rightarrow$ 1.382 THz.
11. $N_{(o\text{-}\mathrm{H_2O})} = 1.4 \times 10^{15}$ cm^{-2}.

CHAPTER 2

1. $\Delta I = I_{ON} - I_{OFF} \Rightarrow (B_\nu(T_{ex}) - B_\nu(T_{BG}))(1 - e^{-\tau})$.
2. Thermal broadening produces a linewidth of $\Delta v_{th} = 0.54$ km/s.
3. $\Delta v_D = 1.41$ km/s.
4. $\tau^{34} \approx 0.86$.
5. a. $N^{34}_{\mathrm{thin}} = 6.9 \times 10^{14}$ cm^{-2}; b. $N_{\mathrm{H_2}} = 9.1 \times 10^{23}$ cm^{-2}; c. $\sim 1 \times 10^3 A_V$; d. Core mass $\approx 4.3M_\odot$
6. $N^{34}_{\mathrm{thin}} \approx 1.2 \times 10^{14}$ cm^{-2}; underestimates N^{34} by a factor of ~6.
7. a. $T_{ex} \approx 18$ K; b. $T_{ex} \approx 10$ K.
8. From Figure 2.11 we have,

$$\begin{aligned} N_{\mathrm{CO}} &= 6.3 \times 10^{16}\ \mathrm{cm}^{-2}\ \Delta V \\ &= 6.3 \times 10^{16}\ \mathrm{cm}^{-2} (2\ \mathrm{km/s}). \\ &= 1.26 \times 10^{17}\ \mathrm{cm}^{-2}. \end{aligned}$$

Chapter 3

1. $T = 48.5$ K.
2. a. $M_d \approx 9.7 \times 10^{30}$ g; b. $M_g \approx 9.7 \times 10^{32}$ g $\Rightarrow 0.49M_\varepsilon$.
3. a. $M_d \approx 1.05 \times 10^{40}$ g $\Rightarrow 5.25 \times 10^6 M_\odot$; b. $M_g \approx 1.05 \times 10^{42}$ g $\Rightarrow 5.25 \times 10^8 M_\odot$
4. a. Gas-dust heating rate: $\Gamma^d_g = 3.58 \times 10^{-16}$ (erg/cm^3 s); b. $(\Gamma^d_g / \Gamma^{cr}_g) = 3.58 \times 10^8 \times$ as effective.
5. In the limit of low optical depth, $\tau \approx 0.3$.

Chapter 4

Programming exercise for the reader.

Chapter 5

1. a. $S_\nu = 1.9 \times 10^{-36}$ (W/m² Hz); b. $W = 2.1 \times 10^{-25}$ W.
2. a.

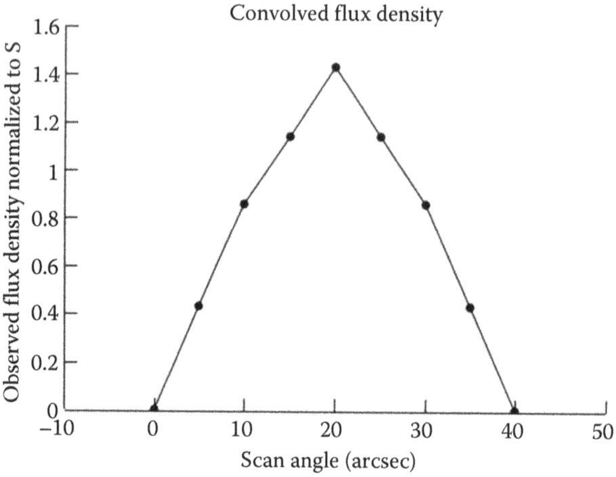

b. $S_\nu^{max} = 1.43S$; c. $T_o = T$, with 1/2 the intensity coming from the central peak and 1/2 from the plateau.

3. $\theta_{FWHM} = 3.6 \times 10^{-5}$ rad ⇒ 7.4″.
4. $\omega_0 = 0.462$ mm.
5. a. Lens focal length is $f = 41.9$ mm; b. Beamwaist at lens is $\omega(41.9\text{ mm}) = 4.0$ mm. Lens should have a diameter, $D \geq 3\omega = 12$ mm; c. (Left for student).
6. Power loss due to reflections is $P_R = 11.7\%$.
   Power loss due to absorption in dielectric is $L_D = 11.4\%$.
   Total loss through lens is 23%.
7. First, find ω at mirror by projecting the beam from the feedhorn.
   The beamwaist at the feedhorn is found to be $\omega_h = 0.465$ mm.
   The lens 3ω diameter is then $D = 46.2$ mm.
   Next, find a value of z_1 from the Cassegrain focus that will project the same beamwaist onto the mirror as the feedhorn does. To do this, first compute the beamwaist at the Cassegrain focus, which for a 14 dB edge taper, is $\omega_{cass} = 3.41$ mm.
   Using the Gaussian beam growth formula, the required distance between ω_{cass} to the mirror is found to be $z_1 = 536$ mm.

Chapter 6

1. The flux density of Jupiter in at the telescope focus is $S_\nu = 7655$ Jy.
2. The required observing time to reach the target rms is 5.8 days.
3. The required local oscillator frequency is $\nu_{LO} = 493.7$ GHz.
4. a. $T_{optic} = 75.1$ K; b. $T_{sys} = 201$ K.

5. $T_{RX} = 1464$ K.
6. Field effect transistor (FET) amplifier. Amplification occurs by modulating the conductivity between the source and drain electrodes by applying the signal to be amplified to the gate electrode.
7. Hot electron bolometer (HEB) receivers provide the lowest noise performance at 1.9 THz. A frequency of 1.9 THz is above the gap frequency of current SIS receivers. Schottky receivers will work at 1.9 THz, but have higher noise and require orders of magnitude more LO power.
8. Due to its high speed and low power/mass, an autocorrelator spectrometer (ACS) is the preferred backend.
9. The spectrometer channel width should be 0.186 MHz.
10. The quantum noise limit of a receiver at 1.9 THz is $T_Q = 91.2$ K.
11. For a junction size of 0.5×0.5 μm, the minimum magnetic flux to suppress the Josephson effect is ~83 Gauss.
12. The Allan time in THz astronomy is a measure of how long one can integrate with a receiver system before the signal noise integrates down slower (by two standard deviations) than expected from the radiometer law.

    With HEB receivers variations in the LO power typically dominate system stability and thereby the Allan time. The receiver should be calibrated at time intervals less than the Allan time.
13. To achieve an order of magnitude improvement in mapping speed requires an array with $N_{pix} = 26$.

Chapter 7

1. $NEP_T = 3.71 \times 10^{-21} (\text{W}/\sqrt{\text{Hz}})$.
2. $NEP = \dfrac{N_{d-a}}{S_E} = 3.75 \times 10^{-16} (\text{W}/\sqrt{\text{Hz}})$.
3. $L_K = 16$ nH.
4. The frequency shift will be 107.3 MHz.
5. The turn-over bandwidth/spectral resolution is $B_{TO} = 102.9$ MHz.
6. The required integration time would be 17.2 s.

Chapter 8

1. a. OTF mapping is particularly well suited for mapping extended sources a few beams across and for large-scale surveys for the following reasons:
      i. Reduced overhead (in time) associated with moving the telescope.
      ii. Uses rapid scanning that helps reduce the effects of variability in atmosphere and detector system.
      iii. One OFF position can be used for many ON's.

   b. Frequency switching is particularly useful when there is no clear off position. It also eliminates overhead associated with moving the telescope.
2. First compute the T_A^* of the Moon; $T_A^* = 287.2$ K. Then compute the physical temperature of the Moon at the wavelength and day of observation; $T_{Moon} = 306.3$ K. The beam efficiency on the Moon is simply the ratio of the two temperatures; $\eta_{Moon} = 0.93$.

3. First compute the T_A^* of Saturn; $T_A^* = 14.2$ K. Then look-up the brightness temperature of Saturn at the wavelength of observation (see Table 8.4); $T_{Sat} \approx 133$ K. The main beam efficiency can then be found from Equation 8.23; $\eta_{mb} = 0.67$.
4. The optical depth at zenith is simply the slope of V_{sky} vs. AM; $\tau_{zenith}^{225} \approx 0.1$.
5. The zenith atmospheric transmission at 690 GHz is 16.6%, while at 230 GHz it is 90%: a lot worse.

Chapter 9

1. $FOV = 11.42''$.
2. The highest angular resolution (i.e., the smallest angular frequencies on the sky to which the interferometer is sensitive) is $\theta_{\rm min} \approx 0.014''$.
3. At 45°, $\theta_{\rm min} \approx 0.02''$.
4. $\Delta S_{\nu}^{\rm rms} = 2.8$ mJy.
5. a. $\theta_{\rm min} \approx 0.37''$; b. $FOV = 74''$; c. First compute the T_{sys} of a single dish under the prescribed atmospheric conditions using Equation 8.31, $T_{sys} \approx 15{,}000$ K. The rms noise level can now be found by substituting into Equation 9.37, $\Delta S_{\nu}^{\rm rms} \approx 27.2$ Jy; d. In space there is no atmosphere to worry about, so the system noise temperature is found to drop to $T_{sys} \approx 1176$ K. The rms noise level then drops to $\Delta S_{\nu}^{\rm rms} \approx 2.17$ Jy. By putting the interferometer above the atmosphere, we gain a factor of ≈ 13 in performance.

APPENDIX 1: TIMELINE OF THz TECHNOLOGY

Contributors (A Few of Many!)	Approximate Year	Development
J. Maxwell[a]	1865	E&M wave theory
S. Langley	1878	Bolometer (platinum strips with lampblack)
D. Hughes	1879	E&M wave demo
H. Hertz[b]	1887	E&M wave demo/publication
E. Branly[c]	1890	Coherer (detector)/publication
J. Bose[d]	1895	Millimeter wave generation/detection
J. Fleming[e]	1904	Vacuum tube diode (patent)
L. DeForest[f]	1906	Vacuum tube triode (patent)
W. Pickard[g]	1906	Silicon detector (patent)
E. Armstrong	1919	Superheterodyne receiver
J. Lilienfeld[h]	1925	Transistor (FET) patent
K. Jansky[i]	1932	Discovery of extraterrestrial radio static
G. Reber[j]	1940	First radio astronomy paper in *Ap. J.*
J. Bardeen, W. Brattain, and W. Shockley	1947	Transistor demonstration
E. Purcell and H. Ewen[k]	1951	First HI detection/receiver
J. Gordon, H. Zeiger, and C. Townes[l]	1954	Maser
F. Low[m]	1961	Gallium-doped germanium bolometer
S. Weinreb[n]	1961	Autocorrelator spectrometer
M. Kinch and Rollin[o]	1963	Hot electron bolometer (HEB) mixer
S. Weinreb et al.[p]	1963	OH detection
R.Wilson, K. Jefferts, and A. Penzias[q]	1970	First CO $J = 1 \rightarrow 0$ detection/receiver
T. Phillips and K. Jefferts[r]	1973	Narrowband HEB receiver for astronomy
Richards et al.[s]/ G. Dolan, T. Phillips, and D. Woody[t]	1979	Superconductor–insulator–superconductor (SIS) mixers
J. Tucker[u]	1979	SIS mixer theory
S. Weinreb[v]	1980	Low noise FET amplifiers
E. Gershenzon et al.[w]	1990	Wideband HEB (phonon-cooled)
E. Prober[x]	1993	Wideband HEB (diffusion cooled)
K. Irwin et al.[y]	1995	Transition edge sensor (TES) with SQUID readout
P. Day et al.[z]	2003	Microwave kinetic inductance detectors (MKIDs)

[a] Maxwell, J., 1865, A dynamical theory of the electromagnetic field, *Philos. Trans. R. Soc. Lond.*,155, 459–512, London.
[b] Hertz, H., 1887, Über sehr schnelle elektrische Schwingungen, *Annals. Phys. Chem.*, 31, 421–448.
[c] Branly, E., 1890. Variations of conductivity under electrical influences, *Minutes of proceedings of the Institution of Civil Engineers*, Vol. 103 by Institution of Civil Engineers (Great Britain) p. 481; (contained in: Académie des Sciences, Paris, vol. cii, p. 78).
[d] Bose: US Patent 755840, Bose, Jagadis Chunder, Detector for electrical disturbances, published September 30, 1901, issued March 29, 1904 (Galena cat-whisker detector).

[e] Fleming: US Patent U.S. 803,684, Fleming, 1905 (Fleming valve).

[f] Pickard: US Patent 836531, Means for receiving intelligence communicated by electric waves, published August 30, 1906, issued November 20, 1905 (silicon diode).

[g] DeForest: U.S. Patent 836,070, Oscillation responsive device (vacuum tube detector—no grid), filed May 1906, issued November 1906.

[h] Lilienfeld: US 1745175, Method and apparatus for controlling electric current first filed in Canada on 1925–10–22, describing a device similar to an FET.

[i] Jansky, K., 1932, Directional studies of atmospherics at high frequencies, *Proc. IRE*, 20, 1920.

[j] Reber, G., 1940, Cosmic static, *Ap. J.*, 91, 621.

[k] Ewen, H. I. and Purcell, E. M., 1951. Observation of a line in the galactic radio spectrum: Radiation from galactic hydrogen at 1420 Mc/sec., *Nature*, 168, 4270.

[l] Gordon, J., Zeiger, H., and Townes, C., 1954. Molecular microwave oscillator and new hyperfine structure in the microwave spectrum of NH, *Phys. Rev.*, 95, 282.

[m] Low, F.J., 1961, Low-temperature germanium bolometer, *J. Opt. Soc. Am.*, 51(11), 1300.

[n] Weinreb, S., 1961. Digital radiometer, *Proc. IEEE*, 49(6), 1099.

[o] Kinch, M. and Rollin, B., 1963, Detection of millimetre and submillimetre wave radiation by free carrier absorption in a semiconductor, *J. Appl. Phys.*, 14, 672.

[p] Weinreb, S., Barrett, A., Meeks, M., and Henry, J., 1963, Low-noise cooled GASFET amplifiers, *Nature*, 200, 829–831.

[q] Wilson, R., Jefferts K., and Penzias, A., 1970, Carbon monoxide in the Orion Nebula, *Ap. J.*, 161, L43–L44.

[r] Phillips, T. G. and Jefferts, K. B., 1973, A low temperature bolometer heterodyne for millimeter wave astronomy, *Rev. Sci. Instrum.*, 44, 1009.

[s] Richards, P., Shen, T., Harris, R., and Lloyd, F., 1979. Quasiparticle heterodyne mixing in SIS tunnel junctions, *Appl. Phys. Lett.*, 34, 345.

[t] Dolan, G. J., Phillips, T. G., and Woody, D. P., 1979, Low-noise 115 GHz mixing in superconducting oxide barrier tunnel junctions, *Appl. Phys. Lett.*, 34, 347.

[u] Tucker, J. R., 1979. Quantum limited detection in tunnel junction mixers, *IEEE J. Quant. Electron.*, 15(11), 1234–1258.

[v] Weinreb, S., 1980, *Electronics Division Internal Report*, No. 202, NRAO.

[w] Gershenzon, E., Gol'tzman, G., Gogidze, I., Gusev, Y., Elant'ev, A., Karasik, B., and Semenov, A., 1990, Millimeter and submillimeter wave range hot electron mixer, *Sov. Phys. Supercond.*, 3, 1582.

[x] Prober, E., 1993, Superconducting terahertz mixer using a transition-edge microbolometer, *Appl. Phys. Lett.*, 62, 2119.

[y] Irwin, K., Nam, S., Cabrera, B., Chugg, B., Park, G., Welty, R., and Martinis, J., 1995, A self-biasing cryogenic particle detector utilizing electrothermal feedback and a SQUID readout. *IEEE Trans. Appl. Supercond.*, 5 (2 pt. 3), 2690–2693.

[z] Day, P., LeDuc, H., Mazin, B., Vayonakis, A., and Zimuidzinas, J., 2003, A broadband superconducting detector suitable for use in large arrays, *Nature*, 425, 817.

APPENDIX 2: MORE THz TRANSITIONS OF ATOMS AND MOLECULES*

TABLE A.1 Important THz Transitions of Atomic Ions

Species	Transition	Frequency (GHz)	Wavelength (μm)
N^+	$^3P_1 \rightarrow {}^3P_0$	1470.3	203.9
C^+	$^2P_{3/2} \rightarrow {}^2P_{1/2}$	1900.536	157.741
N^+	$^3P_2 \rightarrow {}^3P_1$	2459.553	121.889
O^{++}	$^3P_1 \rightarrow {}^3P_0$	3393.045	88.355
N^{++}	$^2P_{3/2} \rightarrow {}^2P_{1/2}$	5230.428	57.317
O^{++}	$^3P_2 \rightarrow {}^3P_1$	5785.712	51.816

TABLE A.2 Important THz Transitions of Neutral Atomic

Species	Transition	Frequency (GHz)	Wavelength (μm)
C^0	$^3P_1 \rightarrow {}^3P_0$	492.162	609.134
C^0	$^3P_2 \rightarrow {}^3P_1$	809.350	370.411
O^0	$^3P_0 \rightarrow {}^3P_1$	2060.061	145.526
Si^0	$^3P_1 \rightarrow {}^3P_0$	2311.751	129.682
Si^0	$^3P_2 \rightarrow {}^3P_1$	4378.194	68.474
O^0	$^3P_1 \rightarrow {}^3P_2$	4744.775	63.184
S^0	$^3P_0 \rightarrow {}^3P_1$	5322.830	56.322

* Data from Melnick, G., 1988, On the road to the large deployable reflector (LDR): The utility of balloon-borne platforms for far-infrared and submillimeter spectroscopy, *International Journal of Infrared and Millimeter Waves*, 9(9), 781.

TABLE A.3 Important THz Transitions of Hydrides

Species	Transition	Frequency (GHz)	Wavelength (μm)
H_2D^+	$J_{K_pK_0} = 1_{10} \to 1_{11}$	372.421	804.982
AlH	$J = 1 \to 0$	377.59	793.96
LiH	$J = 1 \to 0$	443.953	675.280
H_2S	$J_{K_pK_0} = 1_{11} \to 1_{00}$	452.390	662.685
CH	$^2\Pi\ F_1 \to F_2: J = 3/2^+ \to 1/2^-$	532.726	562.751
	$J = 3/2^- \to 1/2^+$	536.761	558.521
H_2O	$1_{10} \to 0_0$	556.936	538.289
NH_3	$J_K = 1_0 \to 0_0$	572.498	523.657
SiH	$^2\Pi\ F_1: J = 3/2^- \to 1/2^+$	624.925	479.726
HCl	$J = 1 \to 0$	625.919	478.964
SiH	$^2\Pi\ F_1: J = 3/2^+ \to 1/2^-$	627.690	477.612
NH	$^3\Sigma^-, N = 1 \to 0, J = 2 \to 1$	974.63	307.60
H_3O^+	$J_K = 0_0^- \to 1_0^+$	984.656	304.464
H_2O	$1_{11} \to 0_{00}$	1113.342	269.273
NH_3	$J_K = 2_0 \to 1_0$	1214.859	246.771
H_2D^+	$J_{K_pK_0} = 1_{01} \to 0_{00}$	1370.204	218.794
SH	$^2\Pi_{3/2}: J = 5/2 \to 3/2$	1382.928	216.781
		1383.257	216.729
CH	$^2\Pi\ F_1: J = 5/2^- \to 3/2^+$	1656.964	180.929
	$J = 5/2^+ \to 3/2^-$	1661.105	180.478
OH	$^2\Pi_{1/2}: J = 3/2^- \to 1/2^+$	1834.760	163.396
	$J = 3/2^+ \to 1/2^-$	1837.853	163.121
CH	$^2\Pi\ F_2: J = 3/2^- \to 1/2^+$	2006.797	149.388
	$J = 5/2^+ \to 3/2^-$	2010.809	149.091
CH_2	$J = 1 \to 0, N_{K_aK_c} = 1_{11} \to 0_{00}$	2344.697	127.860
OH	$^2\Pi_{3/2}\ J = 5/2^+ \to 3/2^-$	2509.963	119.441
	$J = 5/2^- \to 3/2^+$	2514.320	119.234
HD	$J = 1 \to 0$	2675.023	112.071
OH	$^2\Pi_{1/2\to3/2}: J = 5/2^+ \to 3/2^-$	3786.132	79.182
	$J = 5/2^- \to 3/2^+$	3789.216	79.117

APPENDIX 3: COMMONLY USED PHYSICAL AND ASTRONOMICAL QUANTITIES

In THz astronomy, we often find ourselves moving between centimeter-gram-second (CGS) units (e.g., for astronomy and optics) and meter-kilogram-second (MKS) units (e.g., for electromagnetics and engineering). Here, we list often used constants in both systems.

Speed of light:	$c = 2.9979 \times 10^{10}$ cm s^{-1} = 2.9979×10^{8} m s^{-1}
Boltzmann constant:	$k = 1.3805 \times 10^{-16}$ erg K^{-1} = 1.3805×10^{-23} J K^{-1}
Planck constant:	$h = 6.6256 \times 10^{-27}$ erg s = 6.6256×10^{-34} m^2 kg s^{-1}
Electron charge:	$e = 4.8030 \times 10^{-10}$ statC (or esu)
	$= 4.8030 \times 10^{-10}$ g$^{1/2}$ cm$^{3/2}$ s^{-1} = 1.6022×10^{-19} C
Electron mass:	$m_e = 9.1091 \times 10^{-28}$ g = 9.1091×10^{-31} kg
Proton mass:	$m_p = 1.6725 \times 10^{-24}$ g = 1.6725×10^{-27} kg
Atomic mass unit:	amu (or u) = 1.6605×10^{-24} g = 1.6605×10^{-27} kg
Bohr radius:	$a_0 = 5.2918 \times 10^{-9}$ cm = 5.2918×10^{-14} km
Stefan–Boltzmann constant:	$\sigma = 5.670 \times 10^{-5}$ erg cm^{-2} s^{-1} K^{-4} = 5.670×10^{-8} W m^{-2} K^{-4}
Gravitational constant:	$G = 6.6738 \times 10^{-8}$ g^{-1} cm^3 s^{-2} = 6.6738×10^{-11} kg^{-1} m^3 s^{-2}
Mass of Sun:	$M_\odot = 1.9891 \times 10^{33}$ g = 1.9891×10^{30} kg
Radius of Sun:	$R_\odot = 6.955 \times 10^{10}$ cm = 6.955×10^{8} m
Luminosity of Sun:	$L_\odot = 3.846 \times 10^{33}$ ergs s^{-1} = 3.846×10^{26} W
Effective temperature of Sun:	$T_e = 5778$ K

Other quantities:

1 AU = 1.496×10^{13} cm = 1.496×10^{8} km
1 pc = 3.0857×10^{18} cm = 3.0857×10^{13} km
1 yr = 3.1558×10^{7} s
1 debye (D) = 1×10^{-18} statC cm $\approx 3.3356 \times 10^{-30}$ C m
1 eV = 1.6022×10^{-19} J
1 g = 9.8067 m s^{-2}
Impedance of free space: $Z_0 = \eta_0 = 376.730\ \Omega$
Electrical permittivity of free space: $\epsilon_0 = 8.8542 \times 10^{-12}$ F m^{-1} = 8.8542×10^{-12} m^{-3} kg^{-1} s^4 A^2
Electrical permeability of free space: $\mu_0 = 1.2566 \times 10^{-6}$ H m^{-1} = 1.2566×10^{-6} m kg s^{-2} A^{-2}

APPENDIX 4: USEFUL RADIATIVE TRANSFER EXPRESSIONS

Description	Equation	Equation No. in Text
Equation of transfer (distance)	$\frac{dI_\nu}{ds} = j_\nu - k_\nu I_\nu$	Equation 2.1
Equation of transfer (optical depth)	$\frac{dI_\nu}{d\tau_\nu} = I_\nu - S_\nu$	Equation 2.6
Optical depth	$\tau_\nu = \int_0^L k_\nu dz$	Equation 2.8
Emission coefficient	$j_\nu = \frac{h\nu}{4\pi} n_u A_{ul}$	Equation 2.13
Absorption coefficient	$k_\nu = \frac{h\nu}{4\pi}(n_l B_{lu} - n_u B_{ul})$	Equation 2.14
Boltzmann distribution (LTE)	$\frac{n_l}{n_u} = \frac{g_l}{g_u} e^{(h\nu/kT_{ex})}$	Equation 2.16
Planck function (LTE)	$B_\nu = \frac{2h\nu^3}{c^2} \frac{1}{e^{(h\nu/kT_{ex})} - 1}$	Equation 2.19
Solution to equation of transfer (LTE) for on–off observing	$\Delta I_\nu = [B_\nu(T_{ex}) - B_\nu(T_{BG})](1 - e^{-\tau_\nu})$	Equation 2.21
Determining optical depth from ratio of main (M) to isotope (I) line temperatures (LTE)	$R_1 = \frac{T^M}{T^I} = \frac{(1 - e^{-X_I^M \tau_\nu^I})}{(1 - e^{-\tau_\nu^I})}$	Equation 2.38
Excitation temperature (LTE)	$T_{ex} = \left[\frac{k}{h\nu_{ul}} \ln\left[\frac{h\nu_{ul}}{k}\left[\frac{T_R}{f_a(1 - e^{-\tau_\nu})} + J_\nu(T_{BG})\right]^{-1} + 1\right]\right]^{-1}$	Equation 2.40
Column density knowing T_{ex} and τ (LTE)	$N_{thin} = \tau_\nu T_{ex}\left[\frac{c^3 g_u}{8\pi\nu_{ul}^3}\left[\frac{hB}{k}\right] A_{ul}\left(1 - e^{-\frac{h\nu_{ul}}{kT_{ex}}}\right)\right]^{-1} \exp\left(\frac{J}{2}\frac{h\nu_{ul}}{kT_{ex}}\right)\Delta V$	Equation 2.53
Column density in optically thin limit (LTE)	$N_{thin} \approx 4 \times 10^{-5} \frac{\nu^2}{g_u A_{ul} B} e^{\frac{h\nu}{kT_{ex}}\left(1 - \frac{J}{2}\right)} T_{ex} \times II$	Equation 2.65
Critical density	$n_{crit} = \frac{A_{ul}}{\gamma_{ul}}$	Equation 2.74

APPENDIX 5: COMMONLY USED QUASI-OPTICAL EXPRESSIONS

Description	Equation	Equation No.
Gaussian beamwaist size	$\omega(z)^2 = \omega_0^2\left[1+\left(\frac{\lambda z}{\pi\omega_0^2}\right)^2\right]$	Equation 5.27
Power vs. radius in a Gaussian beam	$P(r) = P_0\exp\left(-\frac{2r^2}{\omega(z)^2}\right)$	Equation 5.30
Full-width at half-maximum angle of a Gaussian beam	$\Omega_{FWHM} = 1.18\theta_{\omega_0} = 0.376\frac{\lambda}{\omega_0}$	Equation 5.32
Power coupling between two Gaussian beams	$C_{12} = \frac{4}{[((\omega_{01}/\omega_{02}) + (\omega_{02}/\omega_{01}))^2 + (\lambda/\pi\omega_{01}\omega_{02})^2 d^2]}$	Equation 5.33
Focal length of lens coupling two Gaussian beams	$\frac{1}{f} = \frac{1}{d_1[1+(\pi\omega_{01}^2/\lambda d_1)^2]} + \frac{1}{d_2[1+(\pi\omega_{02}^2/\lambda d_2)^2]}$	Equation 5.34
Power coupled through an aperture	$P_c = 1-\exp\left(-2\left(\frac{r_{optic}}{\omega(z)}\right)^2\right)$	Equation 5.38
Absorption coefficient/unit length through material	$\alpha = \frac{2\pi n\tan\delta}{\lambda_0};\quad n = \sqrt{\varepsilon_R}$	Equations 5.47 and 5.42
Anti-reflection (AR) coatings	$n_{AR} = \sqrt{n_1 n_2};\quad t = \frac{(2\kappa+1)\lambda_D}{4};\quad \lambda_D = \frac{\lambda_0}{n_{AR}}$	Equations 5.51 and 5.52
Fractional power loss through material	$L_d = \left(1-\frac{P}{P_0}\right) = (1-e^{-\alpha d})$	Equation 5.50
Full-width at half-maximum beam size	$\Delta\Omega_{FWHM} = 1.2\frac{\lambda}{D}$	Equation 5.55
Cut-off frequency of rectangular and circular guide	$f_c^{10} = \frac{c}{2a};\quad f_c^{11} = \frac{c}{1.7d}$	Equations 5.83 and 5.84
Gaussian beamwaist of corrugated horn	$\omega_0 = 0.644\,a$	Equation 5.89
Gaussian beamwaist at telescope focus	$\omega_{cass} = 0.22[T_e(\mathrm{dB})]^{1/2}\frac{f_e}{D_p}\lambda$	Equation 5.92

APPENDIX 6: USEFUL HETERODYNE RECEIVER EXPRESSIONS

Description	Equation	Location in Text
Minimum detectable signal	$\Delta T_{\text{rms}} = \sigma_{\text{rms}} = \dfrac{K_s T_{sys}}{\sqrt{\Delta\tau_{\text{int}} B_{pd}}};\quad \sigma_{\text{rms}} \approx \dfrac{1}{3} T_{p-p}$	Equation 6.4
Cascaded noise temperature	$T_{RX} = T_1 + \dfrac{T_2}{G_1} + \dfrac{T_3}{G_1 G_2} + \dfrac{T_n}{G_1 \cdots G_{n-1}}$	Equation 6.14
Noise contribution of component with loss L	$T = T_p(L-1)$	Equation 6.20
Noise contribution of component with noise figure (NF)	$T = \left(10^{\frac{NF_{dB}}{10}} - 1\right) T_0$	Equation 6.2
Watts to dBm conversion	$\text{Power (dBm)} = 10\log_{10}\left(\dfrac{\text{Power (watts)}}{1\times 10^{-3}}\right)$	Equation 6.26
Effective receiver noise temperature	$T_e = \dfrac{T_H - Y \cdot T_C}{Y-1};\quad Y = \dfrac{P_H}{P_C}$	Equation 6.34
Velocity/frequency dispersion	$\Delta v = \dfrac{\Delta\nu}{\nu_0} c$	Equation 6.66
Array advantage	$\varepsilon_{AY} = N_{pix}\left(\dfrac{\eta_{SP}}{\eta_{AY}} \dfrac{T_R^{SP}}{T_R^{AY}}\right)^2$	Equation 6.80
Main beam efficiency	$\eta_{mb} = \beta\gamma = \dfrac{T_A^*(\text{Planet})}{T_{\text{Planet}}} \times \left[1 - \exp\left(\dfrac{-D^2}{\theta^2}\ln 2\right)\right]^{-1}$	Equation 8.23
Deconvolution of beamsize from planet scan	$\theta_{mb} = \sqrt{\left(\theta_{obs}^2 - \dfrac{\ln 2}{2}\theta_{planet}^2\right)}$	Equation 8.25
System noise temperature	$T_{sys} = \dfrac{T_{Rx} + (1 - \alpha e^{-\tau}) T_{amb}}{\alpha e^{-\tau}}$	Equation 8.26
Measuring atmospheric optical depth	$\dfrac{T_{sys}}{T_{amb}} = \dfrac{T_{RX}}{\alpha T_{amb}} + \tau_{zenith}\sec(z)$	Equation 8.32
Atmospheric transmission	$\Gamma_{atm}^{\nu} = e^{-\tau_{zenith}^{\nu}\sec(z)}$	Equation 8.33
Aperture synthesis map rms noise level	$\Delta S_{\nu}^{\text{rms}} = \dfrac{2kT_{sys}}{A_s \eta_Q \sqrt{n_a(n_a-1)\Delta\nu_{IF}\tau_0}} \dfrac{w_{\text{rms}}}{w_{mean}}$	Equation 9.37

APPENDIX 7: DIELECTRIC BEAMSPLITTERS

In many terahertz receiver systems the local oscillator is injected into the mixer, not through a separate port, but along with the input signal from the telescope (see Figure A7.1). This is often accomplished by combining the telescope's signal at frequency, ν_s, and the LO signal at frequency, ν_{LO}, on a thin, dielectric beamsplitter before the mixer. From the beamsplitter the signal and LO travel "hand-in-hand" into the mixer where they are multiplied together to produce a signal with an intermediate frequency, ν_{IF}. What fraction of a signal's power will be transmitted, P_T, or reflected, P_R, by the beamsplitter is determined by the relative permittivity, ε, and thickness, d, of the dielectric, together with the signal's wavelength, λ_0, and polarization. The frequencies of the signal and LO are usually close enough together, that the values of P_T and P_R for each are assumed to be identical. This means one can adjust the ratio of signal to LO power striking the mixer by adjusting the beamsplitter thickness. Usually, one wants to make the beamsplitter as thin as possible to avoid unwanted signal loss. How thin a beamsplitter can be depends on how much LO power is available and how much LO power the mixer needs. At THz frequencies, LO power is often in short supply and values of P_R up to 10% are used, in which case 90% of the incoming sky signal makes it to the mixer. In the Figure A7.1 an absorbing load is used to terminate the unused LO and signal power that passes through or is reflected off of the beamsplitter.

The LOs and mixers used in THz receivers are typically polarized. Care must be taken to arrange them so that they arrive at the beamsplitter with the same polarization. The values of P_T and P_R of the beamsplitter are also dependent on polarization. The reflection coefficients for perpendicular ($\perp$) and parallel ($\|$) polarized signals are given by Goldsmith (1998),

$$r_\perp = \frac{\cos\theta_i - \left(\varepsilon - \sin^2\theta_i\right)^{1/2}}{\cos\theta_i + \left(\varepsilon - \sin^2\theta_i\right)^{1/2}}; \quad r_\| = \frac{\varepsilon\cos\theta_i - \left(\varepsilon - \sin^2\theta_i\right)^{1/2}}{\varepsilon\cos\theta_i + \left(\varepsilon - \sin^2\theta_i\right)^{1/2}}$$

where

θ_i is the angle of incidence (typically 45° for a beamsplitter) and ε the relative permittivity of dielectric

The values of P_T and P_R can then be computed using the following expressions:

$$P_R = \frac{F\sin^2(\delta\phi/2)}{1 + F\sin^2(\delta\phi/2)}; \quad P_T = \frac{1}{1 + F\sin^2(\delta\phi/2)}$$

$$F = \frac{4|r|^2}{\left(1 - |r|^2\right)^2}$$

$$\delta\phi = \frac{4\pi d}{\lambda_0}\left[\varepsilon - \sin^2(\theta_i)\right]^{1/2}$$

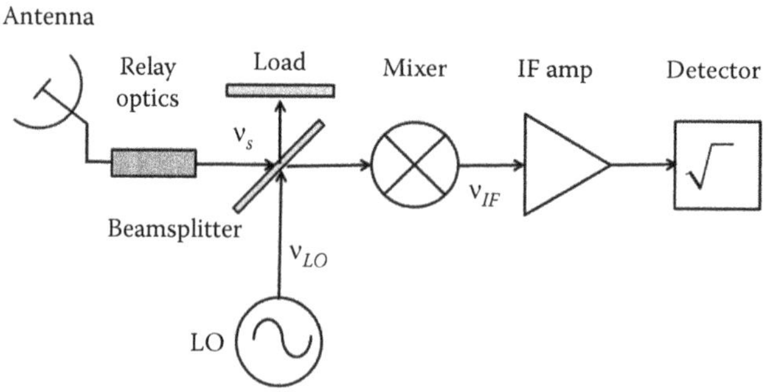

FIGURE A7.1 Heterodyne receiver with beamsplitter LO diplexer.

In the expression for F, the value of r is either $r_\perp$ or $r_\|$, depending on which is being implemented.

For example, at 1.9 THz a 6 μm thick, Mylar beamsplitter will yield a P_R of ~1.2% if the incoming beam is polarized parallel to the dielectric surface or ~9.7% if the field is perpendicularly polarized.

INDEX

J

K

L

M

S

T

Milton Keynes UK
Ingram Content Group UK Ltd.
UKHW052029141024
449569UK00017B/749

9 781138 894648